Jean Guex

Biochronological Correlations

With 156 Figures

Springer-Verlag
Berlin Heidelberg New York London Paris
Tokyo Hong Kong Barcelona Budapest

Professor Dr. JEAN GUEX
Institut de Géologie
UNIL, BFSH 2
CH-1015 Lausanne
Switzerland

Translated by

BARBARA BURKE-HUBBARD
214 University Ave
Ithaca, NY 14850, USA

Title of the original french edition:
Jean Guex: Corrélations biochronologiques et associations unitaires,
Presses Polytechniques Romandes, Cité Universitaire, Centre Midi,
CH-1015 Lausanne, Suisse

ISBN-13: 978-3-540-53937-7 e-ISBN-13: 978-3-642-76583-4
DOI: 10.1007/978-3-642-76583-4

Library of Congress Cataloging-in-Publication Data. Guex, Jean, 1945–[Corrélations biochronologiques et associations unitaires. English] Biochronological correlations / Jean Guex : [translated by Barbara Burke–Hubbard]. p. cm. Translation of: Corrélations biochronologiques et associations unitaires. Includes bibliographical references and index. ISBN 978-3-540-53937-7 1. Stratigraphic correlation. 2. Paleontology. I. Title. QE652.5.G8413 1991 551.7—dc20. 91-20243.

Typesetting: Konrad Triltsch, Graphischer Betrieb, Würzburg
31/3145-543210 – Printed on acid-free paper

Dedicated to the memory of
Dan Patrulius, Carlo Sturani,
and Riccardo Asseretto

Preface

The object of this book is to explain how to create a synthesis of complex biostratigraphic data, and how to extract from such a synthesis a relative time scale based exclusively on the fossil content of sedimentary rocks. Such a time scale can be used to attribute relative ages to isolated fossil-bearing samples.

From a practical point of view, the method described in this book will particularly interest paleontologists and geologists who must construct zonations and establish correlations on the basis of biostratigraphic data that are both plentiful and apparently contradictory.

It is well known that the difficulties involved in constructing biochronologic scales are largely due to the discontinuous nature of the fossil record. We know that the relationships between the first appearances (or disappearances) of different fossil species are rarely constant in stratigraphic sections that are distant from each other. It if often extremely difficult to discover datums or sets of species that are useful in making significant biochronologic correlations on a large scale.

The theoretical model explained here (known as the Unitary Association Method) provides clear solutions to most of these problems.

That method is purely deterministic, as opposed to statistical and probabilistic analytical techniques producing "average" ranges. We demonstrate in Chapter 15 why most of these techniques produce results which are usually not compatible with the original biostratigraphic observations (i.e., the taxonomic contents of the studied samples are not reproduced in the outputs).

The syntheses used here are in the form of a referential (i.e., a system of chronologic reference); the chronologically significant subdivisions of these referentials correspond roughly to the Concurrent Range Zones and to the Oppel Zones of classical stratigraphy. These zones are seen here as discrete (i.e., noncontiguous) units, isolated from each other by separation intervals.

The operations required to construct such zonations are elementary but not always simple. We will begin by analyzing the fundamental properties of biochronologic referentials. This will enable us to assess the validity of stratigraphic correlations based on complex paleontological data. It will also help the reader master the mathematics and algorithms that are indispensable for making sense of the contradictory stratigraphic relationships so often observed among species in different locations.

The ideas presented here were developed in a series of preliminary notes published between 1977 and 1984. To avoid excessive self-citations, we will mention in the text principally the ideas that are due to other authors.

Two different computer programs making it possible to analyze certain complex biostratigraphic data are used a number of time in the present book.

The oldest ones were written by Davaud and published in collaboration with the author (Guex and Davaud 1982, 1984, 1986). The last version of this program will be referred to as "DV-86".

The most recent program, named "BioGraph", was developed on an IBM microcomputer by Savary (Savary and Guex 1991). This very efficient program is described in Chapter 6. It is used to solve most of the difficult problems presented in this book.

JEAN GUEX

Acknowledgments

I wish to thank, first of all, the Swiss National Science Foundation, without whose support this work would not have been possible.

Innumerable discussions with A. Baud, P. Baumgartner, J.-L. Dommergues, L. Edwards, R. Enay, J. Gabilly, H. Hubbard, H. Masson, J. Remane, F. Reuse, G. Scott, D. Taylor, E. T. Tozer, R. Trümpy and D. de Werra indirectly helped me to present these results in a form that I hope will be acceptable to a public of naturalists. These eminent colleagues will find here my sincere thanks for their encouragement and useful criticisms.

I wish in particular to express my gratitude to F. Reuse and H. Hubbard for their invaluable suggestions in revising chapters 3 to 5 and to B. Burke-Hubbard for her excellent translation of the present book.

I am also much indebted and grateful to J. Savary and E. Davaud for their outstanding contribution in computer science, which enabled me to present a detailed analysis of certain examples of application given at the end of this book.

JEAN GUEX

Contents

Chapter 1
Properties of Biochronologic Scales

1.1 Introduction

This book studies the problems of using complex biostratigraphic data to construct biochronologic scales.

Such scales are relative: that is, they describe a sequence of intervals (e.g., zones) whose duration in years is unknown. Several fundamentally distinct categories of relative biochronologic scales exist. We mention here only the most important (Fig. 1.1):

1. Continuous biochronologic scales based on the relative abundances of taxa whose preservation is not selective (e.g., not diagenetically biased) vis-à-vis the different species of the group used. The subdivisions ("Abundance Zones" and "Acme Zones") that make up such scales are contiguous. The Calpionellid zonation introduced by Remane in 1963 belongs to this category.
2. Continuous biochronologic scales whose subdivisions are the intervals separating the first appearance (or extinction) of one species from the first appearance (or extinction) of another. These subdivisions are the "Interval Zones" currently used by micropaleontologists.
3. Discrete biochronologic scales whose subdivisions are characterized by unique and mutually exclusive assemblages of species. These subdivisions are noncontiguous and isolated from each other by intervals of separation. They are similar to "Concurrent Range Zones" and "Oppel Zones".

It is this third group, discrete biochronologic scales, that interests us in this book, for in most situations, the biostratigraphic data are highly discontinuous and cannot be used to establish continuous zonations: discontinuous data clearly generate discrete zonations.

Finding sequences of first appearances (or extinctions) of species within such zones and taking into account the relative abundances of the different taxa used

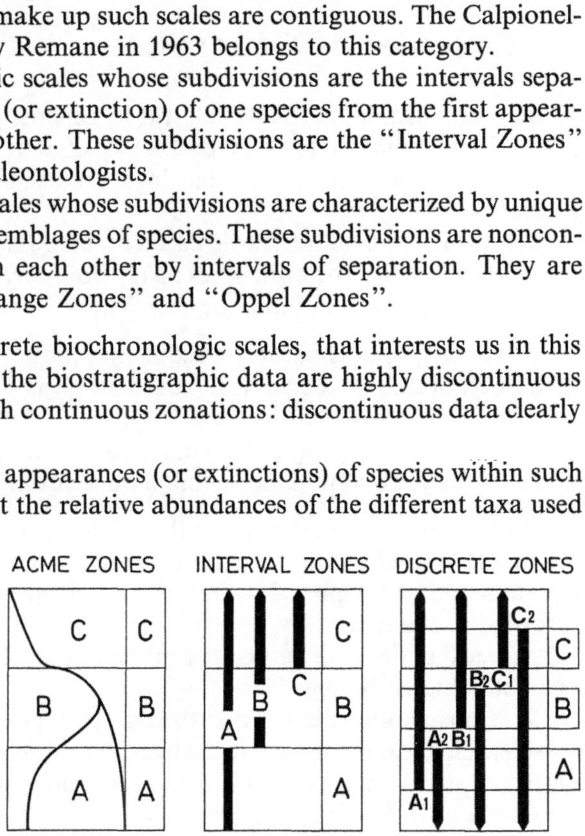

Fig. 1.1. Principal types of zones

to create a system of chronologic reference are subordinate problems: they will be discussed in Chapters 11 and 16.

Throughout the book, we will maintain a strict distinction between "biostratigraphy" and "biochronology".

Biostratigraphy is a general term; it applies to everything that concerns the study of the fossil content of sedimentary rocks (paleontological inventory, study of biofacies, determining ecozones, etc). Biochronology is limited to identifying the taxa in a fossil assemblage that are characteristic of the relative age of the assemblage.

In the following pages, we will also clearly distinguish between two different kinds of range charts:

1. A raw range chart that represents superpositions and coexistences of taxa without regard to the chronologic significance of their stratigraphic relationships will be called a "protoreferential" (Sect. 1.4).
2. When completed by precise additional information about the lateral traceability of its subdivisions, such a range chart will be called a "biochronologic referential" (or referential in short: Sect. 1.5.3). We will see that such information is necessary to distinguish between ecologic versus chronologic control over the stratigraphic distribution of the taxa (Sect. 1.4.2).

1.2 Properties of Discrete Scales

A discrete chronologic scale has more general properties than a continuous chronologic scale. In some respects it is similar to the hardness scale used by mineralogists, with the idea of hardness replaced by that of relative positioning in time.

Constructing a discrete scale based on the fossil content of sedimentary rocks is not always easy. The difficulties result from several important factors independent of taxonomic and sampling problems:

1. The discontinuous nature of sedimentation;
2. Discontinuities in the vertical record of species (selective preservation of material and/or absence of fossilization);
3. Ecologic and biogeographic control over the spatial distribution of taxa;
4. Reworking, contamination, and condensation.

As a result, the data the biostratigrapher must use to construct biochronologic scales are sporadic and discontinuous, and only a small fraction of the total duration and geographic distribution of each species [Fig. 1.2; $D(x)$] is generally recorded in the sediments.

Moreover, the data can be deceptive in that the paleontological record may be contaminated by reworking.

This means that the relative stratigraphic positions observed locally among fossil species commonly are not constant from one place to another. Figure 1.3 gives a typical situation in which species occupy apparently contradictory positions at four different locations.

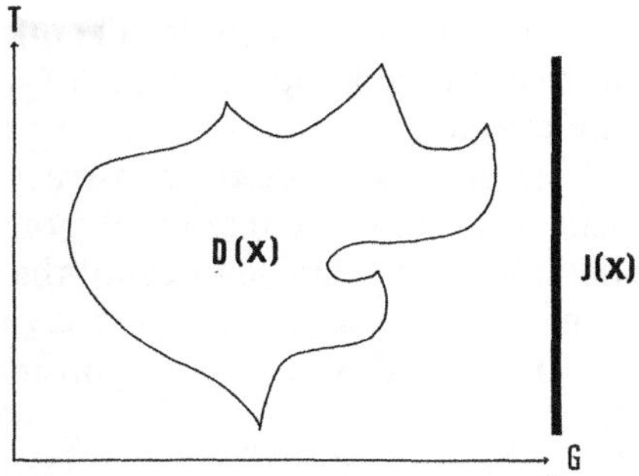

Fig. 1.2. $D(x)$, existence domain of taxon x. $J(x)$, existence interval of taxon x. G, geographic space. T, time (in years)

a

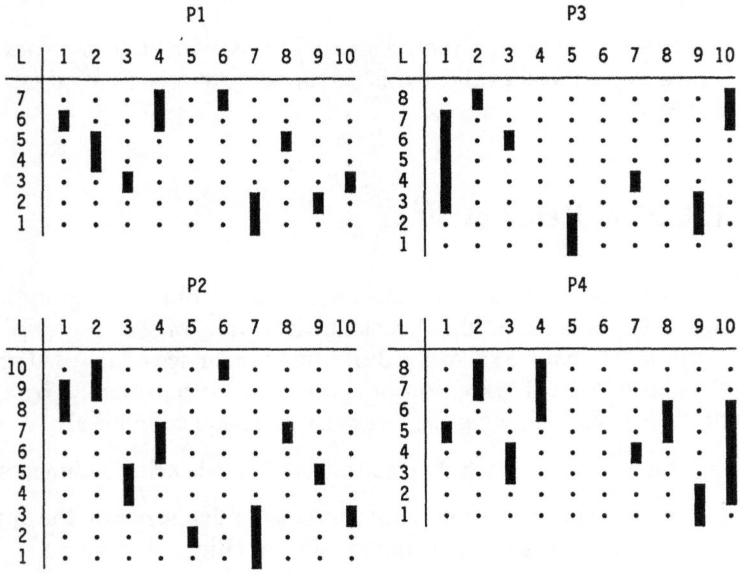

b

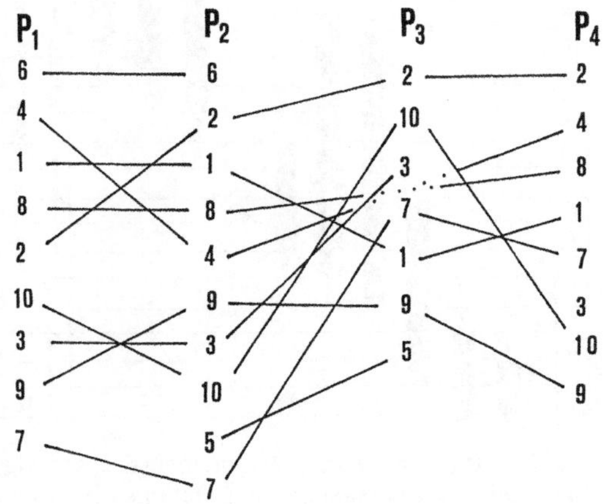

Fig. 1.3. a Local stratigraphic distribution of 10 species (1–10) in four sections $P_1 - P_4$ (L, level). **b** Relative positions of the first local appearances of the 10 species in these sections

This example illustrates the difficulties encountered when trying to establish a reliable chronology from such data. We will study these problems in the following chapters.

Before discussing the construction of a biochronologic referential (i.e. a system of chronologic reference) from such sparse and conflicting biostratigraphic data, we will examine in detail the characteristics of referentials whose subdivisions are defined by the existence spans of the fossil species.

In doing this, it is important to distinguish between the following two problems:

1. That of devising an optimal synthesis from raw biostratigraphic data (resulting in what we will later call a protoreferential) and
2. That of using such a synthesis to find zones that are chronologically meaningful.

These preliminary remarks provide us with a framework for determining which elements of a valid synthesis can serve as a basis for significant chronologic correlations.

1.3 Ideal Referentials

To each species we can associate an interval of time corresponding to its life span: that interval is called the "existence interval" of the species [Fig. 1.2; $J(x)$].

Imagine that we know the duration of existence of all fossil species. In this case the chronologic distribution of species can be represented by an "ideal referential" (Fig. 1.4) showing their respective existence intervals.

One might use such a table to date fossil-bearing sediments in two ways:

1. One might attempt to use the increments that separate the appearances and/or disappearances of the different species (Fig. 1.4; column i); or

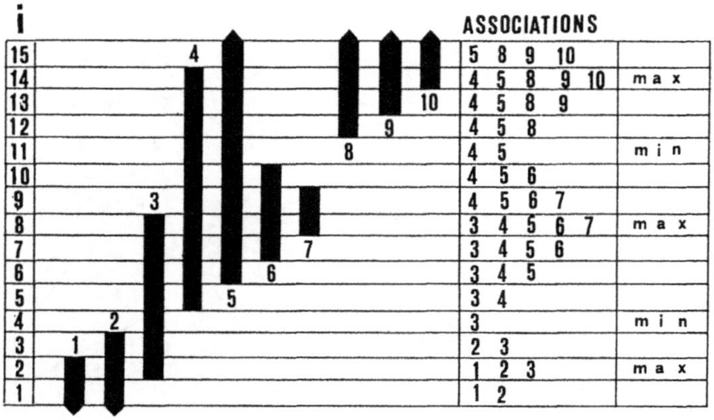

Fig. 1.4. Ideal referential. *I*, Increments separating the first appearances and disappearances of the species. *Max*, maximal intersections of existence intervals. *Min*, separation intervals

2. One might consider only the intersection. . f the existence intervals, reasoning as follows: if a bed contains species x, it was deposited during the existence interval of x; if it contains two species, x and y, it was deposited during the coexistence interval of x and y, and so on. In this approach, the smallest useful intervals are the maximal intersections of existence intervals (Fig. 1.4; max). These subdivisions of minimal duration are separated by separation intervals (Fig. 1.4; min).

Now suppose that a geologist, faced with an actual field problem, has an ideal referential at his disposal, which he wants to use to attribute relative ages to the fossil-bearing beds. He will soon realize that the only useful information he has is information about the intersections of existence intervals.

He cannot, in fact, have any confidence in the relative appearances or disappearances of the species that occur in the vicinity of the separation intervals given in the ideal referential.

Assume for example that three species x, y and z actually appear in the order given in Fig. 1.5.a. In practice, considering the usually poor quality of the fossil record, their apparent order of appearance will often be inverted (Fig. 1.5; I). In this case the information contained in the ideal referential regarding the order of appearances of x, y and z serves only to exemplify the imperfection of the local fossil record.

On the other hand, if the order of appearances observed in a particular stratigraphic section corresponds to that given in the ideal referential, the geologist cannot assume that the first local appearances correspond to the true first appearances of the three species: they could well be displaced in time (Fig. 1.5; D).

In other words, the presence of species x in a sedimentary level tells our geologist only that the level was deposited during the existence interval of species x. Similarly, the presence of species x and y in this level tells him only that it was deposited during the coexistence interval of (x, y), etc.

The only immediately useful chronologic information in the ideal referential is thus information about the intersections of existence intervals of species: for the user of such a table, only the presence of a species is meaningful. Because of this, in the beginning we will not consider the increments given in Fig. 1.4, and we will simplify our ideal referential, so that it gives only an ordered sequence of noncon-

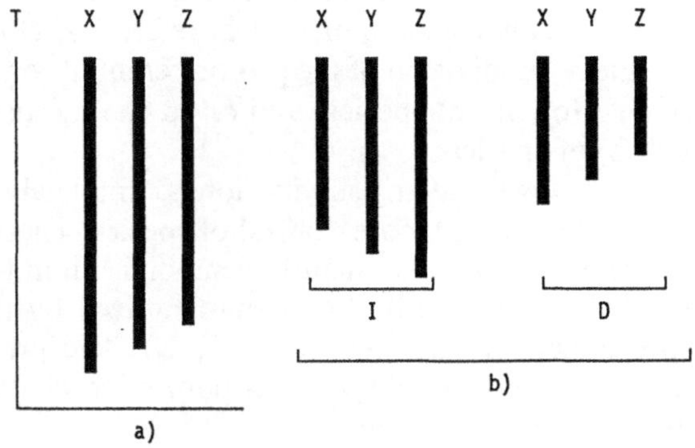

Fig. 1.5. **a** *True* chronologic duration of threee species (x, y, z). T, time (in years). **b** I, inversion of the order of appearances of species x, y and z in local section. D, displacement of the order of appearances of species x, y and z in another section

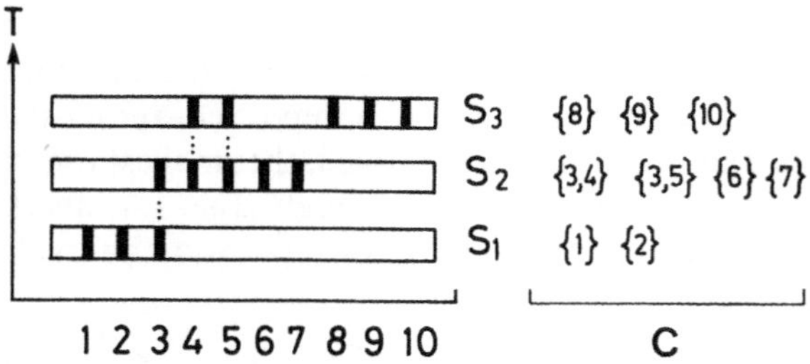

Fig. 1.6. Simplified ideal referential. S_1, S_2, S_3 discrete subdivisions generated by the maximal intersections of the existence intervals of species 1–10 of Fig. 1.4. C, characteristic elements (species and pairs of species) of $S_1 - S_3$

tiguous subdivisions corresponding to the maximal intersections of the species' existence intervals (Fig. 1.6).

The coexistence intervals in this simplified table are all distinct from each other: they correspond to the only associations of species for which uniqueness can be demonstrated using relatively simple criteria.

1.4 Protoreferentials

1.4.1 Synthesis of Actual Biostratigraphic Data

1.4.1.1 Introduction

The stratigraphic literature provides numerous examples of syntheses that faithfully represent the stratigraphic sequences of taxa (range charts). We will show that these range charts are not necessarily meaningful for chronology.

1.4.1.2 Definition

A protoreferential is a discrete range chart giving an exact or optimal summary of all the stratigraphic relationships (i.e., coexistences and/or superpositions) observed among the fossil species in all the areas where they are found. Such a table looks like a simplified ideal referential (Fig. 1.6).

Each subdivision of the protoreferential corresponds to the existence interval of one (or more) species, and/or to the coexistence interval of one (or more) pair(s) of species.

It follows that any subdivision is completely defined by one (or more) species and/or by one (or more) pair(s) of species. These characteristic species and pairs of species allow us to identify a subdivision in fossiliferous strata. For example, subdivision S_2 of Fig. 1.6 is characterized by the species x_6 or x_7, the pair of species (x_3, x_4) and the pair (x_3, x_5). The presence of any one of the above designated individual species or pairs of species in a fossiliferous bed allows us to assign it to subdivision S_2.

1.4.2 Properties of Protoreferentials

A protoreferential does not have the immediate chronologic properties of a simplified ideal referential, and any strict analogy between the two is both incorrect and misleading. In fact, it is easy to demonstrate that the order of the subdivisions of a protoreferential (constructed from raw biostratigraphic data) cannot be interpreted directly as a chronology, even if it is a faithful synthesis of those data.

Consider, for example four stratigraphic sections in which the existence intervals of eight species are partially recorded (Fig. 1.7). The protoreferential of Fig. 1.8 gives an exact representation of the superpositions observed among these species and of the intersections of their apparent existence intervals.

Each subdivision of this protoreferential can be identified in the sections by its characteristic species or pairs of species: the result is given in Fig. 1.9. This figure shows that two subdivisions (S_1 and S_6) can be identified in all the sections. The other four (S_2-S_5) can be identified only in one place and it is not possible to know their relative chronologic positions.

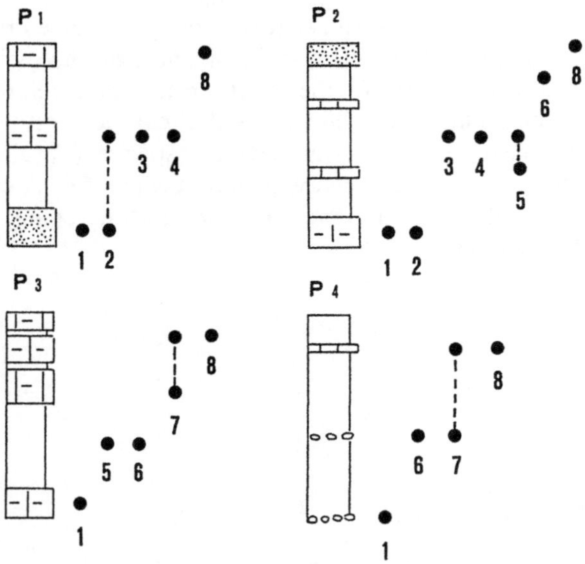

Fig. 1.7. Stratigraphic distribution of eight species (1–8) in four imaginary sections (P_1-P_4)

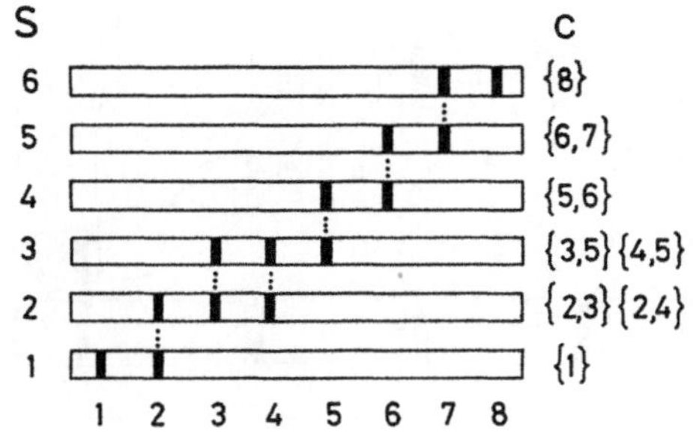

Fig. 1.8. Protoreferential giving an exact representation of the coexistences and superpositions of the eight species of Fig. 1.7. *S*, subdivisions S_1-S_6. *C*, characteristic species/pairs of species of subdivisions S_1-S_6

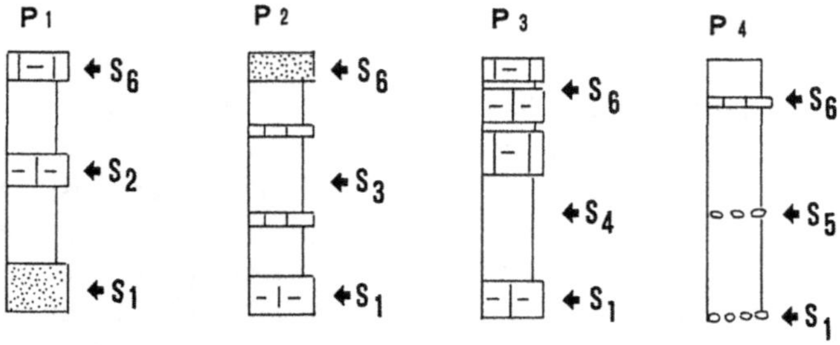

Fig. 1.9. Stratigraphic position in sections P_1-P_4 of the characteristic species of subdivisions S_1-S_6 of the protoreferential in Fig. 1.8

Imagine, now, that we know the actual space-time distribution of the eight species in our four sections and that we project their respective existence domains onto a plane parallel to the time axis (Fig. 1.10).

This figure shows that the characteristic coexistence intervals (i.e., pairs of species) of the subdivisions S_2-S_5 are reversed in sequence compared to the one suggested by the protoreferential.

Thus the apparent sequence of species given by such a protoreferential does not necessarily define a valid chronologic succession, even if it gives a true and complete picture of the local stratigraphic relationships observed among the species. Only the subdivisions that can be identified over vast geographic areas by means of their characteristic species (S_1 and S_6 for example) have any chance of being chronologically meaningful. Such subdivisions are said to be "reproducible" and we will, for the time being, call them "biochronozones" (Sect. 1.5.3).

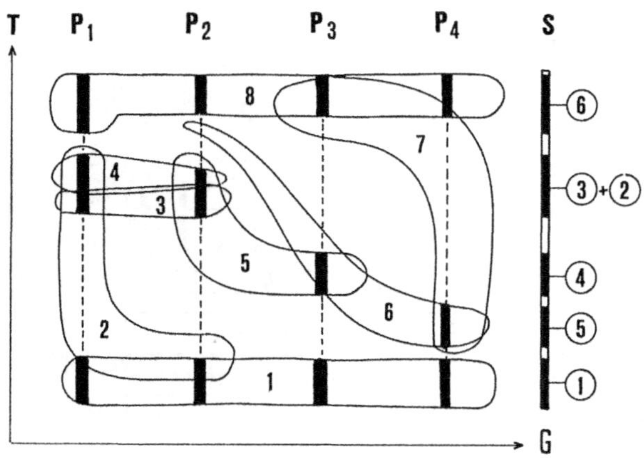

Fig. 1.10. Planar projection of the space-time existence domains of the eight species of Fig. 1.7. The *vertical dotted lines* represent the path of discontinuous sections P_1-P_4 that cross these existence domains and provide a partial record of the existence interval of each of the eight species. The *black rectangles* show the duration and actual position in time of the coexistence intervals characteristic of subdivisions S_1-S_6 of the protoreferential of Fig. 1.8 (*column S*). *T*, time (in years). *G*, geographic space. (Guex 1978 b)

S	P1	P2	P3	P4
6	1	1	1	1
5	0	0	0	1
4	0	0	1	0
3	0	1	0	0
2	1	0	0	0
1	1	1	1	1

Fig. 1.11. Reproducibility matrix associated to the protoreferential of Fig. 1.8. Rows $S_1 - S_6$ represent the six subdivisions of this protoreferential; *columns* $P_1 - P_4$ represent the four sections. The 1's represent the subdivisions that are identified in the different sections (Fig. 1.9)

1.5 Chronologic Interpretation of a Protoreferential

1.5.1 Reproducibility Matrices

To make a quick, visual estimate of the reproducibility of a protoreferential's subdivisions, we construct a double-entry table whose rows correspond to the subdivisions and whose columns correspond to the stratigraphic sections under study. The sequences of subdivisions identified in the different sections are transcribed onto this table, which we call a "reproducibility matrix".

The reproducibility matrix for the protoreferential of Fig. 1.8 is given in Fig. 1.11. It shows that the reproducible subdivisions S_1 and S_6 can immediately be considered as biochronozones: they are defined by the characteristic species given in Fig. 1.8. It also demonstrates that the union of subdivisions $S_2 - S_5$ gives rise to a new subdivision that is itself reproducible.

We will see later that, in a general manner, most chronologically meaningful zones are formed by clustering subdivisions that individually are not reproducible but which are identifiable over a large geographic area when grouped.

1.5.2 Reproducibility and Chronologic Order

Some experts claim that biochronologic scales have no chronologic meaning (Sect. 16.2). This extreme point of view, if valid, would prevent any serious analysis of potential diachroneity in first (and last) local appearances of fossil species (the famous datums). It is worthwhile, then, to devote some time to this "philosophical" question. To do this we will examine once more the relationship between the geometry of the existence domains of species and chronometric time.

Consider an ideal case in which two subdivisions S_1 and S_2 of a referential are universally identifiable by means of their characteristic species, x and y (Fig. 1.12). The superposition of these species is compatible with a chronologic order since this relationship is (by hypothesis) observed in all stratigraphic sections. It is equally clear that the respective existence intervals of these species can overlap in time, even if their existence domains do not intersect in space.

In this case (the worst!) it is still easy to see that, whatever the geometry of the existence domains of x and y, the following relationships are always true:

1. *Inf.* $J(x)$ *is older than inf.* $J(y)$ and
2. *Sup.* $J(x)$ *is older than sup.* $J(y)$,

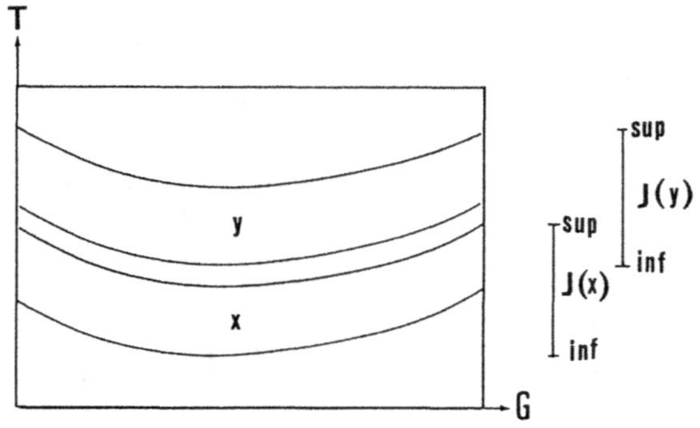

Fig. 1.12. Space-time existence domains of two species, x and y, found everywhere in stratigraphic sequence. These domains do not overlap but the existence intervals of x and y intersect in time

where *inf.* $J(x)$ is the actual time (chronometric) of the first appearance of species x and *sup.* $J(x)$ is the actual time of the extinction of species x.

In other words, a true chronologic order of the first appearances and extinctions of the two species does exist, even if the first appearances and extinctions are diachronous on a regional scale.

In general, then, we can say that at least one part of each characteristic interval of a protoreferential's reproducible subdivisions remains invariant with respect to time (the Reproducibility Principle; Guex 1978 b). Of course we know neither the duration of these intervals, nor their position on the time axis.

This conclusion also holds for the reproducible subdivisions (i.e., biochronozones) formed by the union of the adjacent, nonreproducible subdivisions of a protoreferential.

Finally, we note that a single contradiction in the order of appearance of the species x and y invalidates the conclusion that *inf.* $J(x)$ *is older than inf.* $J(y)$ (Fig. 1.13): a sequence of appearances that is not constant is chronologically meaningless, even if one of the two relationships ($x < y$ or $y < x$) is observed more often than the other. The consequences of this will be discussed in Section 15.3.

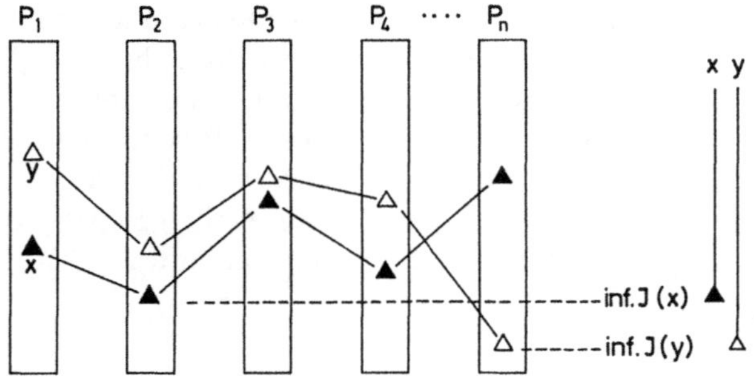

Fig. 1.13. Local, isolated inversion of the appearance order of species x and y. *Dashed lines* indicate the true chronologic appearances of x and y

1.5.3 Biochronologic Referentials and Biochronozones

We saw above (Sect. 1.4.2) that to be useful as a tool for dating, a protoreferential must be accompanied by its reproducibility matrix (i.e., information about the lateral traceability of its subdivisions). In other words, we must consider the combination protoreferential + reproducibility matrix as an entity in itself. We call this entity a biochronologic referential (in short, a "referential").
The reproducible subdivisions (or unions of subdivisions) of a referential are biochronozones. A referential made up of biochronozones is called a "zonal referential".

1.5.4 Instability of Biochronologic Scales

New biostratigraphic investigations often reveal the existence of previously unknown fossil cooccurrences. Such new data will contradict existing biochronologic syntheses and merely reflect the instable nature of biochronologic scales.

An example is given in Fig. 1.14 which shows the case of a sample containing ten species (1–10 in Fig. 1.14.b) that should be datable by means of an "old" scale (i.e., a scale established prior to its collecting; Fig. 1.14.a). Note that there are several incompatible pairs of species in the scale that are observed together in the sample.

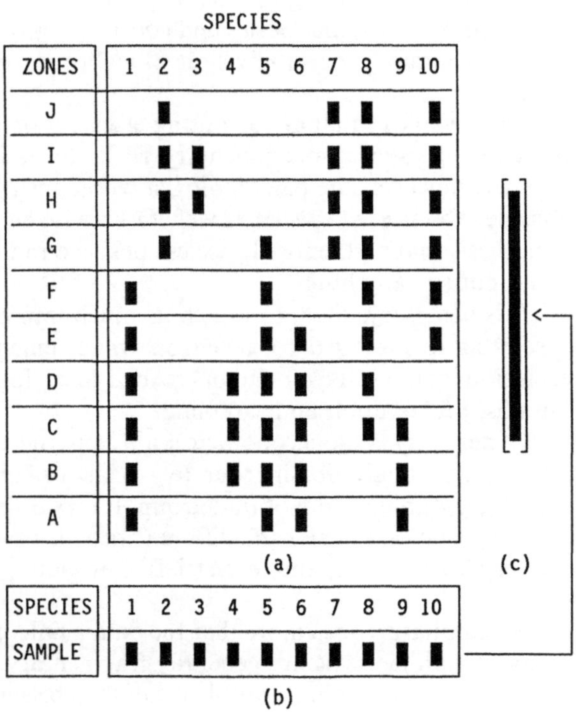

Fig. 1.14 a–c. Example of instable biochronologic scale (see text). **a** biochronologic distribution of ten species (1–10) in ten zones (*A–J*) according to an "old" scale. **b** newly collected sample in which species 1–10 coexist. **c** global uncertainty about the age assignment of the sample

Assuming that the studied sample was not collected from a condensed (or reworked) bed, there are several possible solutions to this problem. The two extremes are:

1. The assemblage belongs to zone C and the ranges of species 2, 3, 7 and 10 must be extended downward in the synthetic range chart; or
2. The assemblage belongs to zone H and the ranges of species 1, 4, 5, 6 and 9 must be extended upward in the synthetic range chart (Fig. 1.14.c).

Finding the sample's accurate biochronologic position depends, in fact, on the global stratigraphic relationships of all the species and this cannot be done without updating the scale.

The formal method described in the following chapters has been specifically devised to facilitate the revision of old biochronologic scales to include such conflicting new data.

1.6 Discrete Biochronologic Scales

Note from the above discussion that we can think of a discrete biochronologic scale as the set of biochronozones of a referential: the characteristic taxa of each of these zones enable us to assign relative ages to the fossiliferous sediments.

Such a scale can always be further refined: in a given state of knowledge it only gives a summary of all the stratigraphic relationships (coexistences and superpositions) observed among the fossil species.

Exclusions of some species and coexistences of others are both equally important in generating the subdivisions (biochronozones) of discrete biochronologic scales.

We saw above that anyone using a scale (even an ideal one) can use only the presence of species in sedimentary rocks to assign a relative age to them. In practice, however, the paleontologist who constructs this scale is still obliged to consider absences of species as well. Otherwise he would be unable to exclude the hypothesis that all the fossil species coexisted in time, which would mean that he could not say anything!

It is necessary, then, to attach as much importance to mutual exclusions of species as to observed coexistences, while remembering that apparent mutual exclusions can result from factors related to ecologic constraints on species distributions, preservation and sampling.

Of necessity, a biochronologic scale constructed from coexistences and exclusions in a particular fossil group (e.g., class or subclass) can be used only within the biogeographic realm of that group. It is essential to construct separate, parallel biochronologic scales for different provinces or ecological environments (at the very least, marine and terrestrial). Correlating these scales will not always be easy.

We emphasize once more that the data a paleontologist must use to construct a biochronologic scale are generally fragmentary, discontinuous and sparse.

The biostratigrapher's problem can thus be summarized as follows. He must:

1. Establish a sequence of mutually exclusive coexistence intervals that provides an accurate picture of the stratigraphic relationships observed locally among the fossil taxa and which has the form of the simplified ideal referential shown in Fig. 1.6.
2. Distinguish, within this picture, the exclusions among taxa due to ecologic or documentary factors from those that are chronologically meaningful; and
3. Determine which anomalies of distribution result from reworking or contamination.

The object of this book is to set forth precise rules for doing this. They follow logically from our description of time as an order relation.

The discussion will be divided into two parts.

First, we will solve a particularly simple biochronologic correlation problem (Chap. 2). This will allow us to present, in an intuitive way, abstract concepts that will be used later. It will also allow us to see what factors make some problems in biochronology difficult.

The second part will discuss these more difficult problems. Their inherent difficulties have to do with the structure of biostratigraphic data and they cannot be resolved without using some formal mathematics. This mathematical underpinning will be given in the simplest possible way in Chapters 3 through 5.

In all of the biochronologic problems given as examples in the following chapters, we assume that the systematics is internally consistent and correct.

Chapter 2
Semiempirical Construction of a Referential

2.1 Introduction

This chapter provides a methodology for constructing a sequence of mutually exclusive existence intervals (or coexistence intervals), and discusses how such a sequence is employed in establishing meaningful biochronologic correlations. The technical details will be examined in the following chapters. Some preliminary definitions are required.

2.2 Definitions

2.2.1 Actual Coexistence and Virtual Coexistence

Two species x and y "actually" coexisted if their coexistence has been observed in at least one sedimentary bed. If, because of ecological factors or inadequate documentation, the coexistence of two species was never observed, but it can be deduced, they are said to have "virtually" coexisted. Two trivial cases of virtual coexistence can be easily recognized:

1. If the vertical distribution of a species in a sedimentary column is discontinuous, we assume that it virtually coexisted with all the species found between its first and last appearance in the column (Fig. 2.1.a).
2. If species x is below species y in one section (i.e., $x < y$), and above it ($y < x$) in another, then we should assume until proven otherwise that x and y virtually coexisted (Fig. 2.1.b).

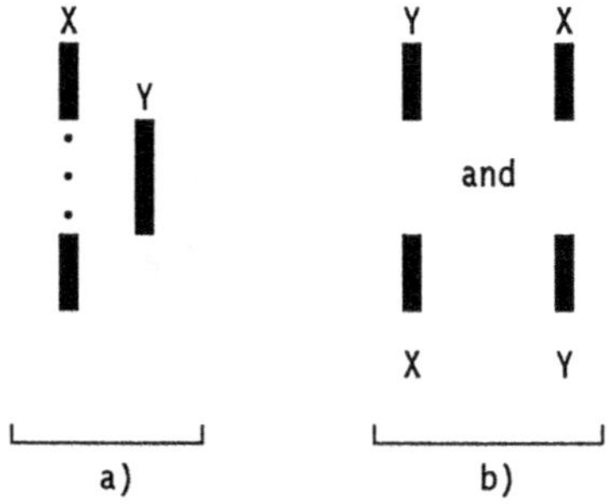

Fig. 2.1 a, b. Virtual coexistence of species x and y. **a** Discontinuous stratigraphic distribution of x in a local section. **b** Inverted stratigraphic superpositions of x and y in two separate sections

2.2.2 Compatibility

Two species are compatible if they actually or virtually coexisted during part of their existence.

2.2.3 Neighborhood

The neighborhood of species x consists of all species compatible with x. It is evident that some species defining the neighborhood are not necessarily compatible with one other (e.g., *Lingula* is compatible with all Mesozoic organisms, including rudists and ceratites, but the latter two are not said to be compatible with each other).

2.2.4 Maximal Set and Unitary Association

A set is "maximal" if it is not contained within a larger set. A "unitary association" (U. A.) is a maximal set of mutually compatible species. An ordered sequence of unitary associations is a protoreferential (Sect. 1.4).

2.2.5 Characteristic Elements of a Unitary Association

Each unitary association (U. A.) corresponds to a coexistence (or existence) interval that is unique. Such an interval is characterized by the species and/or pairs of species that are found exclusively in the U. A. under consideration. These species or pairs of species are the characteristic elements of the association. They make it possible to identify the U. A. in stratigraphic sections (or, inversely, to assign the fossil-bearing beds to a particular unitary association).

An initial brief description of the procedure for identifying unitary associations in fossil-bearing terrains was implicitly given in Figs. 1.7–1.9.

We caution that it must be proven whether or not a species (or pair of species) is characteristic of a biochronologic subdivision. It cannot be an arbitrary choice by the biostratigrapher.

2.2.6 Merging Unitary Associations to Form Biochronozones

In practice it is often necessary to group some unitary associations to increase their reproducibility (Fig. 2.2). This amounts to considering the characteristic elements of the merged unitary associations as having virtually coexisted. The new unitary association that results from this union is a biochronozone that is characterized:

1. By the set of species and pairs of species characteristic of each of the unitary associations that make it up and

2. By a set of new elements consisting of the species and pairs of species that are exclusively found in this union.

A given biochronozone that corresponds to a union of unitary associations is thus defined by the set of characteristic pairs (x, y) where x is any species that made its first appearance in the zone and y is any species that disappeared in the zone (if $x = y$ it is called a characteristic species). The presence of any one of these elements in a bed enables us to identify the zone.

For example, the union formed by subdivisions S_2 and S_3 of Fig. 2.2 is equally well characterized by three pairs: $\{2, 4\}$, which is characteristic of S_2; $\{3, 5\}$, which is characteristic of S_3; and $\{3, 4\}$. The pair $\{3, 4\}$ characterizes neither S_2 nor S_3, but only the union formed by S_2 and S_3, which now make up a biochronozone.

Finally, a union of unitary associations is "strictly identified" in a section if it is located stratigraphically between the levels assigned to the unitary associations above and below it (Fig. 2.2).

2.2.7 Names of Species, Beds and Sections

For the sake of graphical simplicity, we will assign numbers to the species, fossil-bearing beds and sections used in examples. The names corresponding to the numbers are given in the appendices.

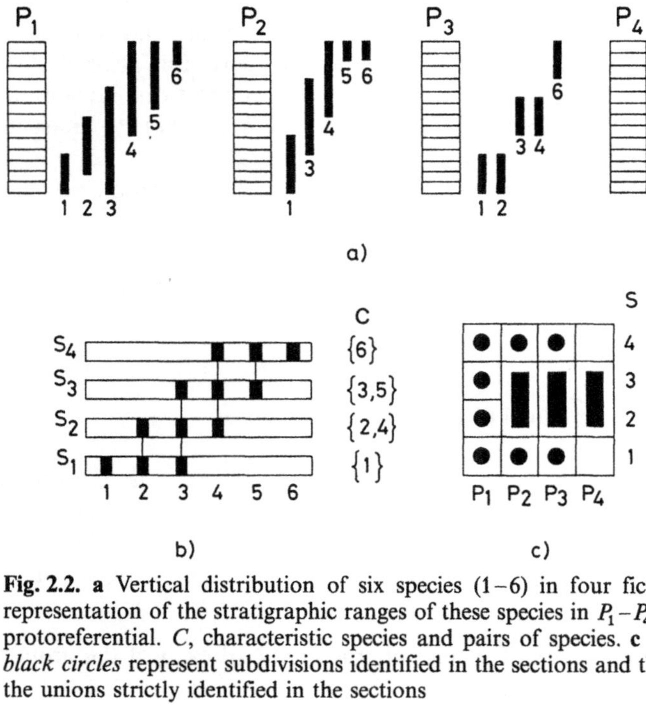

Fig. 2.2. a Vertical distribution of six species (1–6) in four fictitious sections. **b** Synthetic representation of the stratigraphic ranges of these species in P_1–P_4. *S*, subdivisions 1–4 of the protoreferential. *C*, characteristic species and pairs of species. **c** Reproducibility matrix: the *black circles* represent subdivisions identified in the sections and the *black rectangles* represent the unions strictly identified in the sections

2.2.8 Acme

In cases where preservational bias is not prevalent, it can sometimes be useful to study the variations in frequencies and the apparent acme of species (i.e., the stratigraphic intervals in which a taxon is particularly abundant).

In using this kind of information in the unitary association method, we must consider the apparent acme of a species x as a taxon x', distinct from x.

2.2.9 Maximal Horizon

A fossil-bearing bed in an isolated stratigraphic section is a maximal horizon if the set of species that coexist (actually or virtually, see Sect. 2.2.1) in the bed is maximal (NB: in the framework of a study limited to the section).

The maximal horizons of a given section are separated from each other by separation intervals corresponding to the minimal intersections of existence intervals described in Section 1.3 (Fig. 1.4).

In other words, all the beds located between two minimal intersections (separation intervals) are equivalent to the maximal horizon framed by them.

Figure 2.3 gives the local vertical ranges of eight species in a hypothetical section. The superpositional relationships of the species give rise to four distinct maximal horizons ($M_1 - M_4$).

2.2.10 Composite Sections

Whenever lithostratigraphic or structural arguments allow, we must construct composite sections before analyzing the raw biostratigraphic data (Fig. 2.4). This gives the most complete representation possible of the vertical ranges of the

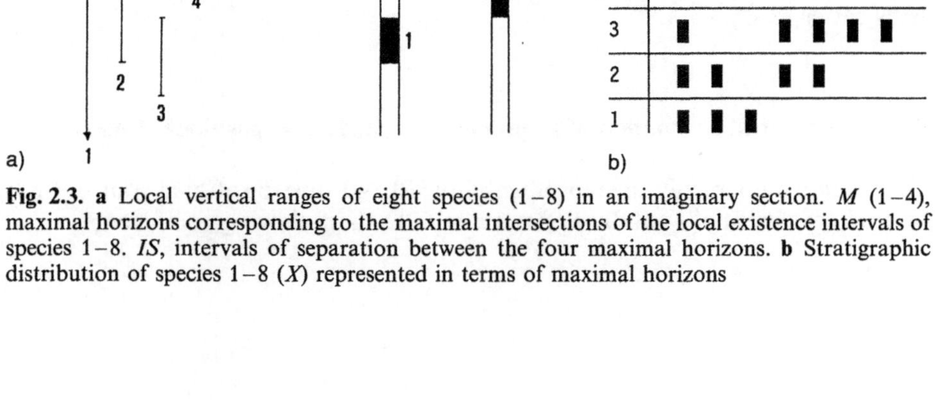

Fig. 2.3. **a** Local vertical ranges of eight species (1–8) in an imaginary section. *M* (1–4), maximal horizons corresponding to the maximal intersections of the local existence intervals of species 1–8. *IS*, intervals of separation between the four maximal horizons. **b** Stratigraphic distribution of species 1–8 (*X*) represented in terms of maximal horizons

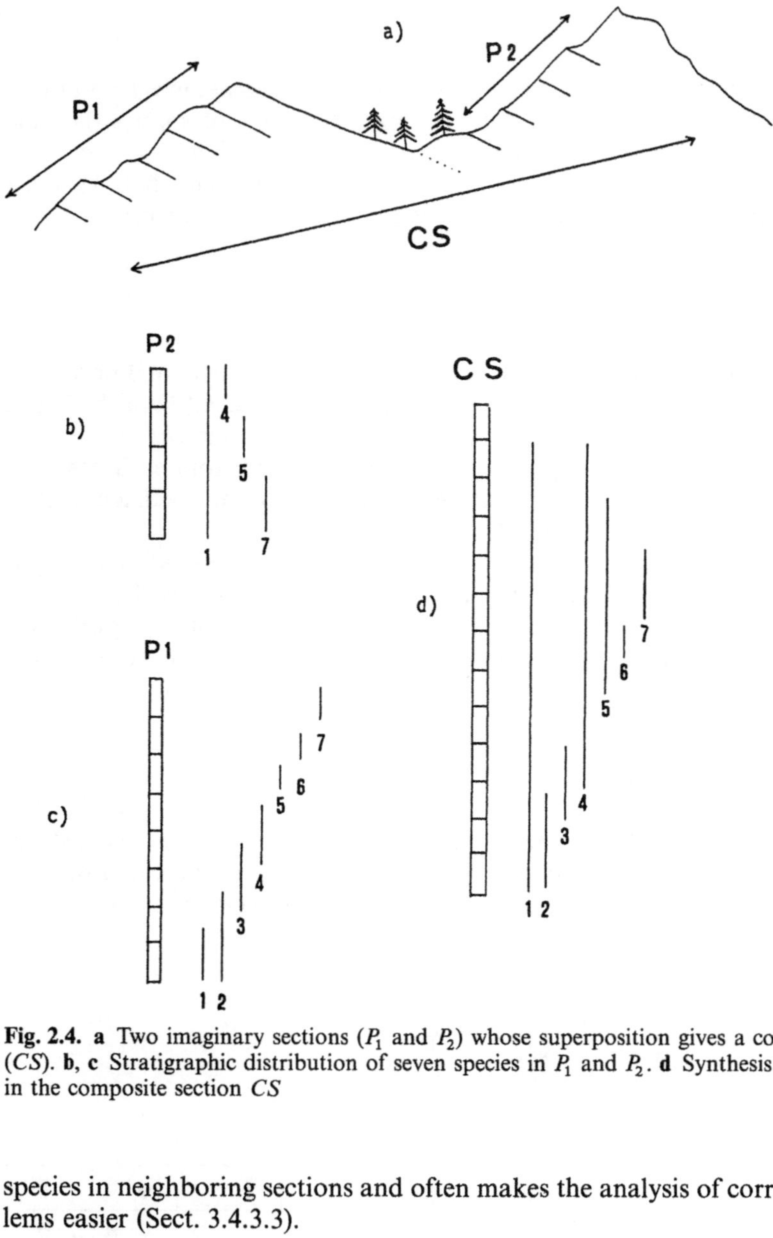

Fig. 2.4. **a** Two imaginary sections (P_1 and P_2) whose superposition gives a composite section (CS). **b, c** Stratigraphic distribution of seven species in P_1 and P_2. **d** Synthesis of these ranges in the composite section CS

species in neighboring sections and often makes the analysis of correlation problems easier (Sect. 3.4.3.3).

2.2.11 Numerical Evaluation of Reproducibility and Superpositional Control

To evaluate numerically the reproducibility (R) and superpositional control (SC) of a protoreferential's subdivisions, we will use the following coefficients:

$$SC_j = SC_j^+ + SC_j^-$$

where SC_j^+ is the number of adjacent subdivisions s_k, identified and located above a subdivision s_j in the reproducibility matrix associated to the protoreferential, and SC_j^- is the number of adjacent subdivisions s_i that are identified and located below the subdivision s_j identified in this reproducibility matrix, and

$$R_j = N_{id}^2/N_{id} + N_{pot}$$

where N_{id} is the number of sections in which a subdivision s_j is identified, and N_{pot} is the number of sections in which it is potentially identifiable (i.e., intercalated between two identified subdivisions but not sampled).
An example is given in Section 2.4.3.

2.3 Method

The first goal of the semiempirical (i.e., manual) method is to find sets of mutually compatible species within each neighborhood.

We start by establishing the neighborhood of each species, based on biostratigraphic observations. The results are entered in a species-species double-entry table.
The numbers corresponding to the various species are placed in the squares along the principal diagonal of this table: these squares are called "name boxes". The squares that are above and below this diagonal and which correspond to pairs of compatible species are marked in black. The others are left empty (Fig. 2.5).

Once this neighborhood table is complete, we permute its rows so that the black squares above the name boxes appear consecutively, with no blanks between them. (The arrangement of the black squares below the name boxes is of no consequence). Next, if such a permutation is possible (we will see later that it is far from always being the case), we move the columns so that the name boxes are again placed along the principal diagonal. This purely mechanical operation reestablishes the table's symmetry.

Next we eliminate the black squares located below the diagonal of the permuted table (this does not destroy any information since the table is symmetrical). The result is that each row corresponds to a group of mutually compatible species (the mathematical explanation for this will be given in Sect. 3.5.2).

The last step is to retain only the rows containing a maximal set (i.e., a set that is not contained within a larger set). If two maximal sets are identical, only one is retained. The remaining sets are unitary associations.

This procedure is illustrated in Fig. 2.5 and an example is studied in Section 2.4.

Once the set of unitary associations has been determined, it is easy to order them by examining the stratigraphic relationships observed locally among their characteristic species and pairs of species. The result is the desired protoreferential. All that remains is to establish the desired biochronologic correlations by identifying the unitary associations in the sections.

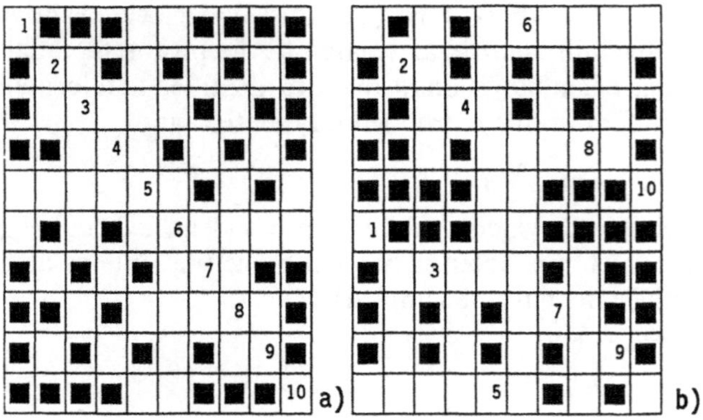

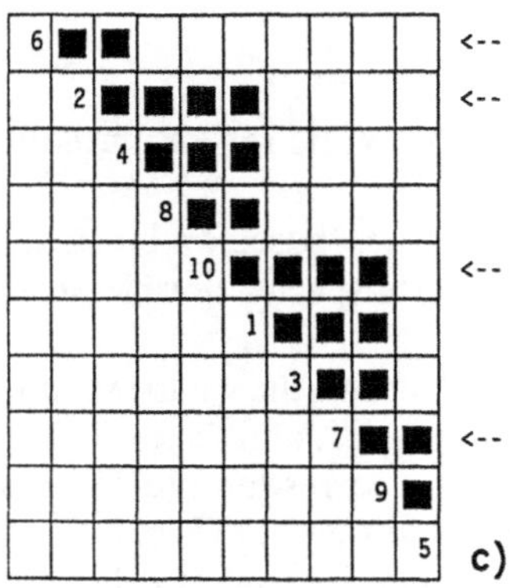

Fig. 2.5. **a** Neighborhood table of 10 species whose local ranges were given in Fig. 1.3. The *black squares* located at the coordinates (2,4) and (4,2) mean that species 2 and 4 are compatible, as one can observe in the section P_4 etc. **b** Permutation of the rows resulting in a consecutive arrangement of *black squares* (compatible pairs) above the *name boxes* (NB: more than one different permutation gives the same result). **c** Permutation of columns reestablishing the table's symmetry; the sets marked by *arrows* are maximal: (6,2,4), (2,4,8,10,1), (10,1,3,7,9) and (7,9,5). **d** Protoreferential. *Column U.A.*: unitary associations. **e** Position of the characteristic elements of the unitary associations of the protoreferential in the different levels (*L*) of sections P_1–P_4

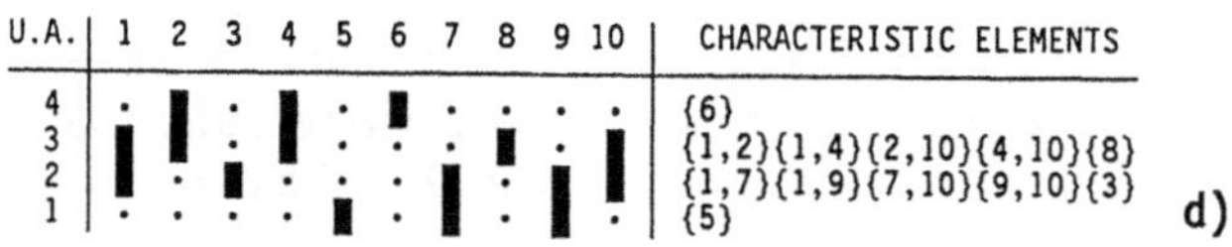

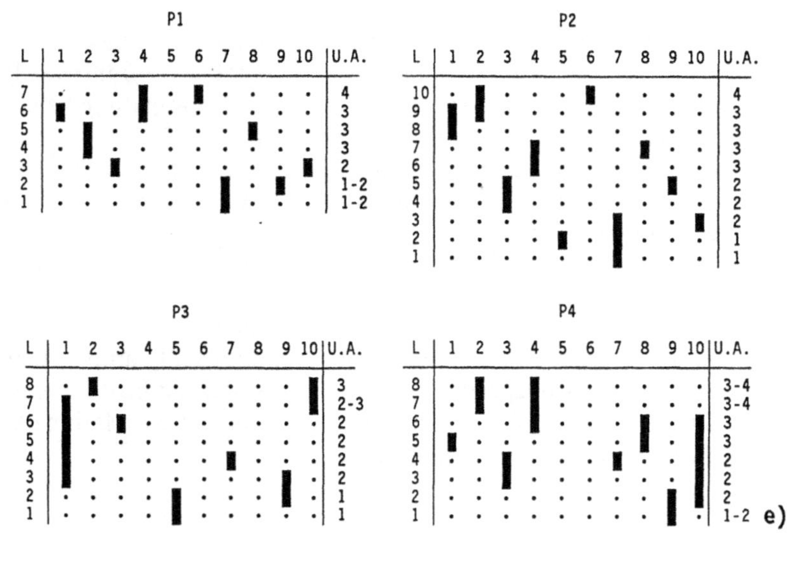

2.4 Example of Application

2.4.1 The Cambrian of the Riley Formation

As an example, we will analyze some of the biostratigraphic data established by Palmer (1954) in eight sections of the Riley Formation in the upper Cambrian of Texas. In this work, made famous in Shaw's *Time in Stratigraphy* (1964), Palmer did not give detailed lithologic descriptions of the sections. Instead he gave very precise information about the vertical ranges of the species he collected.

The local vertical ranges of the 21 most frequent species in these 8 sections are given in Fig. 2.6.

2.4.2 Constructing Unitary Associations

2.4.2.1 Permutation of the Matrix

As described above, we begin by establishing the neighborhood of each species. The result is entered in the table below (Fig. 2.7.a).

By permuting the rows of the table, we obtain a consecutive arrangement of squares with black circles above the name boxes (Fig. 2.7.b). After reestablishing the table's symmetry (Fig. 2.7.c), we eliminate the black squares below the diagonal (Fig. 2.7.d). Now each remaining row corresponds to a set of mutually compatible species. The unitary associations are obtained by retaining only those rows containing a maximal set (Fig. 2.7.d).

To determine the polarity of the sequence of unitary associations and to determine the superpositions of these associations, we simply refer to the original biostratigraphic data (Fig. 2.6) and introduce into the final table (Fig. 2.7.d) information about the local superpositions of the species (denoted $<$ in the figure).

A protoreferential (Fig. 2.8.a) can then be constructed directly from the results obtained in Fig. 2.7.d.

2.4.2.2 Remarks

It is easy to demonstrate that in this particular case, all the possible permutations resulting in a consecutive arrangement of the black squares above the name boxes create an identical order on the set of unitary associations. In addition, except for polarity, which remains unknown, this order satisfies exactly all the stratigraphic superpositional relationships observed locally among the species.

2.4.3 Identifying Unitary Associations

By definition, each unitary association is characterized by one or more species and/or by one or more pair(s) of species (Fig. 2.8.a). We now make the first

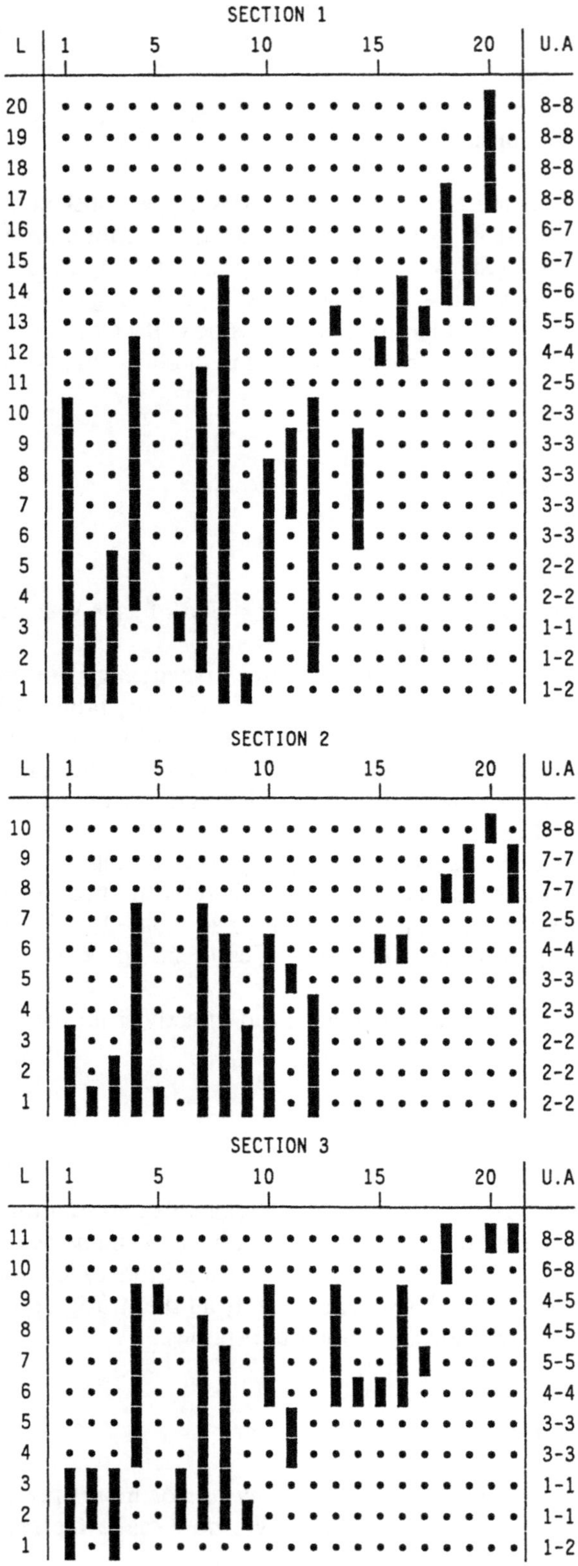

Fig. 2.6. Stratigraphic distribution of 21 species (1–21) in sections 1–8 of Palmer (1954). *L*, fossil-bearing beds. *U.A.*, unitary associations identified in the sections (see Fig. 2.8.a). For names of *numbered* sections, beds and species, see Appendix A.1

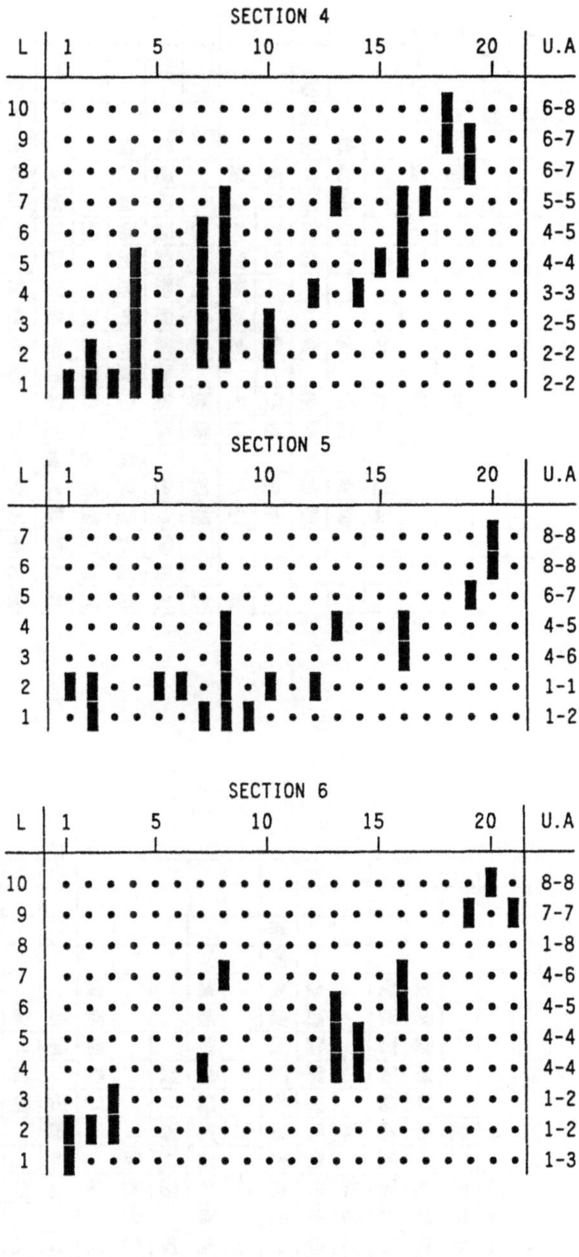

correlation by identifying these associations in the stratigraphic sections. The result is given in column U. A. of Fig. 2.6.

The chronologic significance of each of the eight unitary associations making up our referential depends on its superpositional control and reproducibility.

We begin then by entering in a reproducibility matrix information on the superpositions of the unitary associations identified in the different sections (Fig. 2.6; column U. A.). This is shown in Fig. 2.8.b, which also shows the values

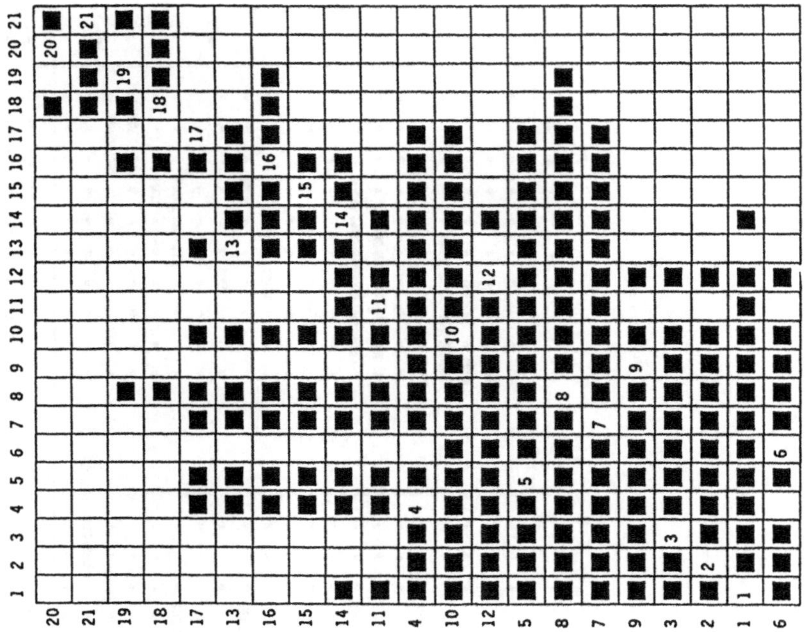

1 2 3 4 5 6 7 8 9 10 11 12 13 14 15 16 17 18 19 20 21
20 21 19 18 17 13 16 15 14 11 4 10 12 5 8 7 9 3 2 1 6

b)

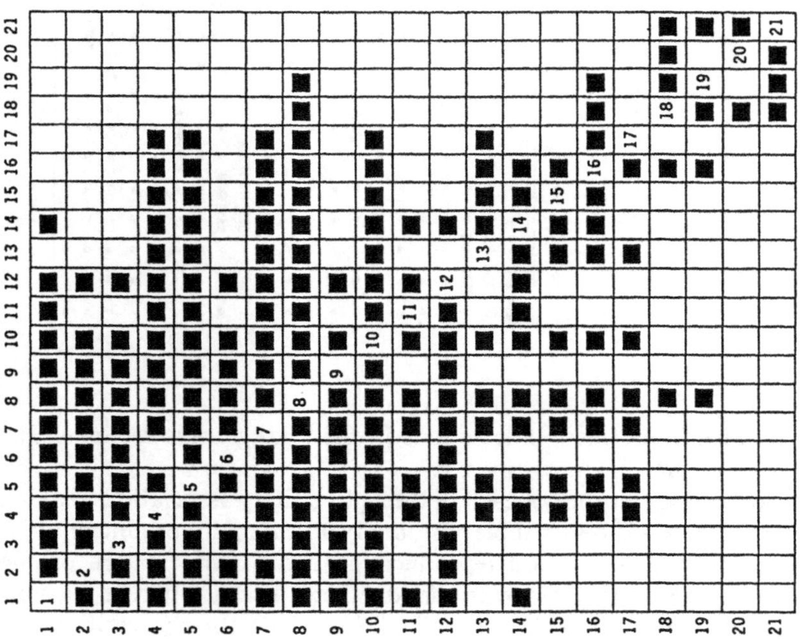

1 2 3 4 5 6 7 8 9 10 11 12 13 14 15 16 17 18 19 20 21
1 2 3 4 5 6 7 8 9 10 11 12 13 14 15 16 17 18 19 20 21

a)

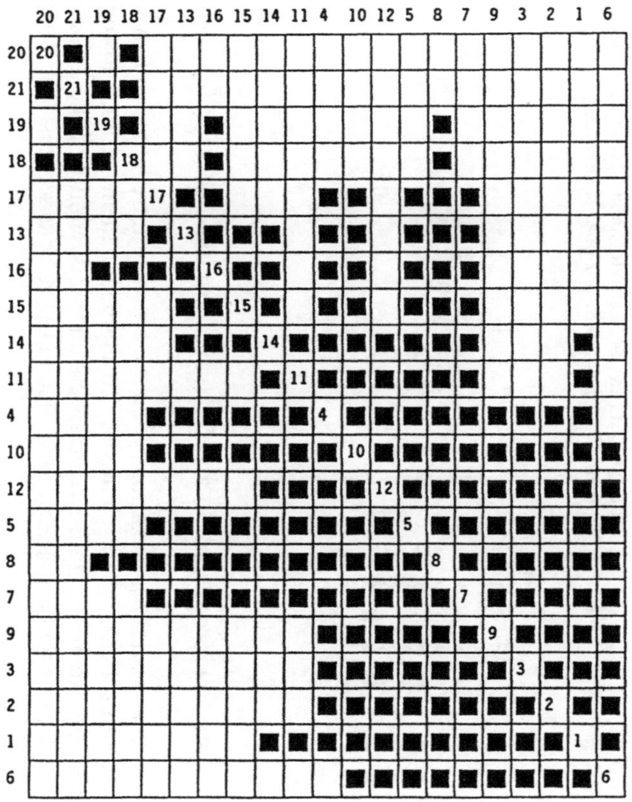

c)

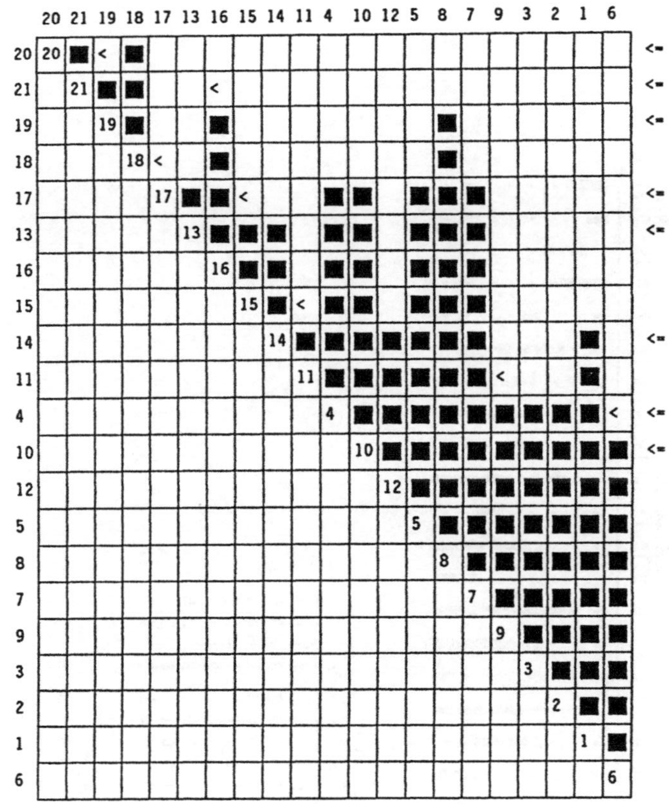

d)

Fig. 2.7. **a** Neighborhoods of species 1–21. **b** Permutation of the rows. **c** Permutation of the columns. **d** Reduced table showing only sets of compatible species: the *arrows* indicate the maximal sets and the *symbol* < gives the superpositional order among the species (for example, species 19 is below species 20); [Guex 1977 (completed)]

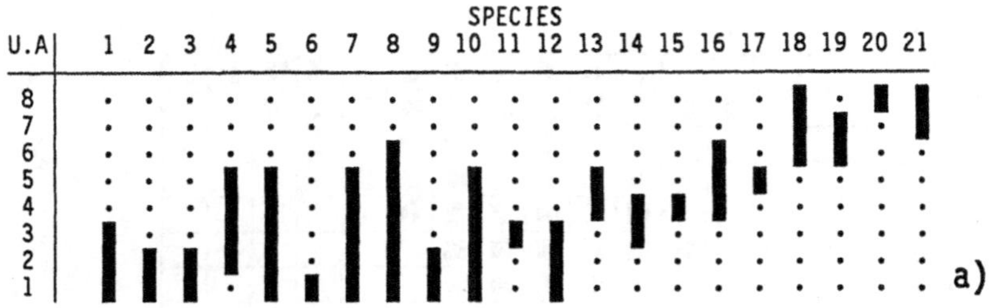

U.A	SECTION 1	2	3	4	5	6	7	8	S.C	R
8	1	1	1	0	1	1	0	1	2	6.0
7	0	1	0	0	0	1	0	0	2	0.8
6	1	0	0	0	0	0	0	0	1	0.1
5	1	0	1	1	0	0	1	0	4	2.0
4	1	1	1	1	0	1	0	1	8	4.5
3	1	1	1	1	0	0	0	1	9	3.6
2	1	1	0	1	0	0	1	1	6	3.6
1	1	0	1	0	1	0	0	1	2	4.0

b)

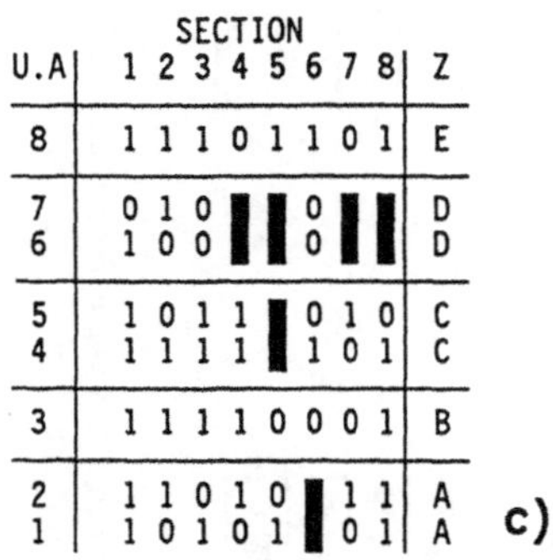

Fig. 2.8. a Protoreferential constructed from Fig. 2.7.d. **b** Reproducibility matrix. *SC*, superpositional control coefficient. *R*, reproducibility. **c** Completed reproducibility matrix. *Black rectangles*: unions of unitary associations that have been strictly identified in the sections. *Column Z*: zones *A–E*. **d** Zonal referential

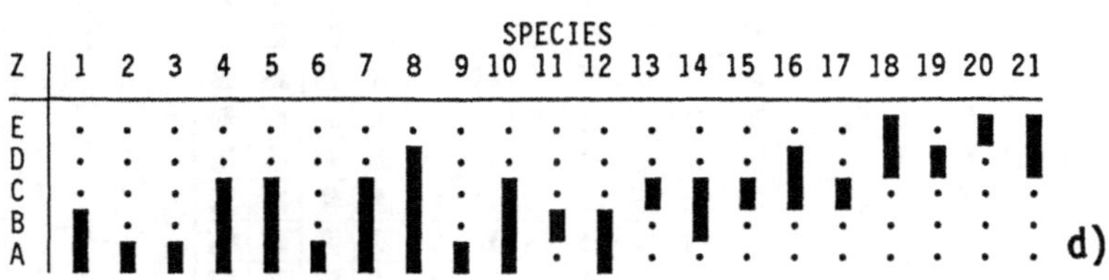

of the coefficients *R* (reproducibility) and *SC* (superpositional control) for each U. A.

We can see that unitary associations 6 and 7 are very unreliable, and that unitary associations 1 and 5 are mediocre. To eliminate them without losing useful information, we now examine the respective reproducibility of the unions of unitary associations that are strictly identified in the various sections (Fig. 2.6; column U. A:).

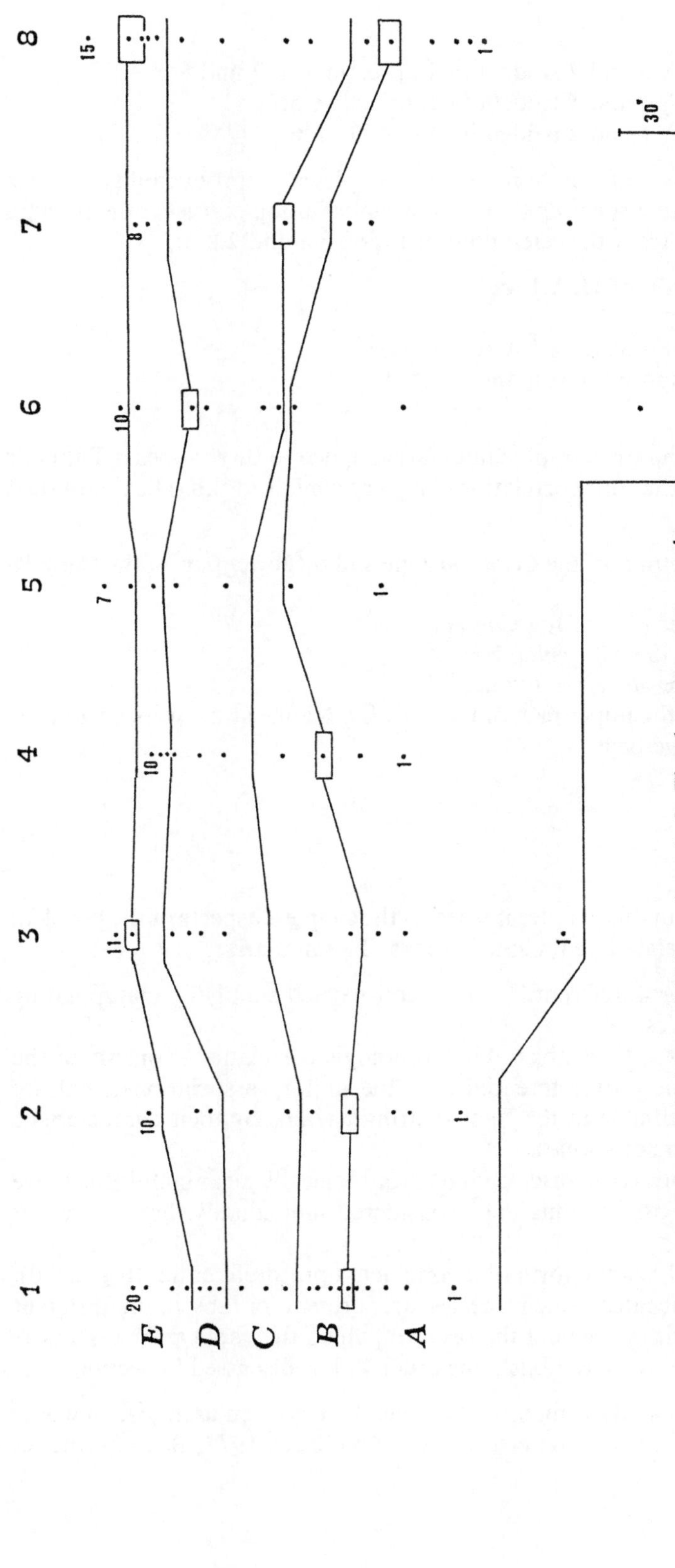

Fig. 2.9. Correlation of Palmer's (1954) sections 1–8. The dots represent samples whose fossil content is given in Fig. 2.6. *A* through *E* are the zones established in Fig. 2.8. Rectangles indicate samples that cannot be assigned precisely to a particular zone

We note that:

1. The union of U. A.6 and 7 is identified in sections 4, 7 and 8;
2. The union of U. A.4 and 5 is identified in section 5;
3. The union of U. A.1 and 2 is identified in section 6.

Transcribing this information in a completed reproducibility matrix (Fig. 2.8.c) immediately suggests that an optimal solution consists of extracting five biochronozones from the referential of Figs.2.8.a and 2.8.b:

1. Zone A = the union of U. A.1 and 2;
2. Zone B = U. A.3;
3. Zone C = the union of U. A.4 and 5;
4. Zone D = the union of U. A.6 and 7;
5. Zone E = U. A.8.

Figure 2.9 gives the stratigraphic intervals assigned to these zones in Palmer's eight sections. Note that our correlations are very similar to those he established empirically:

1. *A* covers the upper part of the *Cedarina* zone and the lower part of the *Coosella* zone;
2. *B* covers the upper part of the *Coosella* zone;
3. *C* corresponds to the *Maryvillia* zone;
4. *D* covers the lower *Aphelaspis* zone;
5. *E* corresponds to the upper part of the *Aphelaspis* zone and covers part of the post-*Aphelaspis* age beds.

2.5 Summary

The example studied in this chapter illustrates the simplest aspects of the problem of using fossils to correlate sedimentary rocks. To summarize:

1. In some cases, a protoreferential can be constructed simply by manipulating neighborhood tables.
2. The framework used to establish biochronologic correlations consists of the subdivisions of this protoreferential (i.e., the unitary associations) that are reproducibly identifiable in the fossil-bearing terrains by their characteristic species and/or pairs of species.
3. Most biochronozones (i.e., discrete and chronologically meaningful zones) are unions of unitary associations that, considered individually, are not reproducible.
4. A zonal referential has the form of a compacted protoreferential (Fig. 2.8.d). Such a table delineates which species are present or absent in different biochronozones without giving the order in which they appear or disappear within the zones. How to establish this order will be discussed in Section 11.8.

The semiempirical analytic method described here is often useful when working with a small number of species (see for example Guex 1977; Baumgartner et al. 1980, or Blieck 1984), but most of the time it cannot be applied in a strict way. Because of the numerous discontinuities in the distribution of fossil species, it is generally not possible to obtain a permutation of the neighborhood table according to the rule described above.

Two examples of the frequently encountered problem of conflicting ranges (Fig. 2.10) demonstrate that the semiempirical approach often is not practicable.

To deal with our biochronologic problem in a general way, we must study the structure of biostratigraphic data in greater detail and employ a certain amount of mathematics.

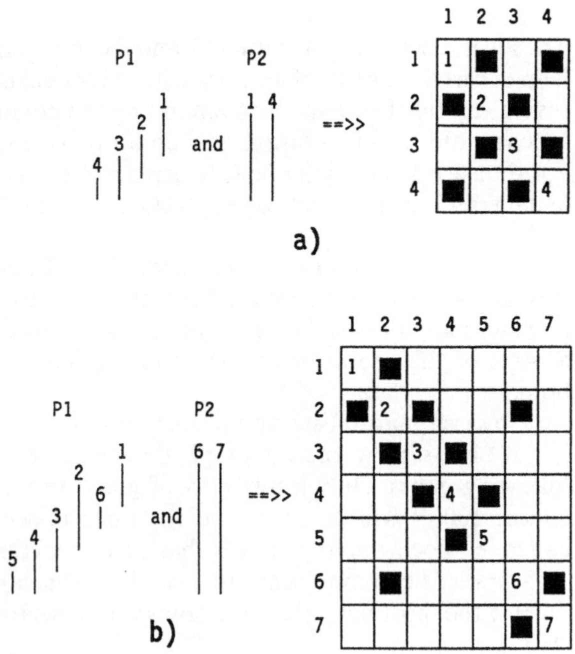

Fig. 2.10 a, b. Two examples of conflicting stratigraphic ranges that cannot be resolved by the manual (i.e., semiempirical) method

Chapter 3
Biochronologic Referentials and Graph Theory

3.1 Representing Stratigraphic Relationships

The ideas discussed in Chapters 1 and 2 can be expressed more precisely with the help of certain aspects of graph theory. This field of applied mathematics provides several useful theorems for overcoming the concrete difficulties so often met in biochronology. In addition, its notation is convenient in describing the algorithms we will use later both to explain the structure of difficult biostratigraphic data, and to construct unitary associations and identify them in fossil-bearing beds.

The basic concepts of graph theory have been made comprehensible to non-specialists in several recent works: Harary (1969), Carre (1979), Golumbic (1980) and especially Roberts (1976, 1978). Stratigraphers interested in the mathematical aspects of the problems discussed in the following chapters can refer to these books.

However, naturalists who dislike formal mathematics can be reassured that it is not necessary to master graph theory in order to understand its use in the following pages. Only a tiny part of graph terminology (about 20 terms out of almost 1000!) will be used, and the main concepts can easily be explained in concrete, biochronologic terms: biochronologic problems have more to do with a naturalist's common sense than with mathematics.

For the most part, the terminology and notation we use are those of Roberts (1976).

In introducing the technical definitions that follow, we once more appeal to the paleontologist's intuition. Imagine that the species whose relationships are being analyzed are represented by a set of points:

1. the stratigraphic superpositions observed among these species can be represented by arrows joining the points. These arrows are, by convention, oriented in the sense $x \rightarrow y$ when species x has been observed below species y;
2. coexistences (observed or virtual) among the species are represented by lines joining the points y and z and indicating that y and z are compatible.

The resulting diagram is called a graph.

3.2 Definitions

3.2.1 The Coexistence Graph G

3.2.1.1 Nonoriented Graph

A nonoriented graph $G = (X, E)$ is composed of a finite set of points $X = \{x_1, x_2, \ldots, x_n\}$ and a family $E = \{e_1, e_2, \ldots, e_m\}$ of unordered pairs of distinct points of X.

Each pair $e = \{x, y\}$ of E is called an edge of the graph, and the points of X are called the vertices of the graph.

The order of the graph is the cardinality of X, i.e., the number of points of X. It is denoted $|X|$.

3.2.1.2 Compatibility and Neighborhood

The edges of G (i.e., the elements of E) represent compatible species. Edges that represent truly associated pairs of species (i.e observed coexistences) are called "real" edges; those that represent virtually associated pairs of species are called "virtual" edges.

Two vertices connected by an edge are neighbors in G. The set of neighbors of a vertex x is denoted $\Gamma(x)$: this set corresponds to the neighborhood of the species x. It contains x. Occasionally we will use the set $\Gamma^*(x) = \Gamma(x) - \{x\}$.

The graph G in Fig. 3.1 represents the coexistences of species 1–8, whose distribution was shown in Fig. 1.7. Such a graph is called a "coexistence graph".

The neighborhood of species 2 (i.e., the set of neighbors of vertex 2 in G) is denoted $\Gamma(2) = \{1, 2, 3, 4\}$.

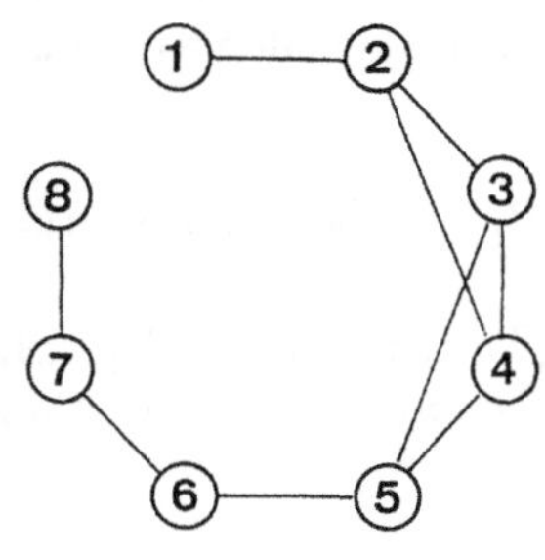

Fig. 3.1. Nonoriented graph G representing the coexistences of the eight species whose stratigraphic distribution is given in Fig. 1.7

3.2.2 The Adjacency Matrix of G

Let $G = (X, E)$ be a nonoriented graph. The adjacency matrix $A = (a_{ij})$ of G is defined by

$$a_{ij} = \begin{cases} 1 \text{ if } x_i \text{ and } x_j \text{ are compatible,} \\ 0 \text{ otherwise.} \end{cases}$$

i \ j	1	2	3	4	5	6	7	8
1	1	1	0	0	0	0	0	0
2	1	1	1	1	0	0	0	0
3	0	1	1	1	1	0	0	0
4	0	1	1	1	1	0	0	0
5	0	0	1	1	1	1	0	0
6	0	0	0	0	1	1	1	0
7	0	0	0	0	0	1	1	1
8	0	0	0	0	0	0	1	1

Fig. 3.2. Adjacency matrix of the graph G of Fig. 3.1. i and j, rows and columns of the matrix

This matrix is equivalent to the neighborhood table discussed in Section 2.3. Figure 3.2 is the adjacency matrix of the graph G of Fig. 3.1.

3.2.3 The Superposition Graph G_s

An oriented graph $G_s = (X, U)$ is a finite set $X = \{x_1, x_2, \ldots, x_n\}$ of vertices and a family $U = \{u_1, u_2, \ldots, u_m\}$ of ordered pairs of distinct vertices of X. Each such pair $u = (x, y)$ is called an "arc" of the graph.

Two incompatible species whose superposition is known form an arc of G_s; by convention the arc (x, y) means that species x lies beneath species y.

The species that have been observed above species x in stratigraphic sections form the set of successors of x: this set is denoted $\Gamma^+(x)$. Similarly, the species that have been observed beneath x form the set of predecessors of x; this set is denoted $\Gamma^-(x)$.

The graph in Fig. 3.3 represents the superpositions of species 1–8 in Fig. 1.7; $\Gamma^-(8) = \{1, 2, 3, 4, 5, 6\}$ is the set of predecessors of species 8 (i.e., vertex 8) and $\Gamma^+(1) = \{3, 4, 5, 6, 7, 8\}$ is the set of successors of species 1.

Such a graph is called a "superposition" graph.

We will use the words "pair" and "couple" interchangeably, unlike some specialists in graph theory for whom pairs are unordered and couples are ordered.

3.2.4 The Biostratigraphic Graph G^*

The semioriented graph $G^* = (X, E + U)$, which combines the edges (or arcs) of G and G_s, is called a biostratigraphic graph. It contains all available information

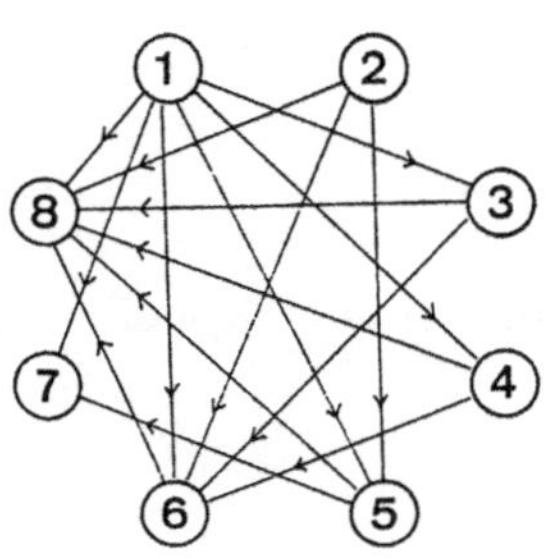

Fig. 3.3. Oriented graph G_s representing the superpositions of the eight species in Fig. 1.7

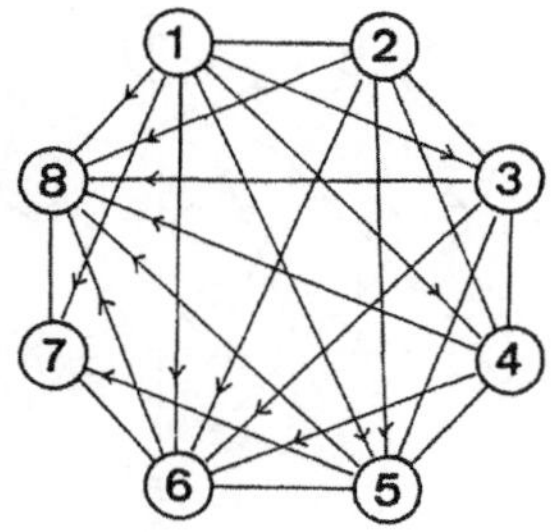

Fig. 3.4. Biostratigraphic graph G^* representing the coexistences and superpositions of the eight species of Fig. 1.7

about the stratigraphic relationships of species: neighborhoods (i.e coexistences) and exclusions with known or unknown stratigraphic relationships.

Those pairs of vertices not connected by an edge or an arc in G^* represent the pairs of species whose stratigraphic relationships are unknown. Fig. 3.4 is the biostratigraphic graph of species 1–8 of Fig. 1.7.

3.2.5 The Adjacency Matrix of G^*

Let $X = \{x_1, \ldots, x_n\}$ be a set of fossil species. The stratigraphic relationships observed among the species of X can also be expressed using a matrix $A = (a_{ij})$ as follows:

$$a_{ij} = \begin{cases} 1 \text{ if } x_i \text{ is in the same level or a lower level than } x_j \\ 0 \text{ otherwise.} \end{cases}$$

This matrix A is the "adjacency matrix" of the biostratigraphic graph G^*; conversely, G^* is the graph associated to A.

The adjacency matrix of the graph in Fig. 3.4 is given in Fig. 3.5

3.2.6 The Complementary Graphs

3.2.6.1 The Graph G^c

If G is a nonoriented graph, then the "complementary graph" G^c of G is the graph whose vertices are the same as the vertices of G, but where an edge connects two vertices precisely if no edge connects the vertices in G.

The complementary graph of the graph of Fig. 3.1 is drawn in Fig. 3.6.a.

i \ j	1	2	3	4	5	6	7	8
1	1	1	1	1	1	1	1	1
2	1	1	1	1	1	1	0	1
3	0	1	1	1	1	1	0	1
4	0	1	1	1	1	1	0	1
5	0	0	1	1	1	1	1	1
6	0	0	0	0	1	1	1	1
7	0	0	0	0	0	1	1	1
8	0	0	0	0	0	0	1	1

Fig. 3.5. Adjacency matrix of the graph G of Fig. 3.4. i and j, rows and columns of the matrix

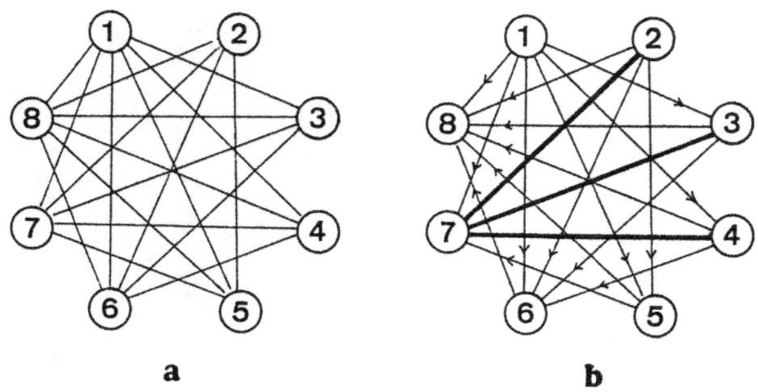

Fig. 3.6. a Graph G^c, complement of the graph G^* of Fig. 3.1. **b** Semioriented graph $G^{c\prime}$, representing the exclusions (superpositions and undetermined relationships) among the eight species of Fig. 1.7

3.2.6.2 The graph $G^{c\prime}$

In our biochronologic problem, the edges of G^c represent any exclusions of species (i.e., superpositions and undetermined stratigraphic relationships). In practice then, we must admit that some of those edges are already oriented: these orientations represent pairs of stratigraphically superposed species.

For reasons that will become clear in Section 3.4.3, we define a semioriented graph $G^{c\prime}$ whose edges represent undetermined stratigraphic relationships and whose arcs represent superpositions given in the graph G_s.

As an example, let us reexamine the coexistences and superpositions of the species whose distribution is described in Fig. 1.7:

1. The coexistences are represented by the graph G of Fig. 3.1; the complement of this graph is drawn in Fig. 3.6.a;
2. the superpositions are represented by the graph G_s of Fig. 3.3: the arcs of this graph induce orientations of the corresponding edges of G^c. By copying these orientations we obtain the semioriented graph $G^{c\prime}$ drawn in Fig. 3.6.b.

3.3 Descriptive and Technical Terms: Definitions

3.3.1 Chain, Circuit and Chord

In a nonoriented graph G, a "chain" is an alternating sequence of distinct vertices and edges beginning and ending in a vertex (Fig. 3.7.a). The length of the chain is the number of its vertices.

A "circuit" is a closed chain (Fig. 3.7.b). A "chord" is an edge connecting two nonconsecutive vertices of a circuit (Fig. 3.7.c). A circuit of length n admitting no chords is denoted Z_n.

3.3.2 Path, Maximal Path, Cycle and Strongly Connected Graph

A "path" in an oriented graph $G_s = (X, U)$ is a sequence of arcs $(u_1, u_2, \ldots, u_n)$, such that the end vertex of each arc coincides with the beginning vertex of the next

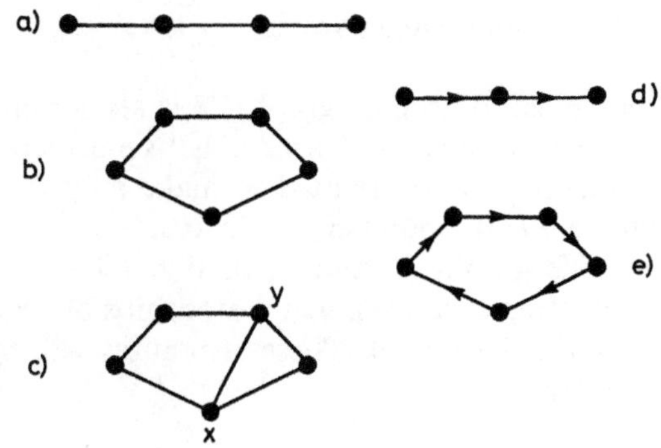

Fig. 3.7. **a** Chain of length 4. **b** Z_5 circuit. **c** Chord: the edge $\{x, y\}$ is a chord of the Z_5 circuit. **d** Path of length 3. **e** C_5 cycle

arc in the sequence. A path is maximal if it is not a subpath of any longer path.

A C_n "cycle" is a closed path, i.e., the endpoint of its last arc coincides with the origin of its first arc (Fig. 3.7.e).

An oriented graph $G_s = (X, U)$ is strongly connected if, for every pair of vertices x and y, there is a path from x to y and a path from y to x.

3.3.3 Subgraphs, Generated Subgraphs and Strong Components

Let $H = (X, F)$ be a graph. A graph $H' = (X', F')$ is a "subgraph" of $H = (X, F)$ if X' is a subset of X and F' is a subset of F.

If H' contains all the edges (resp. arcs) connecting vertices of X', then H' is a "generated subgraph" of H.

The graph H' in Fig. 3.8 is a subgraph of H, and H'' is a generated subgraph of H. Note also that the graphs G and G_s defined in Section 3.2 are subgraphs of our biostratigraphic graph G^*.

A strongly connected subgraph of an oriented graph G_s is called a "strong component" of G_s.

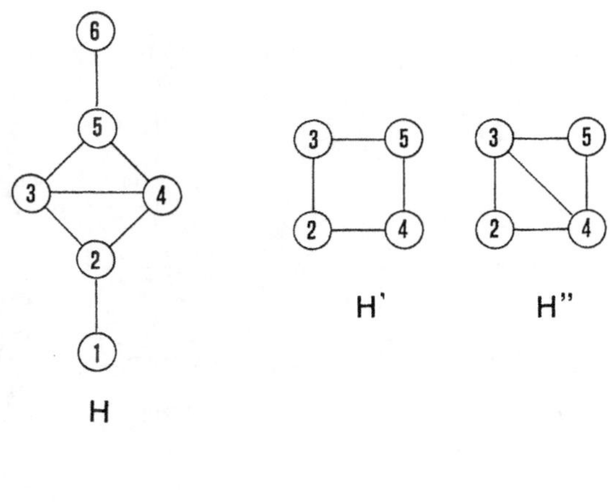

Fig. 3.8. H' is a subgraph of H and H'' is a generated subgraph of H

3.3.4 Semioriented Circuit

Our biostratigraphic graph G^* is semioriented; similarly we will say that a sequence of edges and arcs is a "semioriented circuit" if all the edges in that sequence can be oriented to make a cycle. Such structures will be denoted S_n, where n is the number of vertices.

The graphs S_3 and S_4 in Fig. 3.9 are semioriented circuits: they represent conflicting stratigraphic relationships of species 1, 2 and 3 (for S_3) and species 4, 5, 6 and 7 (for S_4). These structures will be studied in detail in the following chapters.

3.3.5 Orientation of a Graph

3.3.5.1 Transitive Orientation

An oriented graph $G_s = (X, U)$ is "transitive" if whenever $(x, y) \in U$ and $(y, z) \in U$, then $(x, z) \in U$.

Graphs G_1 and G_3 of Fig. 3.10 are transitive, but graphs G_4 and G_5 are not: the arc (x, z) is missing.

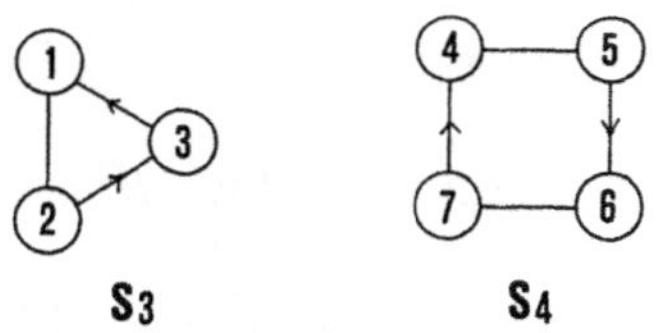

Fig. 3.9. Semioriented S_3 and S_4 circuits

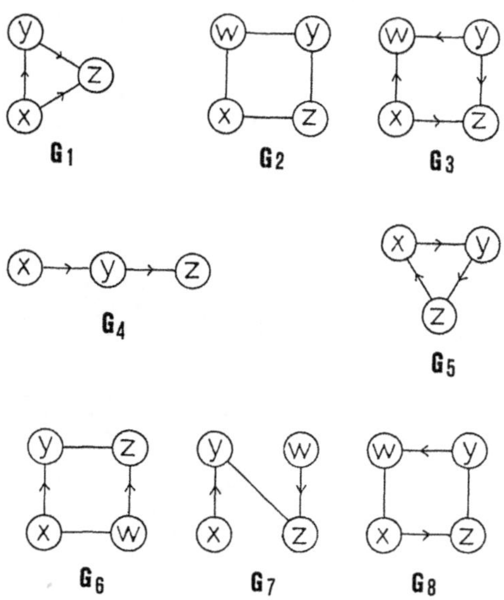

Fig. 3.10. Examples of transitive graphs (G_1 and G_3), nontransitive graphs (G_4 and G_5), transitively orientable graphs (G_2 and G_8) and nontransitively orientable graphs (G_6 and G_7)

3.3.5.2 Transitively Orientable Graph

A nonoriented graph is transitively orientable if and only if an orientation can be chosen for all the edges so that the resulting oriented graph is transitively oriented.

Graph G_2 of Fig. 3.10 is transitively orientable: G_3 is a transitive orientation of G_2.

The properties of transitively orientable graphs are described in detail in Roberts (1976) and Golumbic (1980). The main classical results concerning these graphs are due to Ghouila Houri (1962) and Gallai (1967).

3.3.5.3 Orientability of a Semioriented Graph

A semioriented graph is transitively orientable if its partial orientation can be extended to an orientation making the resulting oriented graph transitively oriented.

3.3.5.4 Examples

Four examples of transitively orientable and nontransitively orientable graphs are given in Figs. 3.6 and 3.10.

1. The edge $\{x, z\}$ of graph G_6 (Fig. 3.10) is lacking; if the edge $\{y, z\}$ of G_6 is oriented in the direction $y \to z$, the graph is not transitive. Since the edge $\{y, w\}$ is also lacking, the opposite orientation of $\{y, z\}$ also fails to make G_6 transitive.
2. As above, the edges $\{x, z\}$ and $\{y, w\}$ of graph G_7 (Fig. 3.10) are both lacking; neither orientation of $\{y, z\}$ makes G_7 transitive.
3. Graph G_8 of Fig. 3.10 is transitively orientable: G_3 is a transitive orientation of G_8.
4. If the edges $\{2, 7\}$, $\{3, 7\}$ and $\{4, 7\}$ are given the orientations (2, 7), (3, 7) and (4, 7), the graph $G^{c\prime}$ of Fig. 3.6.b becomes transitively oriented.

3.3.6 Clique, Maximal Clique, and Unitary Association

A "clique" of G is a collection of vertices, each of which is a neighbor of every other. A clique is "maximal" if it is contained in no larger clique. Maximal cliques will be denoted k.

The graph of Fig. 3.11 is made up of two maximal cliques: $\{1, 2, 3, 4\}$ and $\{1, 2, 3, 5\}$. The three vertices $\{1, 2, 3\}$ form a clique which is not maximal, since it is contained in both maximal cliques above.

Strictly speaking, a maximal clique has only real edges. A "unitary association", however, is a maximal clique of which some edges may be virtual. We will use the same notation for maximal cliques and for unitary associations; the reason for this will become clear in Section 3.5.

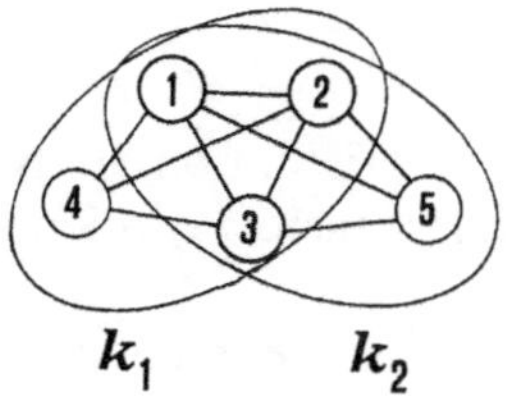

Fig. 3.11. Two maximal cliques (k_1 and k_2)

S6	0	0	0	0	0	0	1	1
S5	0	0	0	0	0	1	1	0
S4	0	0	0	0	1	1	0	0
S3	0	0	1	1	1	0	0	0
S2	0	1	1	1	0	0	0	0
S1	1	1	0	0	0	0	0	0
X =	1	2	3	4	5	6	7	8

Fig. 3.12. Subset-vertex incidence matrix. The sets $S_1 - S_6$ represent the maximal cliques of the graph G of Fig. 3.1, and X represents the set of the eight vertices of this graph. This matrix corresponds precisely to the protoreferential of Fig. 1.8

3.3.7 Incidence Matrix and Maximal Clique Matrix

Let $X = \{x_1, x_2, \ldots, x_n\}$ be a set of n vertices and $S = \{s_1, \ldots, s_m\}$ be a family of m subsets of X. The "subset-vertex incidence matrix" is the m by n matrix (m rows and n columns) whose $(i, j)^{\text{th}}$ entry (i^{th} row, j^{th} column) is 1 if vertex x_i belongs to subset s_j and 0 otherwise.

The subset-vertex incidence matrix in which S represents the maximal cliques of a graph G and X represents the vertices of G is called the "maximal clique matrix" of G.

Fig. 3.12 is a subset-vertex incidence matrix where the subsets $S_1, \ldots, S_6$ are precisely the maximal cliques of the graph G in Fig. 3.1. This matrix is the maximal clique matrix of G.

3.3.8 Consecutive 1's Property

A matrix composed of 1's and 0's has the "consecutive 1's property" if its rows can be permuted so that the 1's in each column have no 0's between them. The matrix in Fig. 3.12 has this property.

3.3.9 Triangular Matrix

A matrix is upper (or lower) triangular if all the entries beneath (or above) its main diagonal are 0. The matrix in Fig. 3.13 is lower triangular.

3.3.10 Triangulated Graph

A nonoriented graph is "triangulated" if all its circuits of length greater than 3 have a chord (Figs. 3.7.c and 3.14).

i \ j	1	2	3	4	5	6	7	8
1	1	0	0	0	0	0	0	0
2	1	1	0	0	0	0	0	0
3	1	1	1	0	0	0	0	0
4	0	1	1	1	0	0	0	0
5	0	1	1	1	1	0	0	0
6	0	0	1	1	1	1	0	0
7	0	0	0	0	1	1	1	0
8	0	0	0	0	0	0	1	1

Fig. 3.13. Lower triangular matrix. *i* and *j*, rows and columns of the matrix

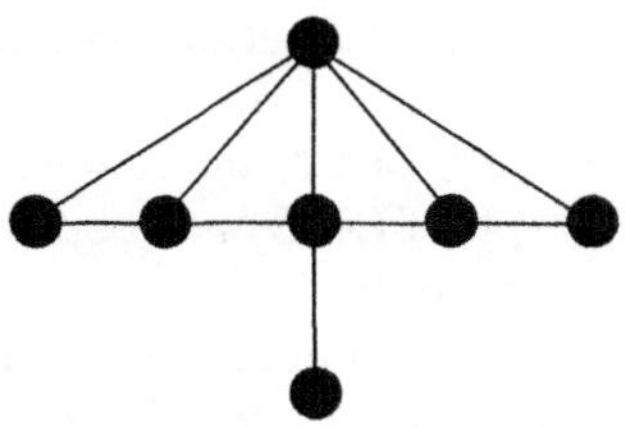

Fig. 3.14. Triangulated graph

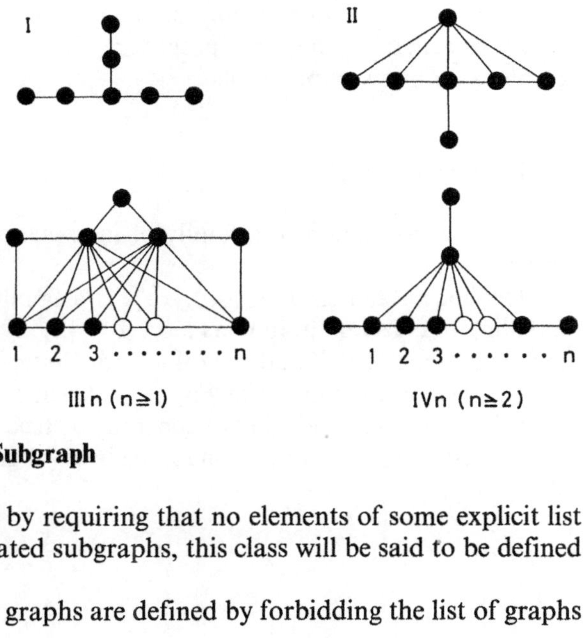

Fig. 3.15. Asteroidal triples of Lekkerkerker and Boland. *I–IV*, original notation used by Lekkerkerker and Boland (1962)

3.3.11 Forbidden Generated Subgraph

If a class of graphs is defined by requiring that no elements of some explicit list of subgraphs appear as generated subgraphs, this class will be said to be defined by "forbidden subgraphs".

For instance, triangulated graphs are defined by forbidding the list of graphs Z_n, $n > 3$.

3.3.12 Asteroidal Triple

Vertices x, y, z of a nonoriented graph G form an "asteroidal triple" if there exists a chain C_1 between x and y, a chain C_2 between x and z, and a chain C_3 between y and z, such that there is no edge between z and C_1, y and C_2 or x and C_3.

Figure 3.15 shows examples of asteroidal triples.

3.3.13 Interval

Technically, an "interval" is a line segment, which we will take to be closed, i.e., containing its own extremities.

In our problem, we will be concerned with the time intervals corresponding to the life span of species, already called "existence intervals". We should not expect an existence interval to be entirely recorded in a single section, or the record of an existence interval in different sections to be synchronous.

3.3.14 Intersection Graph.

Let S be a set, and $A = \{A_1, \ldots, A_n\}$ be a family of subsets of S. The intersection graph of (A, S) has one vertex x_i for each subset A_i, and an edge joining x_i to x_j if and only if $A_i \cap A_j \neq \emptyset$.

For instance, to Fig. 1.7 we can associate a set S consisting of four distinct time axes T_1, T_2, T_3, T_4, one for each section, and subsets $A_1, \ldots, A_8$ of S, $A_i \cap T_k$ being the observed (and partially recorded) existence interval of the ith species in the kth section. Figure 3.1 is the intersection graph of this family.

More generally, the same procedure allows us to interpret any coexistence graph as an intersection graph.

3.3.15 Interval Graph

The intersection graph of a family of intervals on a line is called an "interval graph".

This means that if $J = \{J_1, \ldots, J_n\}$ is a family of intervals, then the corresponding intersection graph has a set $X = \{x_1, \ldots, x_n\}$ of vertices, with an edge connecting x_i and x_j if and only if J_i and J_j intersect.

For instance, the graph in Fig. 3.1 is an interval graph. This is not obvious from the definition, and will not generally be true of coexistence graphs. However, it is easy to verify that this one is the interval graph associated to the intervals of Fig. 1.8.

3.4 Interval Graphs

3.4.1 Main Characterizations

The literature contains several characterizations of interval graphs. Three theorems are particularly adapted to linking graph theory with our biochronologic problems; they are due to Fulkerson and Gross (1965), Gilmore and Hoffman (1964) and Lekkerkerker and Boland (1962).

3.4.2 Fulkerson and Gross (1965)

3.4.2.1 Theorem

A graph G is an interval graph if and only if its maximal clique matrix has the consecutive 1's property.

3.4.2.2 Discussion

This statement concerns the relationship between protoreferentials and the maximal clique matrix of an interval graph (Figs. 3.1 and 3.12).

For example, the protoreferential given in Fig. 2.8.a is a maximal clique matrix with the consecutive 1's property, and its associated graph is an interval graph.

In a first approximation, constructing a protoreferential is the same problem as transforming a coexistence graph (representing the local intersections of partially recorded existence intervals of species), into an interval graph.

3.4.3 Gilmore and Hoffman (1964)

3.4.3.1 Theorem

A graph G is an interval graph if and only if (1) G has no Z_4 as generated subgraph; (2) the complementary graph G^c admits a transitive orientation.

3.4.3.2 Discussion

We first observe that the theorem characterizes interval graphs by forbidden generated subgraphs.

Clearly, a Z_4 circuit does not represent a sequence of intervals on a line (Fig. 3.16.a). More generally the reader will easily realize that a graph containing chordless Z_n circuits ($n > 4$) as generated subgraphs does not represent such a sequence.

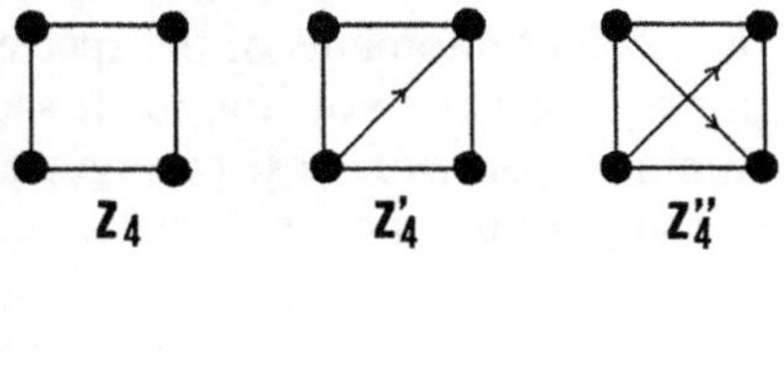

Fig. 3.16. **a** The Z_4 circuit represents intervals drawn on a circle. **b** Z_4, Z'_4, Z''_4. Three types of generated Z_4 subgraphs of biostratigraphic graphs

Figure 3.16.b shows three different kinds of circuits of length 4 that appear frequently as generated subgraphs in a biostratigraphic graph. The interpretation of such structures will be discussed in Section 4.2.

Property (2) of the theorem can be understood as follows: if two intervals of an oriented line do not intersect, then one lies above the other; this induces an orientation on the complementary graph that is clearly transitive. One can verify easily that the edges of a graph can be transitively oriented if and only if this graph has no nontransitively orientable generated subgraphs. In other words, the generated subgraphs of G^c that are not transitively orientable are (by definition) forbidden configurations in the complement of an interval graph.

The practical consequences of the second condition of this theorem are not evident and should be discussed in some detail.

We begin by recalling that in the problem studied here, the edges of G^c represent mutual exclusions of species. These exclusions have two possible origins:

1. The stratigraphic relationship between two species (x and y) is undetermined, for the species have never been found together in the same section.
2. x and y are only observed in stratigraphic sequence (they are never found in the same bed).

In the case of our particular problem, this means that some orientations of edges of G^c are imposed by the stratigraphic observations that must be interpreted. These orientations are those of the arcs of G_s (and of G^*) that represent observed superpositions of the species. It is this constraint that led us to define the semioriented graph $G^{c\prime}$ (Sect. 3.2.6.2), whose edges represent undetermined stratigraphic relationships and whose arcs represent superpositions.

In fact, our problem of transforming G into an interval graph is a double problem: not only must we determine whether G^c is transitively orientable but also (and most important) we must determine whether the (forced) orientations of $G^{c\prime}$ can be completed to make a transitive orientation.

Taking into account what we saw above, this means that we must also detect the generated subgraphs of $G^{c\prime}$ that are not transitively orientable (and of course those that are not transitive). These subgraphs are the forbidden configurations of our graph $G^{c\prime}$: they correspond precisely to the graphs representing conflicting (i.e.,cyclic) stratigraphic relationships of fossil species. The most common are the C_3 cycles and the semioriented S_3 and S_4 circuits of the biostratigraphic graph G^* (Fig. 3.17).

3.4.3.3 Composite Sections

Last, we note that the existence of these conflicting structures is partially due to the fact that the local stratigraphic sections record only a very small portion of the existence domains of the species. Figure 3.18 shows in a simple way that the prior construction of composite sections (Sect. 2.2.10) enables us to immediately eliminate some of these configurations. The biostratigraphic graph of the relationships of the species whose ranges are sketched in Fig. 2.4 admits six generated

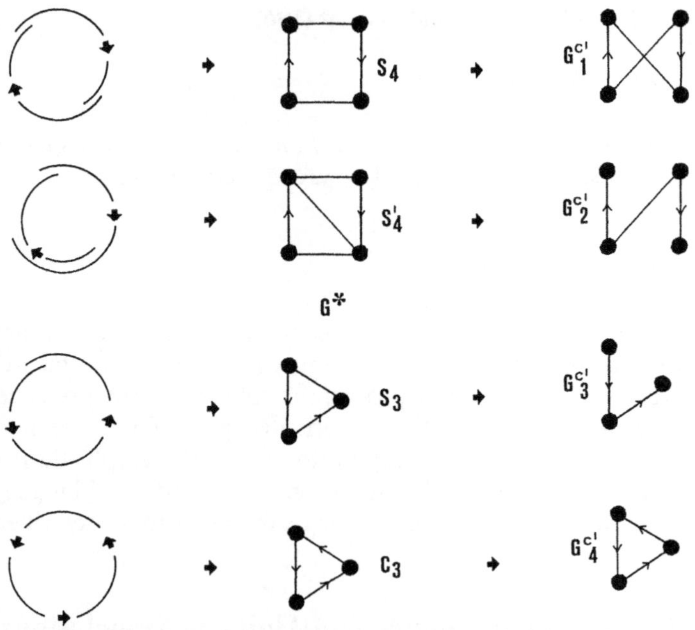

Fig. 3.17. Relationship between the forbidden configurations of a biostratigraphic graph G^* (semioriented S_3, S_4, and S_4' circuits and C_3 cycles) and the nontransitive subgraphs (or, non-transitively orientable subgraphs) of $G^{c'}$. The edges of $G_1^{c'}$ and $G_2^{c'}$ cannot be oriented in such a way as to obtain a transitive graph. The graphs $G_3^{c'}$ and $G_4^{c'}$ are not transitive

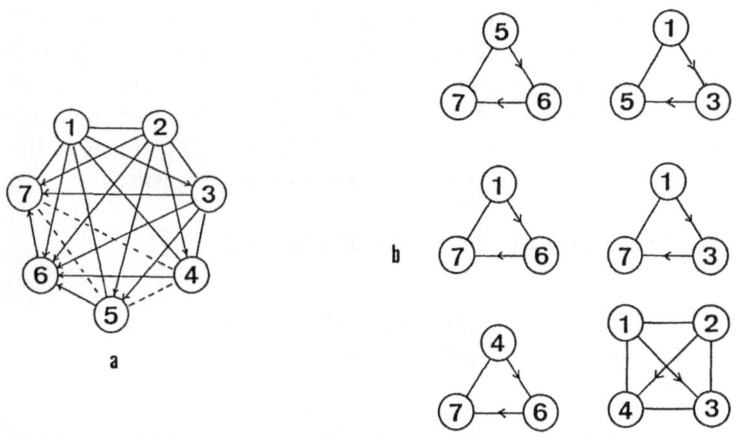

Fig. 3.18. a Graph G^* representing the coexistences and superpositions among the species whose distribution is given in Fig. 2.4 (sections P_1 and P_2). The *dotted lines* represent initial virtual coexistences. **b** Forbidden configurations of the same graph

subgraphs that represent conflicting stratigraphic relationships. These forbidden configurations do not appear if we use geological arguments, such as marker lithologic levels or marker biostratigraphic horizons. Such arguments allow the superposition of sections P_1 and P_2 of Figs. 2.4.b and 2.4.c in order to work on a unique composite section (Fig. 2.4.d).

3.4.4 Lekkerkerker and Boland (1962)

3.4.4.1 Theorem

A graph G is an interval graph if and only if it is triangulated and does not admit asteroidal triples (Fig. 3.15) as generated subgraphs.

3.4.4.2 Discussion

This theorem is very useful since it enables us to limit drastically the number of forbidden configurations that must be detected in a biostratigraphic graph G^*. However, as far as our problem is concerned, the presence of asteroidal triples (Fig. 3.15) in G (i.e., the nonoriented part of G^*) has no consequence for the transformation of this graph into an interval graph: these structures, generated by species whose distribution is limited to restricted biogeographic domains, are automatically destroyed by one of the algorithms described in Section 5.4.3.

3.5 Maximal Cliques and Unitary Associations

3.5.1 Helly's Property

It is important to keep in mind that the uniqueness of a coexistence interval of n species (in other words, a maximal clique with n vertices) can be established based on fragmentary biostratigraphic data coming from a great many localities (at most $(n^2 - n)/2$ localities in which only a pair of species is found each time).

This fundamental property of cliques whose vertices represent intervals is called "Helly's property". Berge (1973, p. 352) states it as follows: "If a family of intervals does not contain two disjoint intervals, then a point exists that is common to all of them."

We can prove this proposition thus (Fig. 3.19):

Let

$J = \{J_1, J_2, \ldots, J_n\}$ be a family of intervals;

x be the set of points of the interval Ji

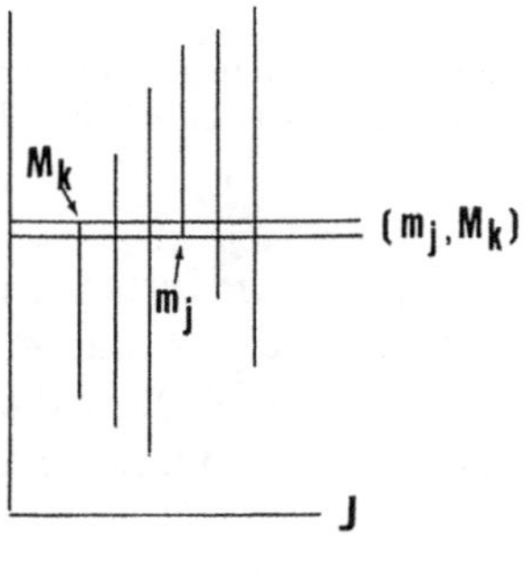

Fig. 3.19. The interval $[m_j, M_k]$ is the intersection of the intervals of the family J

m_i and M_i be the minimum and maximum of J_i so that

$J_i = [m_i, M_i] = \{x \mid m_i \leqq x \leqq M_i \text{ for all } i\}$;

J_k be the interval with the smallest M_k;

J_j be the interval with the largest m_j.

By hypothesis, $J_j \cap J_k \neq \emptyset \rightarrow m_i \leqq M_k$ for all i. In other words we have: $m_i \leqq M_k$.

Consequently: $[m_j, M_k] = \bigcap [m_i, M_i] \neq \emptyset$.

It follows from this assertion that, in an interval graph, a maximal clique characterizes a unique interval. This interval of minimal duration is the intersection of all the intervals making up the maximal clique.

In our problem, a maximal clique that has virtual edges may contain disjoint intervals. For this reason we must distinguish between the terms maximal clique and unitary association. It is clear that in general, a unitary association is a union of maximal cliques.

In practice, unitary associations and maximal cliques have identical properties. Thus we will use the same notation for these two objects in all the set-theoretical calculations (complements, unions and intersections) we will perform later.

3.5.2 Triangular Matrices, Consecutive 1's, and Cliques

3.5.2.1 Theorem

If a vertex-vertex incidence matrix is triangular and has, vertically, the consecutive 1's property, then its rows are cliques.

3.5.2.2 Proof

Let $A = (a_{ij})$ be a triangular matrix with

$$a_{ij} = \begin{cases} 1 \text{ if } x_i \text{ is a neighbor of } x_j \text{ and if } i \leq j, \\ 0 \text{ otherwise.} \end{cases}$$

Let $\Gamma_h(x_i) = \{x_j \mid a_{ij} = 1\}$

Let $\Gamma_v(x_i) = \{x_j \mid a_{ji} = 1\}$

The property of consecutive 1's means that for all x and y such that $\Gamma_v(x) \cap \Gamma_v(y) \neq \emptyset$,

either $x \in \Gamma_v(y)$,

or $y \in \Gamma_v(x)$.

We wish to prove that $\Gamma_h(x_i)$ is a clique. To do this we must prove that all x and $y \in \Gamma_h(x_i)$ are neighbors.

If x and $y \in \Gamma_h(x_i)$, then $\Gamma_v(x)$ and $\Gamma_v(y)$ contain x_i

Thus $\Gamma_v(x) \cap \Gamma_v(y) \neq \emptyset$ implies that, because of the consecutive 1's property

either $x \in \Gamma_v(y)$,
or $y \in \Gamma_v(x)$.

In both cases x and y are neighbors.

3.5.2.3 Remark

It is obvious that the *maximal clique-vertices* incidence matrix that corresponds to the triangular matrix considered here also has the consecutive 1's property. Therefore the graph associated to it is an interval graph. The *manual* method of searching for unitary associations that we described in Section 2.3 is based on the above theorem.

3.6 Summary of Methodology

3.6.1 The Problem

The biochronologic problem discussed in the first two chapters can be briefly restated in terms of graphs.

Imagine a utopian situation in which:

1. The sedimentation is continuous in time and space;
2. The fossil record in the sedimentary rocks is perfect: all species that existed at a given time are fossilized everywhere;
3. All the species are ubiquitous throughout their respective existence.

In such an ideal case the coexistence graph G of the fossil species would correspond exactly to an interval graph. The maximal clique matrix of this graph would correspond to a simplified ideal referential (Fig. 1.6), which could be used directly as a stratigraphic clock.

The reality, of course, is completely different. Many coexistences of synchronous species cannot be observed in the sedimentary record because of the discontinuous nature of sedimentation, the fragmentary character of the fossil record, and ecological constraints limiting the biogeographic distribution of species.

As a result, a graph representing observed coexistences of fossil species is usually not an interval graph. The conditions that must be met for it to be one (Sects. 3.4.2, 3.4.3) are generally not satisfied and the graph will usually admit forbidden generated subgraphs, even when it represents restricted biostratigraphic data.

The constraints inherent in the fossil record also mean that the maximal clique matrix of a graph constructed from real data (in other words, a protoreferential) cannot be interpreted directly as a chronologic sequence of associations of species, even if the graph is an interval graph.

This was implicitly proven above (Sect. 1.4) with the help of Figs. 1.8–1.10: the graph representing the coexistences of the eight species whose local stratigraphic relationships are given in Fig. 1.7 is an interval graph (Fig. 3.1). However, the maximal clique matrix of this interval graph does not represent a chronology because the true position in time of the subdivisions is partially the inverse of the order suggested by the protoreferential.

Yet we have seen that the superpositional order of the reproducible subdivisions of this protoreferential can reasonably be called a chronologic order (Sect. 1.5.2).

This means that the problem of constructing a protoreferential from sparse and fragmentary biostratigraphic information is virtually solved in all cases in which an interval graph can be constructed from a biostratigraphic graph.

In the next two chapters we will discuss certain criteria that make it possible to transform G into a graph whose properties are approximately those of an interval graph. This technical discussion is fairly long. To assist the reader, we will begin with a simplified summary of the method.

3.6.2. Protoreferentials and Interval Graphs

The purpose of the operations described in Chapter 4 is to establish a protoreferential that is an exact or optimal synthesis of complex biostratigraphic data. Such a synthesis is given in the form of a rectangular table whose rows represent groups of species (unitary associations) and whose columns correspond to species. The existence of a species in one of the groups is denoted by a 1 in the corresponding row and its absence is denoted by a 0. (Fig. 3.12). In graph terminology, such a table is a maximal clique matrix that has the consecutive 1's property.

To such a matrix is associated an interval graph (Sect. 3.3.15). For a graph to be an interval graph, it cannot have certain types of generated subgraphs. These forbidden configurations are listed in Section 3.4.

The problem of constructing a protoreferential is thus easily stated in terms of graphs: let G^* be a biostratigraphic graph representing the stratigraphic relationships observed among the fossil species. How can G (i.e., the nonoriented part of G^*) be transformed into an interval graph? The answer is obvious. One must eliminate, within G^*, the forbidden generated subgraphs that represent conflicting stratigraphic relationships. We do this by replacing certain arcs of G^* by virtual edges.

3.6.3 Biostratigraphic Graphs and Interval Graphs

The generated subgraphs of the biostratigraphic graph G^* that represent conflicting stratigraphic relationships of the different species correspond precisely to the configurations that are forbidden in an interval graph. These configurations will be examined in detail in Section 4.2 and we will see that they can be eliminated easily by adding virtual edges to G: these fictitious edges represent virtual coexistences of species with contradictory stratigraphic relationships. Adding such edges is generally done at the expense of certain arcs that represent superpositions of species. It should thus be done in such a way as to destroy a minimum of arcs.

Two different techniques of doing this are described in Section 4.4.2 and Chapter 6. The graph generated by replacing arcs of G by virtual edges is denoted G'.

3.6.4 Unitary Associations, Graph G_k and Protoreferentials

An algorithm for constructing the unitary associations of G' is described in Section 5.2. We then construct an oriented graph G_k whose vertices correspond to unitary associations and whose arcs represent the superpositional relationships of these unitary associations (Sect. 5.3).

To construct G_k (i.e., to find its arcs) we examine the stratigraphic relationships observed among the species. Let k_i and k_j be two distinct unitary associations, let x be a species of k_i not found in k_j and let y be a species of k_j not found in k_i. We deduce the superposition of k_i and k_j from the superposition of x and y (in other words, we join k_i and k_j by an arc whose orientation is that of the arc (x, y) in G^*).

In general the graphs G_k include several distinct *maximal paths*: a simple criterion for reducing G_k to a unique path is described in Section 5.4.3. After

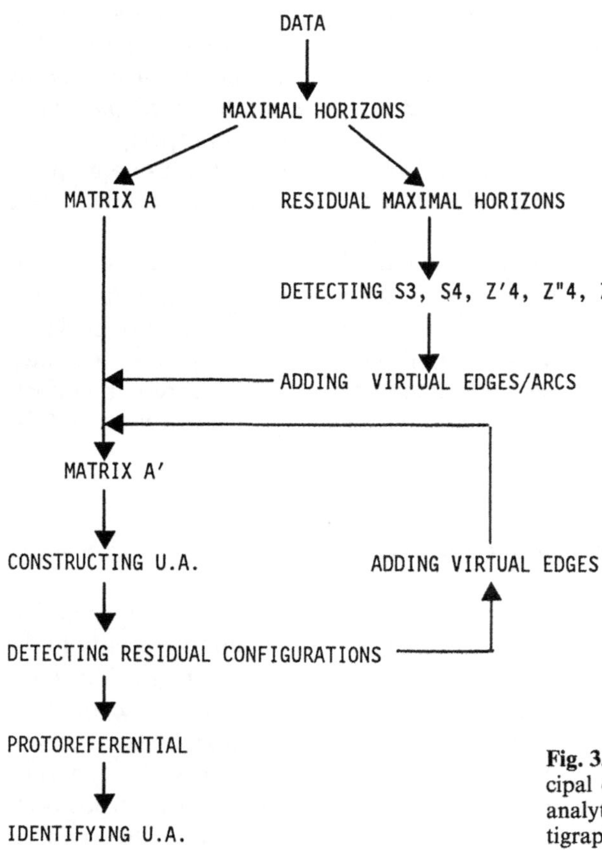

Fig. 3.20. Flow chart showing the principal operations performed during the analytic treatment of complex biostratigraphic data

doing this, we transcribe in a *U. A.-species* incidence matrix the species contained in the unitary associations whose superpositional order has been established. This matrix corresponds to the desired protoreferential.

3.6.5 Correlations

Biochronologic correlations are made with the help of the characteristic elements (species and pairs of species) of the different unitary associations. The chronologic reliability of the U. A. (or unions of U. A.) depends on their reproducibility. The criteria by which reliable zonations can be established from a referential (i.e., a protoreferential and reproducibility matrix) are discussed in Section 5.6.

3.6.6 Summary

A flow chart of the procedures described in the following chapters is given in Fig. 3.20. It illustrates the relationships among the operations performed on the biostratigraphic graph and the resulting synthesis.

3.7 Summary of Notations

$A-B$:	difference between sets A and B
$A \cup B$:	union of sets A and B
$A \cap B$:	intersection of sets A and B
$A \subseteq B$:	A is a subset of B
$X \in A$:	x is an element of A
$x \notin A$:	x is not an element of A
$\|A\|$:	cardinality of A (= number of elements of A)
$\emptyset$:	empty set
$G = (X, E)$:	nonoriented graph (coexistences)
$G_s = (X, U)$:	oriented graph (superpositions)
$G^* = (X, E + U)$:	biostratigraphic graph
U :	set of arcs of G_s and G^*
E:	set of edges of G and G^*
X:	set of vertices of G, G_s and G^*
$\{x, y\}$:	edges of G and G^*
(x, y):	arcs of G_s and G^*
G^c:	complement of G
$G^{c\prime}$:	partial orientation of G^c
$\Gamma(x)$:	set of neighbors of x in G (neighborhood)
$\Gamma^*(x)$:	$\Gamma(x) - \{x\}$
$\Gamma^+(x)$:	set of successors of x in G_s and G^*
$\Gamma^-(x)$:	set of predecessors of x in G_s and G^*

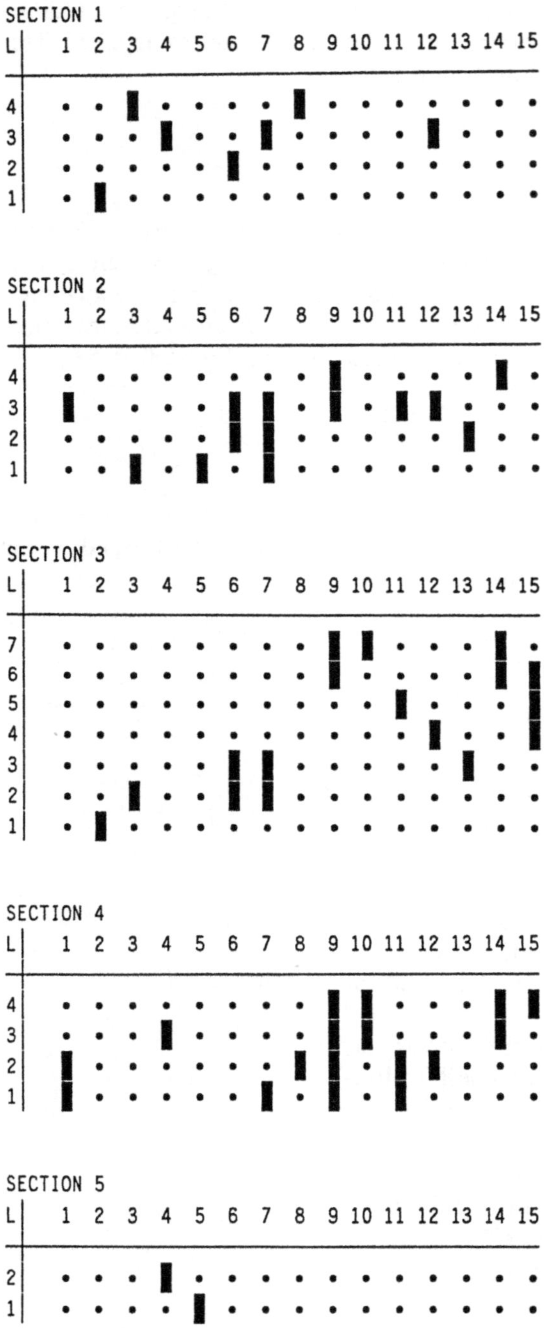

Fig. 3.21. Stratigraphic distribution of 15 alveolinid species (1–15) in 11 sections. *B*, beds (maximal horizons). For names of sections, beds (*L*) and species, see Appendix B.1

SECTION 6

L	1	2	3	4	5	6	7	8	9	10	11	12	13	14	15
4	•	•	•	•	•	•	•	•	•	█	•	•	•	█	•
3	•	•	•	█	•	•	•	•	█	•	•	•	•	•	•
2	█	•	•	•	•	•	█	•	•	•	•	•	•	•	•
1	•	•	•	•	•	•	█	•	•	•	•	•	█	•	•

SECTION 7

L	1	2	3	4	5	6	7	8	9	10	11	12	13	14	15
4	•	•	•	•	•	•	█	•	•	•	•	•	█	•	█
3	•	•	█	•	•	•	█	•	•	•	•	•	•	•	•
2	•	•	•	•	█	•	•	•	•	•	•	•	•	•	•
1	•	█	•	•	•	•	•	•	•	•	•	•	•	•	•

SECTION 8

L	1	2	3	4	5	6	7	8	9	10	11	12	13	14	15
6	•	•	•	•	•	•	•	•	•	█	•	•	•	█	•
5	•	•	•	•	•	•	•	█	•	•	•	•	•	█	•
4	•	•	•	•	•	•	•	•	•	•	•	•	█	•	•
3	•	•	•	•	•	█	•	•	•	•	•	•	•	•	•
2	•	•	•	•	•	•	█	•	•	•	•	•	•	•	•
1	•	█	•	•	█	•	•	•	•	•	•	•	•	•	•

SECTION 9

L	1	2	3	4	5	6	7	8	9	10	11	12	13	14	15
2	•	•	•	•	•	█	•	•	•	█	•	█	•	•	•
1	•	•	•	•	•	•	•	•	•	█	•	•	•	•	█

SECTION 10

L	1	2	3	4	5	6	7	8	9	10	11	12	13	14	15
3	•	•	•	•	•	•	•	█	•	•	•	•	•	•	•
2	•	•	•	•	•	•	•	•	•	█	•	•	•	•	•
1	█	•	•	•	•	█	•	•	•	•	█	█	•	•	•

SECTION 11

L	1	2	3	4	5	6	7	8	9	10	11	12	13	14	15
1	•	•	•	•	•	•	•	•	█	█	•	•	•	█	•

Z_n:	circuit of length n without chord in G and G^*
C_n:	cycle of length n in G_s and G^*
S_n :	semioriented circuit of length n in G^*
$\rightarrow$:	implies that
$\leftrightarrow$:	if and only if
$\mid$:	such that
$\exists$:	there exists
$\neq$:	different from

3.8 A Concrete Problem: the Ilerdian Alveolinid from Yugoslavia

To illustrate the logical steps described in the following chapters, we will now analyze a concrete problem that is significantly more difficult than the one studied in Section 2.4. It concerns data published by Drobne (1977) on the stratigraphic distribution of the Ilerdian alveolinids from Yugoslavia.

We will simplify the problem by limiting our discussion to analyzing the relationships of the 15 species of alveolinid with the largest geographical distribution in the territory studied by Drobne.

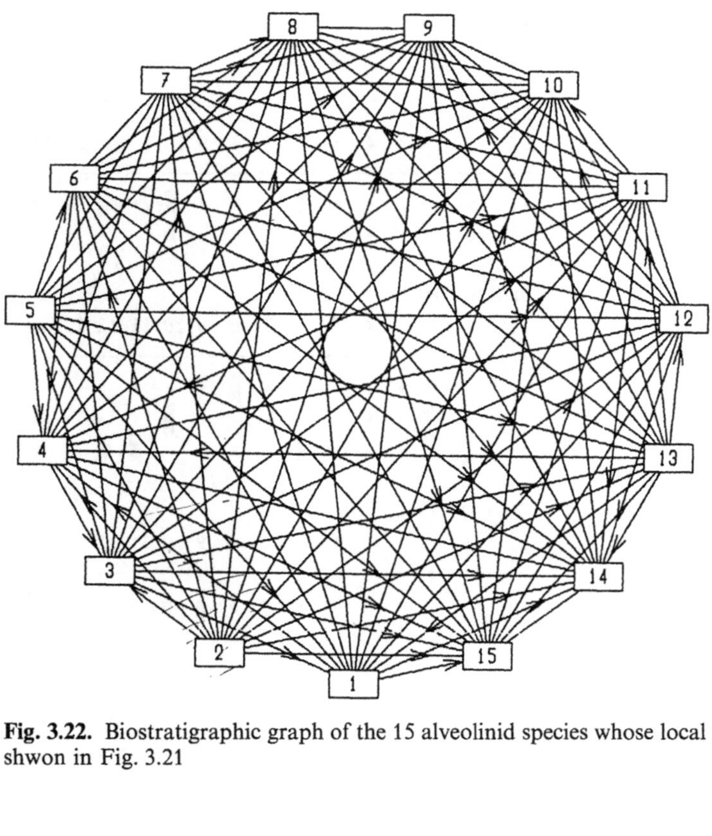

Fig. 3.22. Biostratigraphic graph of the 15 alveolinid species whose local vertical distribution is shwon in Fig. 3.21

We have omitted from our list (Appendix B.1) identifications considered uncertain by Drobne (1977).

The local stratigraphic distribution of the 15 species is illustrated in Fig. 3.21. This table shows only associations and superpositions observed locally among the species: only the maximal horizons (Sect. 2.2.9) generated by relationships among the species are considered.

Level 4 of section 1 (Drobne's sample Fh-5) contains a reworked fauna. This simple fact makes the problem particularly interesting as it gives us an opportunity to establish a set of biochronologic criteria for detecting reworking. The graph G^* representing the stratigraphic relationships among these alveolinid is illustrated in Fig. 3.22.

Chapter 4
Transforming a Biostratigraphic Graph

4.1 Introduction

The characterization of interval graphs given by Gilmore and Hoffman (Sect. 3.4.3) immediately suggests that a biostratigraphic graph can be transformed into an interval graph by destroying the forbidden generated subgraphs found in it. This can be done in two ways:

1. Removing certain vertices of G^*: this simplistic approach suppresses information whose potential biochronologic significance is not known a priori (Sect. 15.4.2). We will use this method only to detect reworking.
2. Adding virtual edges to the biostratigraphic graph. This is the approach we will use here, for it has the advantage of taking into consideration all the stratigraphic relationships observed within a group of species.

Before examining the criteria that will be used to eliminate the forbidden generated subgraphs, we note that these structures can result from reworking or contamination as well as from insufficient documentation.

Therefore if the content of a sedimentary bed is clearly reworked (e.g., there is obvious resedimentation or physical deterioration of the fossils), that bed will not be considered in our analysis of the biostratigraphic data. Similarly, we will not consider fossil associations observed in condensed beds (phosphorites, hard grounds, etc).

In other words, we begin by assuming that:

1. The observed fossil associations represent chronologically true coexistences;
2. The observed stratigraphic superpositions are not due to reworking or downworking.

The problem of detecting reworking using purely biochronologic criteria will be discussed in Chapter 7.

4.2 Searching for Virtual Coexistences

4.2.1. Forbidden Configurations of G^*

4.2.1.1 Principal Categories

Two categories of generated subgraphs of G^* that correspond to forbidden configurations in an interval graph are particularly common: they are the circuits of length 4 (Z_4, Z'_4, and Z''_4) and the semioriented S_3 and S_4 circuits.

In the following paragraphs we will begin by studying the virtual coexistences that can be deduced from these structures.

4.2.1.2 Z_4 Circuits

The Z_4 circuits of a biostratigraphic graph can take three distinct forms:

1. The simple form in which we do not know the superpositional relationships of the two pairs of mutually exclusive species creating the circuit. In this case we can conclude that one or both of the pairs virtually coexisted. This assertion is proven by Fig. 4.1 (Z_4).
2. The intermediary form in which one of the superpositional relationships is known. If this relationship corresponds to a true chronologic relationship, then we can conclude that the pair of mutually exclusive species for which the stratigraphic relationship is unknown virtually coexisted. This is proven by Fig. 4.1 (Z'_4).
3. The complete form in which the circuit admits two arcs representing the contradictory stratigraphic relationships of the species. These conflicting circuits can have two meanings:
   a) the two pairs of mutually exclusive species implicated in the circuit virtually coexisted; or
   b) one of the observed superpositions represents a true chronologic relationship and the other (resulting from insufficient documentation or from the discontinuous biostratigraphic distribution of one of the species) should be interpreted as a virtual coexistence (Fig. 4.1; Z''_4). In either case, adding a single virtual edge to the Z_4 circuit destroys this forbidden configuration.

4.2.1.3 Semioriented S_3 and S_4 Circuits

Consider three species x_1, x_2 and x_3 whose stratigraphic relationships are represented by an S_3 in G^*. If the three species are found in sequence in some localities but the respective existence domains of x_1 and x_3 intersect in others, then either x_1 or x_3 virtually coexisted with x_2 (Fig. 4.2).

We can also come across situations where two pairs of species $\{x_1, x_4\}$ and $\{x_2, x_3\}$ represent real coexistences although x_1 is observed below x_2 and x_3 is observed below x_4 (Fig. 4.2).

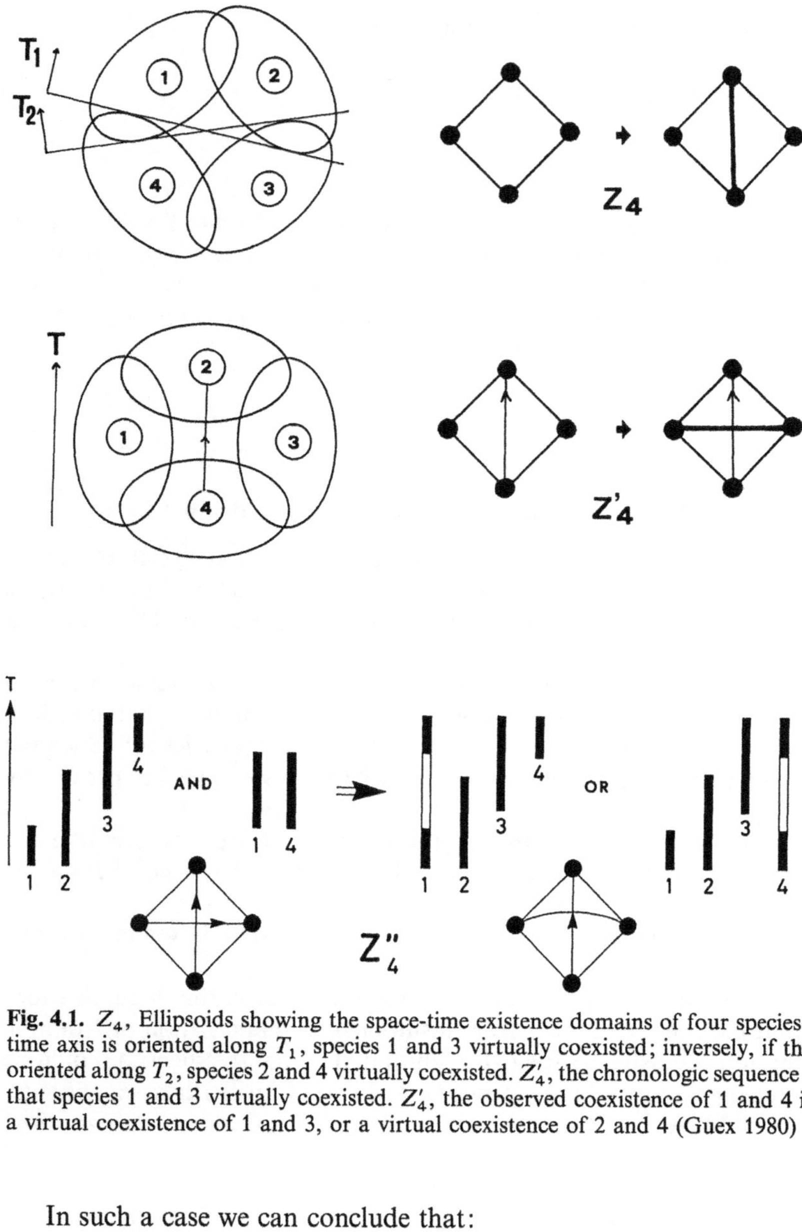

Fig. 4.1. Z_4, Ellipsoids showing the space-time existence domains of four species (1–4). If the time axis is oriented along T_1, species 1 and 3 virtually coexisted; inversely, if the time axis is oriented along T_2, species 2 and 4 virtually coexisted. Z_4', the chronologic sequence 4 → 2 implies that species 1 and 3 virtually coexisted. Z_4', the observed coexistence of 1 and 4 implies either a virtual coexistence of 1 and 3, or a virtual coexistence of 2 and 4 (Guex 1980)

In such a case we can conclude that:

1. If the superposition $x_1 \rightarrow x_2$ corresponds to a true chronologic relationship, then x_4 virtually coexisted with x_3;
2. If the superposition $x_3 \rightarrow x_4$ corresponds to a true chronologic relationship, then x_1 virtually coexisted with x_2.

In this case too, replacing just one of the arcs of a S_3 or S_4 configuration by a virtual edge is sufficient to destroy the forbidden subgraph: generally speaking, it is obvious that when two superpositions conflict, at least one must be false.

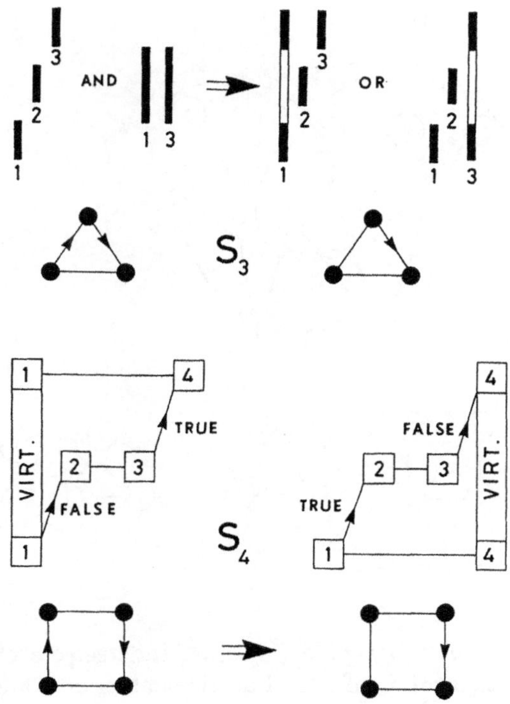

Fig. 4.2. Virtual coexistences that can be deduced from the configurations S_3 and S_4

4.3 Detecting Circuits and Cycles of G^*

4.3.1 Preliminary Remark

The structures examined here were defined in Section 3.3, and the properties of the Z_4 circuits and of the semioriented S_3 and S_4 circuits were discussed in the preceding section.

4.3.2 Detection of S_3, S_4, S_4' and Z_4

4.3.2.1 Method

Let $S_{ij} = \Gamma^+(x_i) \cup \Gamma^+(x_j)$ and $P_{ij} = \Gamma^-(x_i) \cup \Gamma^-(x_j)$.

For each edge $\{x_i, x_j\}$ of G, we construct

$$C_{ij} = P_{ij} \cap S_{ij}\,.$$

If $C_{ij} \neq \emptyset$, then the edge $\{x_i, x_j\}$ is implicated in one (or several) S_3.

Next we search for the sets

$$A_{ij} = S_{ij} - C_{ij} \quad \text{and} \quad B_{ij} = P_{ij} - C_{ij}\,.$$

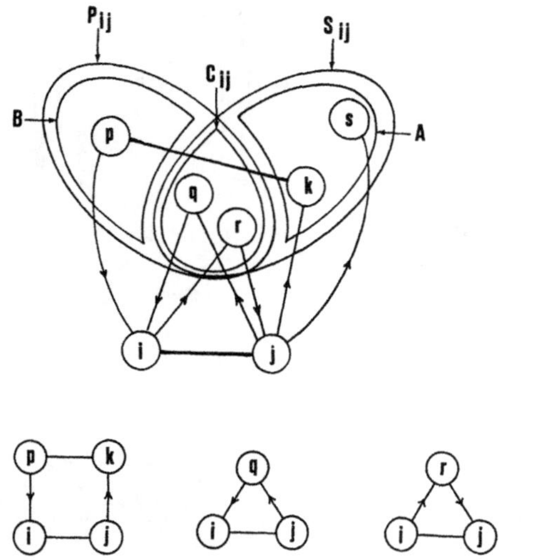

Fig. 4.3. Detection of semioriented S_3 and S_4 circuits (see text). *i*, *j*, *k*, *p*, etc. denote the different vertices of the circuits

A subgraph S_4, S_4' or Z_4'' is detected each time that an edge exists between an element x_k of A and an element x_p of B and neither x_k nor x_p forms an S_3 with $\{x_i, x_j\}$.

A schematization of this procedure (due to Davaud) is shown in Fig. 4.3.

4.3.2.2 Technical Remarks

S_4 is a subgraph of S_4' and of Z_4'': in practice, the detection of subgraphs of the S_4 type necessarily leads to the detection of the generated S_4' and Z_4'' subgraphs.

Another method for detecting generated subgraphs of G^* is given in Chapter 6.

4.3.3 Detection of Z_4 and Z_4'

4.3.3.1 Z_4 Circuit

For each pair (x_i, x_j) in G^* such that $a_{ij} = a_{ji} = 0$ in the matrix A (i.e., the adjacency matrix *of* G^*) we construct

$$I_{ij} = \Gamma(x_i) \cap \Gamma(x_j).$$

If $|I_{ij}| > 1$, we denote $\{(x_k, x_p)\}$ the pair or pairs of species of I_{ij}. If $a_{kp} = a_{pk} = 0$, then we have detected one or more Z_4.

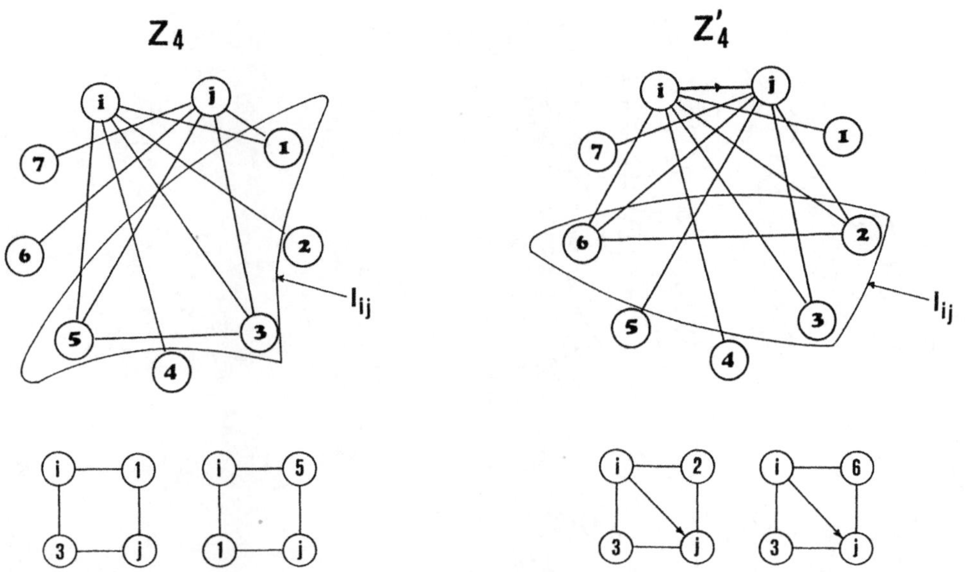

Fig. 4.4. Detection of Z_4 and Z'_4 circuits. Z_4, $I_{ij} = \{1, 3, 5\}$; $a_{kp} = a_{pk} = 0$ for (1, 3) and (1, 5). Z'_4, $I_{ij} = \{2, 3, 6\}$; $a_{kp} = a_{kp} = 0$ for $\{2, 3\}$ and $\{3, 6\}$ (see text)

4.3.3.2 Z'_4 Circuit

For each pair (x_i, x_j) in G^* such that $a_{ij} = 1$ and $a_{ji} = 0$ in the matrix A, we do the same operation as above.

If $|I_{ij}| > 1$, we denote $\{(x_k, x_p)\}$ the pair or pairs of species of I_{ij}. If $a_{kp} = a_{pk} = 0$, then we have detected one or more Z'_4.

Figure 4.4 illustrates this procedure.

4.3.4 Detection of C_3 Cycles

For each arc (x_i, x_j) and (x_k, x_i) of G^*, we first construct

$$\Gamma^-(x_i) \cap \Gamma^+(x_j) = A_{ij}\,.$$

If the arc (x_k, x_i) has not yet been examined, we next construct

$$\Gamma^+(x_i) \cap \Gamma^-(x_k) = B_{ij}\,.$$

If $A_{ij} \neq \emptyset$ and/or if $B_{ij} \neq \emptyset$, then we have detected one or more C_3 cycles. Figure 4.5 illustrates this operation.

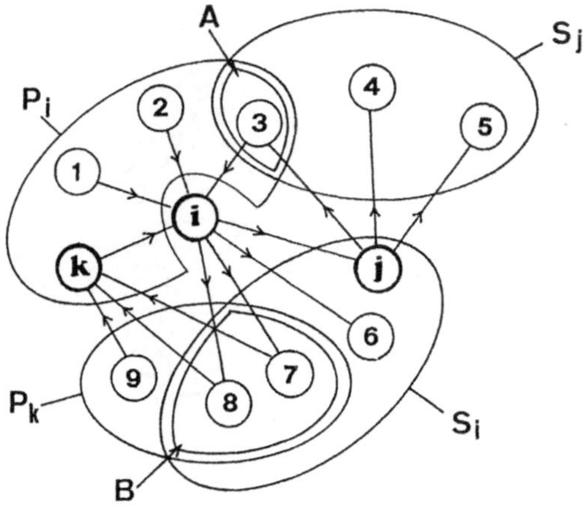

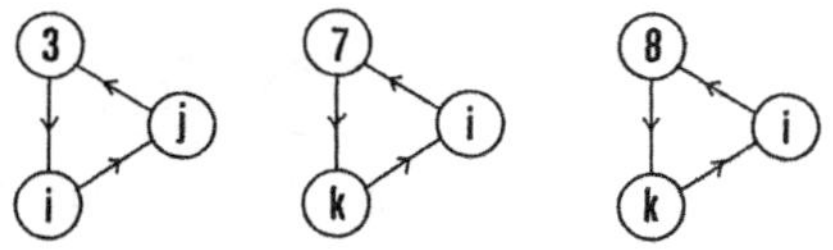

Fig. 4.5. Detection of C_3 cycles: $P_i = \Gamma^-(x_i)$ and $S_i = \Gamma^+(x_i)$ (see text)

4.3.5 A Real Case: the Graph of the Alveolinid

Detecting and listing all the subgraphs discussed here requires a computer when the biostratigraphic graph is even moderately complicated. As an example, we have reproduced in Fig. 4.6 all the forbidden configurations of the graph G^* of our standard example of ilerdian alveolinids (Fig. 3.22): this graph is small but it still admits 19 Z_4'' circuits, 14 semioriented S_3 circuits, and three C_3 cycles.

4.4 Adding Virtual Edges to G

4.4.1 Complexity of the Problem

A graph representing the stratigraphic relationships of 100 species can admit several thousand distinct and imbricated generated S_3, S_4, and Z_4 subgraphs, even when the biostratigraphic observations are abundant and of good quality. In the example of paleogene nannoplankton that we will study below (Chap. 11), there are more than 3800. Yet this is a biochronologic problem with particularly excellent basic data.

Faced with the complexity of the imbrications among the configurations that must be destroyed in order to transform the nonoriented part of G^* (i.e., the graph G) into an interval graph, we must make certain choices: systematically replacing the arcs implicated in the generated S_3, S_4, and Z_4 subgraphs by edges

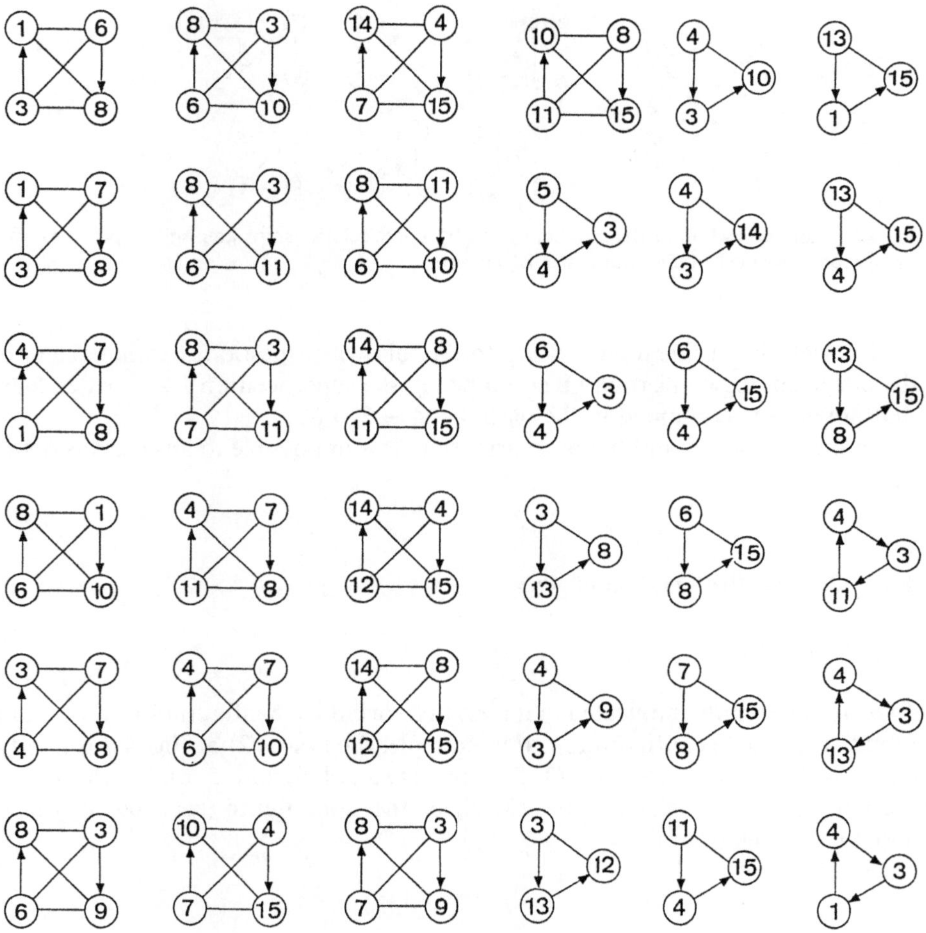

Fig. 4.6. Forbidden configurations of the alveolinid graph (see Fig. 3.22)

can cause a serious loss of chronologic information, even if we make an effort to preserve the arcs that represent reproducible superpositional relationships.

Figure 4.7 illustrates in a simple manner the dilemma so often encountered: it shows a set of six imbricated Z_4'' circuits involving six species (1–6). Let us assume that the stratigraphic relationship $1 \rightarrow 2$ [i.e., arc (1,2) in the Z_4''] has a reproducibility equal to the sum of the observed stratigraphic superpositions $3 \rightarrow 4$, $3 \rightarrow 5$, $3 \rightarrow 6$, $4 \rightarrow 5$, $4 \rightarrow 6$, $5 \rightarrow 6$.

If we preserve the arc (1, 2) on the assumption that it represents a true chronologic relationship (see above), then we must replace the arcs (3, 4), (3, 5), (3, 6), (4, 5), (4, 6), and (5, 6) by virtual edges in order to destroy the six imbricated Z_4'' (Fig. 4.7).

However, these structures can just as well be destroyed by replacing the single arc (1, 2) by an edge: this would preserve the other six arcs and the six observed stratigraphic superpositions that they represent.

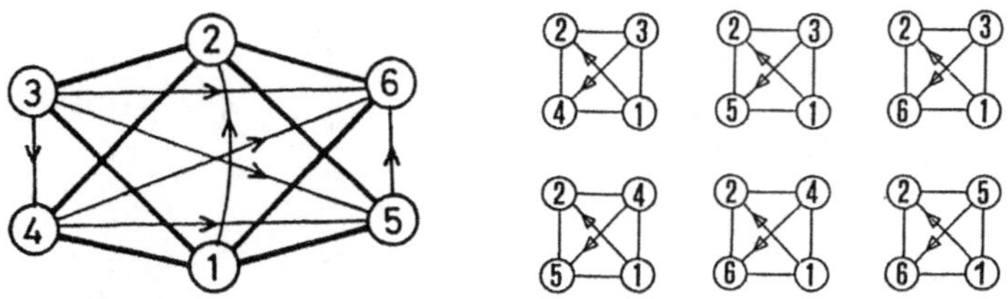

Fig. 4.7. Representation of the conflicting stratigraphic relationships among six species (1–6); this graph is formed of six imbricated Z_4'' circuits

In brief, the problem boils down to the following question: to what extent is the biochronologic information contained in the superposition $1 \rightarrow 2$ better than that potentially contained in the path $3 \rightarrow 4 \rightarrow 5 \rightarrow 6$?

In the absence of additional information, it is impossible to answer this question.

4.4.2 Optimal Preservation of Arcs

4.4.2.1 Definition

Two arcs are jointly implicated in imbricated forbidden configurations if they are found in the same configuration. For example, the arc (1, 2) of Fig. 4.7 is jointly implicated with the arcs (3, 4), (3, 5), (3, 6), (4, 5), (4, 6) and (5, 6), but those arcs are not jointly implicated in the six Z_4'' of the same figure (each belongs to a different configuration).

4.4.2.2 Optimization Procedure

An optimization procedure for dealing with the above problem was developed by Davaud (see Guex and Davaud 1982).

To begin, all the generated S_3, S_4 and Z_4'' subgraphs (Sect. 4.3) are detected, and the pairs of arcs jointly implicated in these structures are recorded in an arc-arc matrix $T = (t_{ij})$ where:

$$t_{ij} = \begin{cases} 1 \text{ if the arc } i \text{ is jointly implicated with the arc } j \text{ in a forbidden} \\ \quad \text{generated subgraph,} \\ 0 \text{ if not.} \end{cases}$$

Next, the arcs implicated in forbidden subgraphs are selectively replaced by virtual edges. Two criteria are used to destroy a minimum of arcs (in other words, to preserve a maximum of information on the species' superpositional relationships) while eliminating as many forbidden configurations as possible:

1. The most frequently implicated arcs are progressively replaced by virtual edges, the matrix T being readjusted each time.

2. When two arcs are implicated the same number of times, the least reproducible (i.e., the one representing the stratigraphic relationship observed the least frequently) is destroyed first.

At the end of this procedure, all the configurations detected during the first iteration are destroyed, but the new (virtual) edges that were substituted for the destroyed arcs can themselves create new forbidden configurations. Thus it is necessary to repeat the above operation until it no longer creates forbidden configurations.

4.4.2.3 Example

The initial generated Z_4'' and S_3 subgraphs of the graph G^* corresponding to the alveolinids was given in Fig. 4.6. The arc-arc matrix that allows us to optimize the replacement of jointly implicated arcs in these configurations by virtual edges is illustrated in Fig. 4.8.

During the first iteration, the arcs (7, 8), (6, 8), (4, 15), (4, 3), (8, 15), (3, 13), (1, 15), and (6, 4) are progressively replaced (in that order) by edges. The reader will easily verify that all the forbidden subgraphs of the first generation are destroyed during this first step. The virtual edges created as a result give rise to forbidden configurations of the second and third generations (Fig. 4.9.a, b).

ARC																											
1←3	1:	0	1	1	0	0	0	0	0	0	0	0	0	0	0	0	0	0	0	0	0	0	0	0	0	0	0
6→8	2:	1	0	0	1	0	0	0	0	0	0	1	1	1	0	0	0	0	0	0	0	1	1	0	0	0	0
7→8	3:	1	0	0	0	1	1	0	0	0	0	1	0	1	0	0	1	0	0	0	0	0	1	0	0	0	0
1→10	4:	0	1	0	0	0	0	0	0	0	0	0	0	0	0	0	0	0	0	0	0	0	0	0	0	0	0
1→4	5:	0	0	1	0	0	0	0	0	0	0	0	0	0	0	0	0	0	0	0	0	0	0	0	0	0	0
3←4	6:	0	0	1	0	0	0	1	1	0	0	1	1	0	0	0	0	0	1	0	0	0	0	0	0	0	0
4←5	7:	0	0	0	0	0	1	0	0	0	0	0	0	0	0	0	0	0	0	0	0	0	0	0	0	0	0
4←6	8:	0	0	0	0	0	1	0	0	0	0	0	0	0	0	1	0	1	0	0	0	0	0	0	0	0	0
3→13	9:	0	0	0	0	0	0	0	0	0	1	0	0	0	1	0	0	0	0	0	0	0	0	0	0	0	0
8←13	10:	0	0	0	0	0	0	0	0	1	0	0	0	0	0	0	0	0	0	0	0	0	1	0	0	0	0
3→9	11:	0	1	1	0	0	1	0	0	0	0	0	0	0	0	0	0	0	0	0	0	0	0	0	0	0	0
3→10	12:	0	1	0	0	0	1	0	0	0	0	0	0	0	0	0	0	0	0	0	0	0	0	0	0	0	0
3→11	13:	0	1	1	0	0	0	0	0	0	0	0	0	0	0	0	0	0	0	0	0	0	0	0	0	0	0
12←13	14:	0	0	0	0	0	0	0	0	1	0	0	0	0	0	0	0	0	0	0	0	0	0	0	0	0	0
7→10	15:	0	0	0	0	0	0	0	1	0	0	0	0	0	0	0	0	1	0	0	0	0	0	0	0	0	0
4←11	16:	0	0	1	0	0	0	0	0	0	0	0	0	0	0	0	0	1	0	0	0	0	0	0	0	0	0
4→15	17:	0	0	0	0	0	0	0	1	0	0	0	0	0	0	1	1	0	0	1	1	0	0	0	0	0	1
3→14	18:	0	0	0	0	0	1	0	0	0	0	0	0	0	0	0	0	0	0	0	0	0	0	0	0	0	0
7→14	19:	0	0	0	0	0	0	0	0	0	0	0	0	0	0	0	0	1	0	0	0	0	0	0	0	0	0
12→14	20:	0	0	0	0	0	0	0	0	0	0	0	0	0	0	0	0	1	0	0	0	0	1	0	0	0	0
10←11	21:	0	1	0	0	0	0	0	0	0	0	0	0	0	0	0	0	0	0	0	0	0	1	0	0	0	0
8→15	22:	0	1	1	0	0	0	0	0	0	1	0	0	0	0	0	0	0	0	0	1	1	0	1	0	0	0
11→14	23:	0	0	0	0	0	0	0	0	0	0	0	0	0	0	0	0	0	0	0	0	0	1	0	0	0	0
1←13	24:	0	0	0	0	0	0	0	0	0	0	0	0	0	0	0	0	0	0	0	0	0	0	0	0	1	0
1→15	25:	0	0	0	0	0	0	0	0	0	0	0	0	0	0	0	0	0	0	0	0	0	0	0	1	0	0
4←13	26:	0	0	0	0	0	0	0	0	0	0	0	0	0	0	0	0	1	0	0	0	0	0	0	0	0	0

Fig. 4.8. Matrix $T = (t_{ij})$ of the arcs jointly implicated in the forbidden configurations S_3, S_4 and Z_4'' illustrated in Fig. 4.6

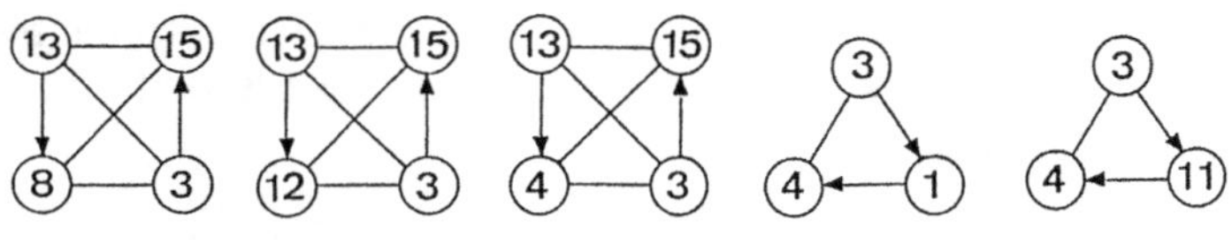

a)

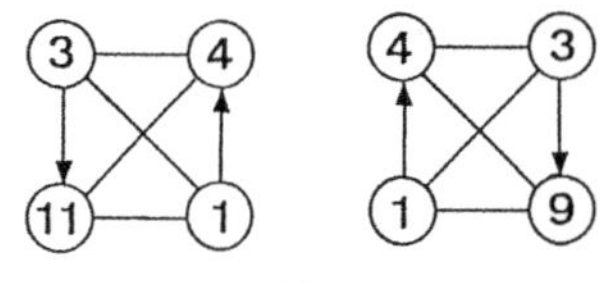

b)

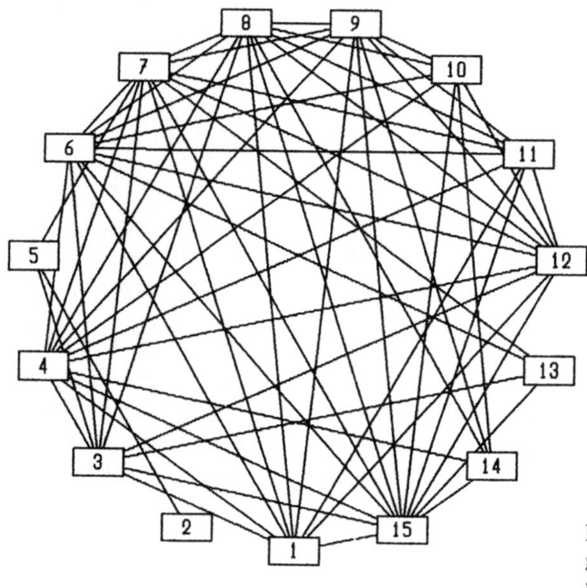

c)

Fig. 4.9. a Z_4'' and S_3 of the second generation. **b** Z_4'' of the third generation. **c** Graph G' (see text)

During the second iteration, the arcs (3, 15), (11, 4), and (3, 1) are destroyed. In the third and final iteration the arc (1, 4) is replaced by an edge to transform the initial nonoriented graph G into a graph G' which, in this particular case, is an interval graph (Fig. 4.9.c).

4.4.2.4 Remark

Applying the preceding algorithm is sometimes difficult, for the number of generated S_3, S_4 and Z_4 subgraphs can grow exponentially with the complexity of the problem [see Moon and Moser (1965) for the mathematical aspect of this question, remembering that the worst cases are essentially constructions of Z_4]. The matrix T of the DV-86 program has a limited size and the memory available for the optimal preservation of arcs is easily swamped when too difficult biochronologic problems are analyzed.

In such cases we must use another optimization procedure, which has the advantage that it requires a relatively limited memory, whatever the complexity of the problem (Guex 1988). This algorithm is described in Chapter 6: it considers exclusively the reproducibility of arcs (i.e., the number of sections in which a superposition of two species is observed) as the criterion for optimization and serves as the basis for a new computer program (BioGraph) written by Savary.

Chapter 5
Protoreferentials, Referentials, and Correlations

5.1 Introduction

Transforming a biostratigraphic graph G^* by adding virtual edges according to the procedures just described gives rise to a new graph G' (the partial nonoriented graph of the transformed G^*). This new graph is not always an interval graph: G' can still admit Z_n circuits ($n > 4$) and asteroidal triples (Fig. 3.15).

We will see, incidentally, that these structures do not have a pernicious effect on the construction of referentials, for they will be automatically eliminated when certain operations are performed later.

In addition, experience shows that these transformations of G^* make it possible to eliminate almost all the contradictory stratigraphic relationships that make interpreting complex biostratigraphic data difficult (or even impossible). Thus we can construct unitary associations directly from G'.

In summary, the operations that will now be described take place in the following order:

1. Construction of unitary associations of the graph G';
2. Seriation of the unitary associations and construction of a protoreferential;
3. Interpretation of the referential and correlations;
4. Search for biochronozones.

5.2 Construction of Unitary Associations

5.2.1 Algorithm

The problem of constructing the unitary associations of G' is technically identical to the problem of constructing the maximal cliques of any nonoriented graph.

Several algorithms that can be used to solve this problem have been published in the recent literature (see Johnston 1976; Hansen 1980). The one we constructed (Guex 1980), is very close to that of Bierstone (in Auguston and Minker 1970). It proceeds by induction and the general idea is as follows: knowing the set of maximal cliques of a subgraph G_n of order n, we can determine the set of maximal cliques of the graph G_{n+1} obtained by adding to it a vertex and the edges of G_n incident to this vertex (G_n is thus a subgraph of G_{n+1}).

Consider a subgraph $G_n = (X_n, E)$ where $X_n = \{x_1, \ldots, x_n\}$ is the set of vertices.

$\{x_{n+1}\}$ will be a vertex added to the subgraph G_n.
$\Gamma^*(x_{n+1})$ will be the set of neighbors of $\{x_{n+1}\}$ in X_n.
$K_n = \{\{k_n\}\}$ will be the set of maximal cliques of G_n at the nth step.
C_{n+1} will be a set of cliques (not necessarily maximal) at the step $n + 1$.
K_{n+1} will be the set of maximal cliques of G_{n+1} at the step $n + 1$.

It is obvious that if:

$\{k_n \cap \Gamma^*(x_{n+1})\} = \emptyset$, then k_n and $\{x_{n+1}\}$ are two distinct cliques of G_{n+1}.
$\{k_n \cap \Gamma^*(x_{n+1})\} = k_n$, then $\{k_n \cup \{x_{n+1}\}\}$ is a maximal clique of G_{n+1}.
$\{k_n \cap \Gamma^*(x_{n+1})\} \neq \emptyset$ and $\neq k_n$, then k_n is a maximal clique of G_{n+1} and $\{(k_n \cap \Gamma^*(x_{n+1}) \cup \{x_{n+1}\}\}$ is a clique of G_{n+1}.

We can thus write

$$C_{n+1} = \{k_n, (k_n \cap \Gamma^*(x_{n+1}) \cup \{x_{n+1}\} \mid k_n \in K_n\}.$$

We denote as R the operation of eliminating all the nonmaximal cliques of G_{n+1}. Thus we can write:

$$K_{n+1} = R(C_{n+1}).$$

The algorithm is iterative: in the beginnning we set

$$X_1 = \{x_1\},\ K_1 = \{x_1\} \quad \text{and} \quad \Gamma^*(x_2) = \{x_1\} \text{ or empty}.$$

5.2.2 Example

The nonoriented graph of Fig. 5.1 represents the real and virtual coexistences of the alveolinid 1–8 of our standard example (see the graph $G^{*\prime}$ of Fig. 4.9.c).

The progressive construction of the cliques (C) and the maximal cliques (K) established at each step of the above algorithm (step 1, 2, ..., 8) is given in Table 5.1

The complete solution for the 15 vertices of the graph $G^{*\prime}$ of Fig. 4.9.c is the following:

$k_1 = \{2, 5\}$
$k_2 = \{3, 5, 7\}$
$k_3 = \{4, 6, 8, 9, 10, 12, 15\}$
$k_4 = \{1, 4, 6, 7, 8, 9, 11, 12, 15\}$

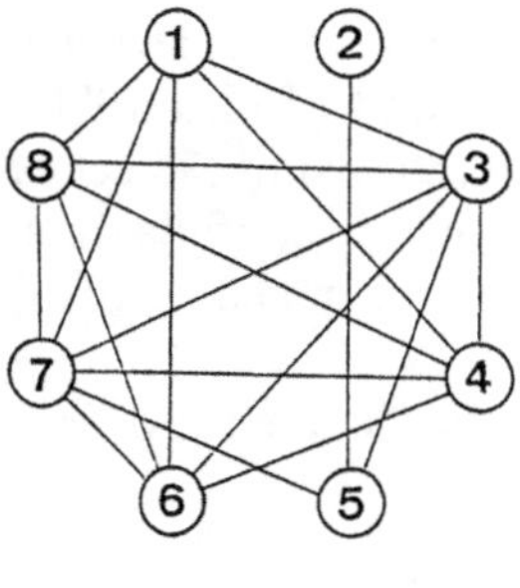

Fig. 5.1. Graph representing the coexistences (observed and virtual) among the alveolinid 1–8 of the graph G' of Fig. 5.9.c

Table 5.1. Construction of the maximal cliques

K_1	= {1}
C_2	= {1} {2}
$R(C_2)$	= {1} {2}
K_2	= $R(C_2)$
C_3	= {1} {2} {1,3}
$R(C_3)$	= {2} {1,3}
K_3	= $R(C_3)$
C_4	= {2} {1,3} {1,3,4}
$R(C_4)$	= {2} {1,3,4}
K_4	= $R(C_4)$
C_5	= {2} {1,3,4} {2,5} {3,5}
$R(C_5)$	= {1,3,4} {2,5} {3,5}
K_5	= $R(C_5)$
C_6	= {1,3,4} {2,5} {3,5} {1,3,4,6} {3,6}
$R(C_6)$	= {2,5} {3,5} {1,3,4,6}
K_6	= $R(C_6)$
C_7	= {2,5} {3,5} {1,3,4,6} {5,7} {3,5,7} {1,3,4,6,7}
$R(C_7)$	= {2,5} {3,5,7} {1,3,4,6,7}
K_7	= $R(C_7)$
C_8	= {2,5} {3,5,7} {1,3,4,6,7} {3,7,8} {1,3,4,6,7,8}
$R(C_8)$	= {2,5} {3,5,7} {1,3,4,6,7,8}

$$k_5 = \{3, 6, 7, 13, 15\}$$
$$k_6 = \{4, 8, 9, 10, 14, 15\}$$
$$k_7 = \{1, 3, 4, 6, 7, 8, 12, 15\}$$

5.3 Seriation of Unitary Associations

5.3.1 Method

The simplest way to establish a partial order on the set of unitary associations is to use information about the stratigraphic relationships observed among the species. This information is given by the graph G_s' (the oriented part of the biostratigraphic graph transformed by adding virtual edges).

For each pair of unitary associations (k_i, k_j), we perform the following operation:
First we construct the sets

$$A_{ij} = k_i - k_j \quad \text{and} \quad B_{ij} = k_j - k_i\,,$$

where A_{ij} is the set of species of k_i that are not in k_j and where B_{ij} is the set of species of k_j that are not in k_i.

The superpositional relationship between k_i and k_j is known if an arc exists between an element of A_{ij} and an element of B_{ij}.

Next we introduce the known relationships of the different pairs of unitary associations in a matrix $B = (b_{ij})$ where:

$$b_{ij} = \begin{cases} 1 \text{ if } k_i \text{ is found below } k_j, \\ 0 \text{ otherwise.} \end{cases}$$

The oriented graph associated to this matrix is denoted G_k. It represents all the information on the stratigraphic relationships established among the unitary associations. We now use this graph to construct a referential.

5.3.2 Example

In our alveolinid problem, the superpositional relationships of the species are given by the oriented part of the transformed biostratigraphic graph of Fig. 4.9.c.

The method of constructing the graph G_k described here is summarized in Table 5.2. This table gives the species content of the sets A_{ij} and B_{ij} for each pair (k_i, k_j) of unitary associations, as well as the superpositional relationships (arcs) between some elements of these sets (NB: a single arc is sufficient to establish the relationship between k_i and k_j).

The relationships among the unitary associations k_1 to k_7 that we can deduce from this table are represented in the matrix B and its associated graph G_k (Fig. 5.2).

Table 5.2. Construction of the graph G_k

(k_i,k_j)	A	B	Arc	Relation
(k_1,k_2)	{2}	{3,7}	(2,3)	k_1-->k_2
(k_1,k_3)	k_1	k_3	(2,4)	k_1-->k_3
(k_1,k_4)	k_1	k_4	(2,4)	k_1-->k_4
(k_1,k_5)	k_1	k_5	(2,3)	k_1-->k_5
(k_1,k_6)	k_1	k_6	(2,4)	k_1-->k_6
(k_1,k_7)	k_1	k_7	(2,3)	k_1-->k_7
(k_2,k_3)	k_2	k_3	(5,4)	k_2-->k_3
(k_2,k_4)	k_2	k_4	(5,4)	k_2-->k_4
(k_2,k_5)	{5}	{6,13,15}	(5,6)	k_2-->k_5
(k_2,k_6)	k_2	k_6	(5,4)	k_2-->k_6
(k_2,k_7)	{5}	{1,4,6,8,12,15}	(5,4)	k_2-->k_7
(k_3,k_4)	{10}	{1,7,11}	(7,10)	k_4-->k_3
(k_3,k_5)	{4,8,9,10,12}	{3,7,13}	(7,10)	k_5-->k_3
(k_3,k_6)	{6,12}	{14}	(6,14)	k_3-->k_6
(k_3,k_7)	{9,10}	{1,3,7}	(7,10)	k_7-->k_3
(k_4,k_5)	{1,4,8,9,11,12}	{3,13,15}	(3,11)	k_5-->k_4
(k_4,k_6)	{1,6,7,11,12}	{10,14}	(6,14)	k_4-->k_6
(k_4,k_7)	{9,11}	{3}	(3,11)	k_7-->k_4
(k_5,k_6)	{3,6,7,13}	{4,8,9,10,14}	(6,14)	k_5-->k_6
(k_5,k_7)	{13}	{1,4,8,12}	(13,1)	k_5-->k_7
(k_6,k_7)	{4,9,10,14}	{1,3,6,7,12}	(7,10)	k_7-->k_6

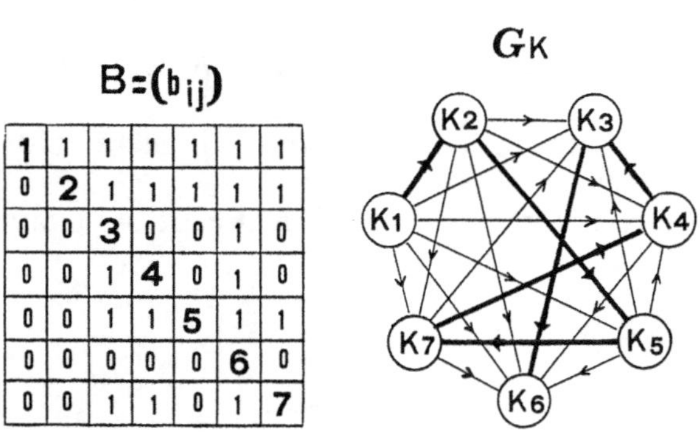

Fig. 5.2. Matrix B giving the superpositions among the U.A.'s k_1–k_7 constructed in Section 5.2. The graph G_k associated to this matrix

5.4 Graph G_k and Protoreferential

5.4.1 Preliminary Remark

The U. A.-species matrix associated to a graph G_k is (by implicit definition) a protoreferential if G_k is made up of a unique path. But G_k is usually made up of several distinct maximal paths, more or less imbricated, and it can also admit isolated vertices (Fig. 5.5). Before we can establish a protoreferential we must detect these maximal paths and reduce them to a unique path.

Experience shows that in reasonably well documented cases, these graphs have only a small number of vertices (about 30 for problems dealing with 100 species) and a very limited number of maximal paths.

To list these paths, we use the following algorithm.

5.4.2 Detecting the Maximal Paths of G_k

5.4.2.1 Method

A path of G_k can be characterized by the list of its vertices. We call such a list a vector. The length of this vector equals the number of elements it contains.

Let $K = \{\{k\}\}$ be the set of unitary associations whose superpositional relationships are represented by G_k.

Let $V = \{\{v\}\}$ be a (variable) set of vectors of elements of K, all the vectors the same length, and each representing a partial path of G_k.

Initialization: V = the set of vectors of length 1 that consist of the elements $k_1 \in K$ such that $\Gamma^+(k_1) = \emptyset$.

Algorithm: at each step we associate to each element $v = (k_1, k_2, \ldots, k_n)$ of V the set of vectors $(k_1, k_2, \ldots, k_n, k_{n+1})$ such that k_{n+1} is an immediate predecessor of k_n (in other words: $\Gamma^+(k_{n+1}) \cap \Gamma^-(k_n) = \emptyset$).

If k_n has no immediate predecessor, we write $(k_1, k_2, \ldots, k_n)$ which is then a maximal path of G_k, and we eliminate it from V.

5.4.2.2 Example

Vertex 6 of the graph G_k of Fig. 5.2 has no successor: we call it k_1. The immediate predecessor of k_1 is vertex 3: we call it k_2. The immediate predecessor of k_2 is vertex 4: we call it k_3, etc.

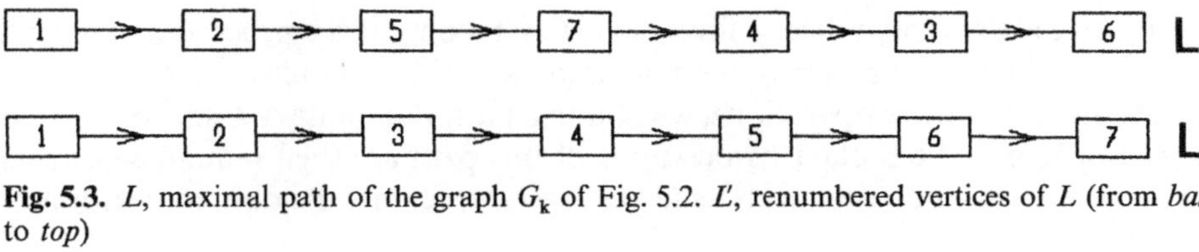

Fig. 5.3. *L*, maximal path of the graph G_k of Fig. 5.2. *L'*, renumbered vertices of *L* (from *base* to *top*)

		SPECIES														
		1	2	3	4	5	6	7	8	9	10	11	12	13	14	15
	7	0	0	0	1	0	0	0	1	1	1	0	0	0	1	1
	6	0	0	0	1	0	1	0	1	1	1	0	1	0	0	1
	5	1	0	0	1	0	1	1	1	1	0	1	1	0	0	1
U.A.	4	1	0	1	1	0	1	1	1	0	0	0	1	0	0	1
	3	0	0	1	0	0	1	1	0	0	0	0	0	1	0	1
	2	0	0	1	0	1	0	1	0	0	0	0	0	0	0	0
	1	0	1	0	0	1	0	0	0	0	0	0	0	0	0	0

Fig. 5.4. Protoreferential (U.A.-species incidence matrix) associated to the graph L' of Fig. 5.3

The result of these operations is given in Fig. 5.3: the graph G_k here is made up of a unique path, denoted as L (note that the renumbered vertices in L' are labeled from base to top in ascending order). The U. A.-species matrix associated to it corresponds to the desired protoreferential (Fig. 5.4).

5.4.3 Reduction of G_k to a Unique Path

5.4.3.1 Method

When G_k is made up of several imbricated or disconnected maximal paths, we must transform it into a graph with a unique path. We do this in the following way.

Let L be the longest of the maximal paths of G_k (if several L's are possible, we choose the one with the most reproducible unitary associations).

Let k'_p, $p \in (1, \ldots, p_L)$ be the U. A. in L and k''_q, $q \in (1, \ldots, q_L)$ be the others.

Now let M_L be the maximal cliques matrix associated to L. If this matrix does not have the consecutive 1's property (Sect. 3.3.8), we give it that property by substituting 1's for any 0's that are between the 1's in the columns. We then eliminate any cliques that are no longer maximal at the end of this operation.

To reduce G_k to a unique path, we take advantage of the fact that the maximal cliques matrix associated to its longest modified path now has the same properties as a protoreferential. The position that the unitary associations k''_q must occupy in this referential can be determined by using the following rule:

For a given k''_q and for each k'_p we calculate

$$W = \max |k'_p \cap k''_q| .$$

If there is a *unique* index p for which W is maximum, then we replace the corresponding k'_p by $k'_p \cup k''_q$.

If there are several indices p for which W is maximum, we proceed to a second iteration (n^{th}) after comparing the remaining set of the couples (k'_p, k''_q).

At the end of these operations, we obtain a path L' of unitary associations that is well-ordered. The unitary associations of this path are then renumbered from 1 to n: the U. A.-species matrix associated to L' represents the desired protoreferential.

5.4.3.2 Example

The example studied here anticipates a concrete problem that will be discussed later (Sect. 5.6.3) in a different context.

Let k_1 to k_7 be the following unitary associations:

$k_1 = \{2, 5\}$;
$k_2 = \{3, 5, 7\}$;
$k_3 = \{3, 6, 7, 13, 15\}$;
$k_4 = \{1, 3, 4, 6, 7, 8, 12, 15\}$;
$k_5 = \{1, 4, 6, 7, 8, 9, 11, 12, 15\}$;
$k_6 = \{4, 6, 8, 9, 10, 12, 15\}$;
$k_7 = \{4, 8, 9, 10, 14, 15\}$.

Suppose that the graph representing the superpositional relationships of these unitary associations is that of Fig. 5.5.a and take the sequence $k_2 \to k_3 \to k_5 \to k_7$ as the maximal path L of reference (with the convention that $k_2 = k'_1$, $k_3 = k'_2$, $k_5 = k'_3$, and $k_7 = k'_4$). The different values of W for the intersections $k'_p \cap k''_q$ are given in Fig. 5.5.b.

These values were established as follows:

$k_6 \cap k_2 = \emptyset \quad \to W = 0;$
$k_6 \cap k_3 = \{6, 15\} \quad \to W = 2;$
$k_6 \cap k_5 = \{4, 6, 8, 9, 12, 15\} \to W = 6;$
$k_6 \cap k_7 = \{6, 8, 9, 10, 15\} \quad \to W = 5$ etc.

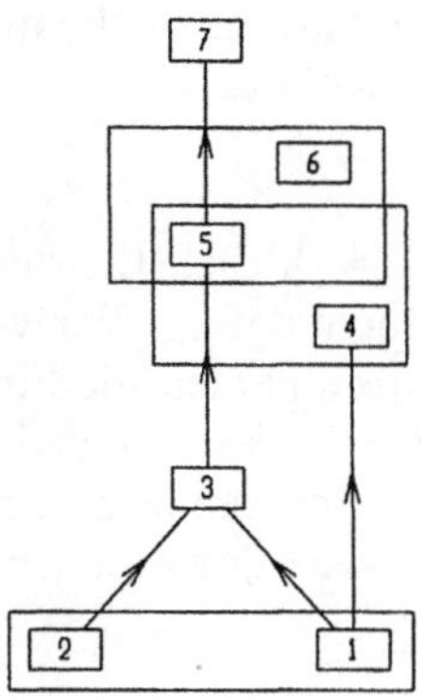

a)

	k_2	k_3	k_5	k_7	
k_6	0	2	6	5	-> $k_6 \cup k_5$
k_4	2	4	7	3	-> $k_4 \cup k_5$
k_1	0	0	0	0	-> $k_1 \cup k_2$

values of W

b)

SPECIES

U.A.	1	2	3	4	5	6	7	8	9	10	11	12	13	14	15
4:	0	0	0	1	0	0	0	1	1	1	0	0	0	1	1
3:	1	0	1	1	0	1	1	1	1	1	1	1	0	0	1
2:	0	0	1	0	0	1	1	0	0	0	0	0	1	0	1
1:	0	1	1	0	1	0	1	0	0	0	0	0	0	0	0

c)

Fig. 5.5. a Graph G_k with several distinct maximal paths: the rectangles surround the U.A.'s which must be merged to transform G_k into a unique path. **b** Values of W for the intersections $k'_p \cap k''_q$ ($k'_p = k_2, k_3, k_5, k_7$ and $k''_q = k_1, k_4, k_6$). **c** Protoreferential resulting from the unions of U.A.'s

The rule stated above thus leads us to merge k_6 and k_5; k_4 and k_5; and k_1 and k_2.

The protoreferential corresponding to the unique path of the modified graph G_k is given in Fig. 5.5.c.

5.4.3.3 Technical Remark

When G_k is made up of disconnected maximal paths, some of which contain only U. A.'s that are older (or younger) than those in L, it is necessary to treat these unitary associations separately, or else they will be completely consumed by the first (or last) U. A. of L'. This very special situation (rare in real cases) can be avoided by constructing composite sections before processing the raw data (Sect. 2.2.10).

5.5 Identifying Unitary Associations

5.5.1 Method

The presence of a characteristic element of a unitary association in a fossil-bearing bed can be detected by checking whether the species content of that bed is a subset of the U. A.

To do this, we proceed as follows:

Let X_{ij} be the species content of the bed n_i in the section p_j, and $K = \{k_m, m = 1, \ldots, p\}$ be the set of unitary associations of the graph G'.

For each bed n_i and each section p_j we calculate:

$$Y = |X_{ij} - k_m| \quad \text{for } m = 1 \text{ to } p.$$

If there is a unique index m for which $Y = 0$, then the bed n_i of section p_j contains at least one of the characteristic elements of k_m. We will assign this bed to U. A. k_m and we will say that this association is identified in the bed.

If there are several indices m for which $Y = 0$, then the bed does not contain a characteristic element of one particular U. A. but can be assigned to any one of them (in other words, to the set of unitary associations for which $Y = 0$).

5.5.2 Example

Consider the set $K = \{k_1, k_2, \ldots, k_7\}$ of the unitary associations constructed in Section 5.2.2 and shown in Fig. 5.4.

The varying values of $Y = |X_{ij} - k_m|$ for the 15 samples of Drobne's first three sections (P_1 to P_3; Fig. 3.21) are given in Table 5.3.

These values were calculated as follows:

1. Bed 1 of section 1 contains exclusively species 2, which exists only in U. A. k_1: this bed is thus assigned to U. A. k_1.

Table 5.3. Identification of unitary associations (see text)

P_i	n_j	X_{ij}	Y_1	Y_2	Y_3	Y_4	Y_5	Y_6	Y_7	U.A.
1	1	{2}	0	1	1	1	1	1	1	k_1
1	2	{6}	1	1	0	0	0	0	1	k_3-k_6
1	3	{4,7,12}	3	2	2	0	0	1	2	k_4-k_5
1	4	{3,8}	2	2	1	0	1	1	1	k_4
2	1	{3,5,7}	2	0	1	1	2	3	3	k_2
2	2	{6,7,13}	3	2	0	1	1	2	3	k_3
2	3	{1,6,7,9,11,12}	6	5	4	2	0	3	5	k_5
2	4	{9,14}	2	2	2	2	1	1	0	k_7
3	1	{2}	0	1	1	1	1	1	1	k_1
3	2	{3,6,7}	3	1	0	0	1	2	3	k_3-k_4
3	3	{6,7,13}	3	2	0	1	1	2	3	k_3
3	4	{12,15}	2	2	1	0	0	0	1	k_4-k_6
3	5	{11,15}	2	2	1	1	0	1	1	k_5
3	6	{9,14,15}	3	3	2	2	1	1	0	k_7
3	7	{9,10,14}	3	3	3	3	2	1	0	k_7

2. Bed 4 of section 3 contains only species 12 and 15, which coexist in U. A. k_4, k_5 and k_6: this bed can thus be assigned to any one of these three unitary associations but to no other, etc.

The final result of this procedure is illustrated in the correlation table of Fig. 5.6 and the corresponding reproducibility matrix.

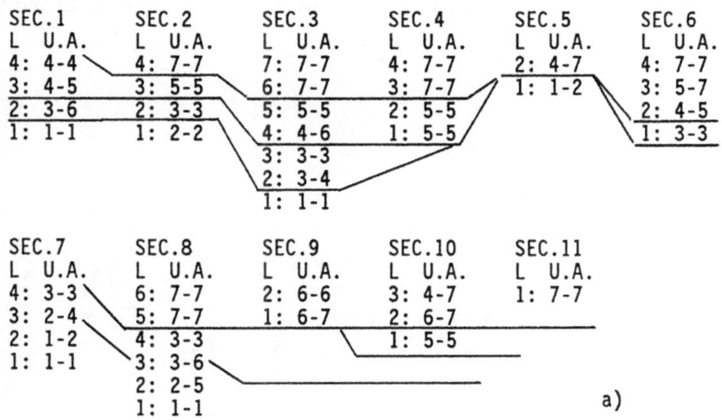

Fig. 5.6. a Correlation of Drobne's (1977) sections 1–11, generated by the protoreferential of Fig. 5.4. **b** Reproducibility matrix associated to this protoreferential. The unions of U.A.'s strictly identified in the sections are indicated by *black rectangles*

The correlation lines connecting the different sections correspond to boundaries between the reproducible biochronologic subdivisions (zones A to D). The establishment of these zones will be studied in greater detail in Section 5.6.

5.5.3 Virtual Pairs of Characteristic Species

The fact that unitary associations are maximal cliques in which some edges are virtual can have the consequence that some U. A. are exclusively characterized by virtual pairs. These associations can obviously not be identified anywhere and have no influence on the correlations.

Figure 5.7 shows how the procedure of replacing jointly implicated arcs in the forbidden configurations by virtual edges (Sect. 4.4.2) sometimes leads to such constructions. The imaginary biostratigraphic graph of Fig. 5.7.a contains two imbrications of S_3 graphs (Fig. 5.7.b) that can be destroyed by replacing the arcs (6, 5) and (4, 3) by virtual edges. The maximal clique matrix (Fig. 5.7.d) of the graph $G^{*\prime}$ (i.e., modified G^*; see Fig. 5.7.c) represents a range chart (protoreferential) in which the subdivisions k_2 and k_4 are characterized by the virtual pairs $\{3, 4\}$ and $\{5, 6\}$.

Note that any method generating a range chart which fails to represent such virtual pairs is incorrect. As a matter of fact, ignoring subdivisions k_2 and k_4 would imply absurd correlations of the stratigraphic sections where the species' superpositions $4 \rightarrow 3$ and $6 \rightarrow 5$ are observed (i.e., the respective positions of subdivisions k_1/k_3 and k_3/k_5 would be reversed in these sections).

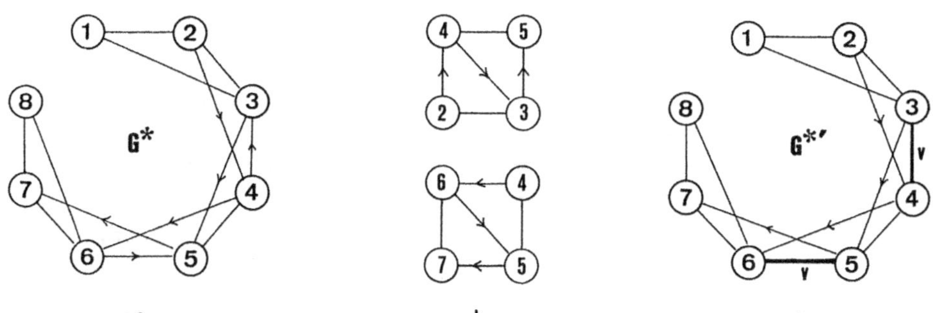

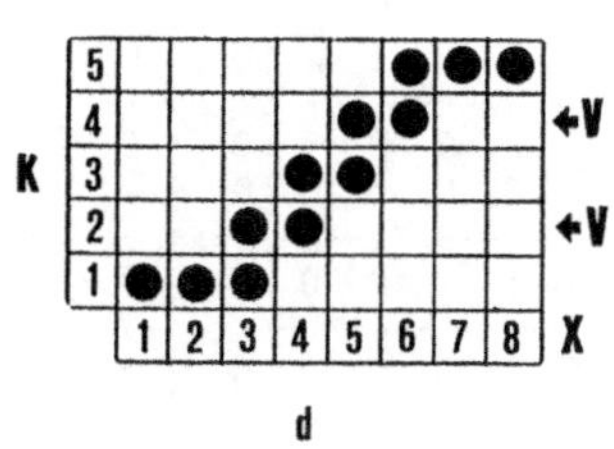

Fig. 5.7. a Fictitious biostratigraphic graph G^*. **b** Two pairs of imbricated S_3 circuits in G^*. **c** $G^{*\prime}$, graph G^* modified by adding the virtual edges $\{3, 4\}$ and $\{5, 6\}$. **d** X, species 1–8 of the graph G^*. K, unitary associations k_1–k_5 of the graph $G^{*\prime}$. V, virtual pairs ($\{3, 4\}$ and $\{5, 6\}$) characteristic of the unitary associations k_2 and k_4, which are not identifiable anywhere

5.6 Optimal Biochronologic Subdivisions

5.6.1 Preliminary Remark

Let us recall that a biochronozone corresponds either to a unitary association that is reproducibly identifiable, or to a union of several unitary associations that are individually not very reproducible, but which are reproducible when combined.

We will examine here three techniques for establishing biochronozones from initial sequences of unitary associations (Sect. 5.3).

5.6.2 Empirical Unions of Nonreproducible U. A.

5.6.2.1 Method

At the end of the preceding operation, all the fossil-bearing beds studied were assigned: (1) to a single U. A. if they contained a characteristic element of that U. A.; (2) to several U. A. if they contained only characteristic elements of a union of adjacent unitary associations.

The basic principle that must be applied to eliminate unitary associations that are not chronologically significant was briefly described earlier (Sect. 2.4.3). In any case, we try to unite the nonreproducible unitary associations so that the resulting union of unitary associations will itself be reproducible.

To do this we use: (1) the information in the reproducibility matrix of the referential of the problem being studied, and (2) information on the unions of unitary associations that are strictly identified in the fossil-bearing beds of the different sections (Sect. 2.2.6).

Of the immense diversity of forms that reproducibility matrices can take, we will look at three situations that occur fairly frequently. These typical cases are illustrated in Fig. 5.8, as follows:

1. Figure 5.8.a: the union of unitary associations 2, 3 and 4 gives rise to a reproducible biochronologic subdivision located between U. A.1 and U. A.5 (which are reproducible).
2. Figure 5.8.b: if the union of U. A.2 and U. A.3 is strictly identified in the sections, we merge them; if this union is not identified there, we merge U. A.2 with U. A.1, and U. A.3 with U. A.4.
3. Figure 5.8.c: if the union of U. A.2 and U. A.3 is not identified in sections P_2 and P_3, we merge U. A.2 with U. A.1, and U. A.3 with U. A.4. If it is identified in sections P_2 and P_3 we merge U. A.2 with U. A.3.

5.6.2.2 Remark

If two possible unions give equally good reproducibility and superpositional control (as could be the case in Fig. 5.8.b), we group unitary associations on the basis of the similarity of their faunal content.

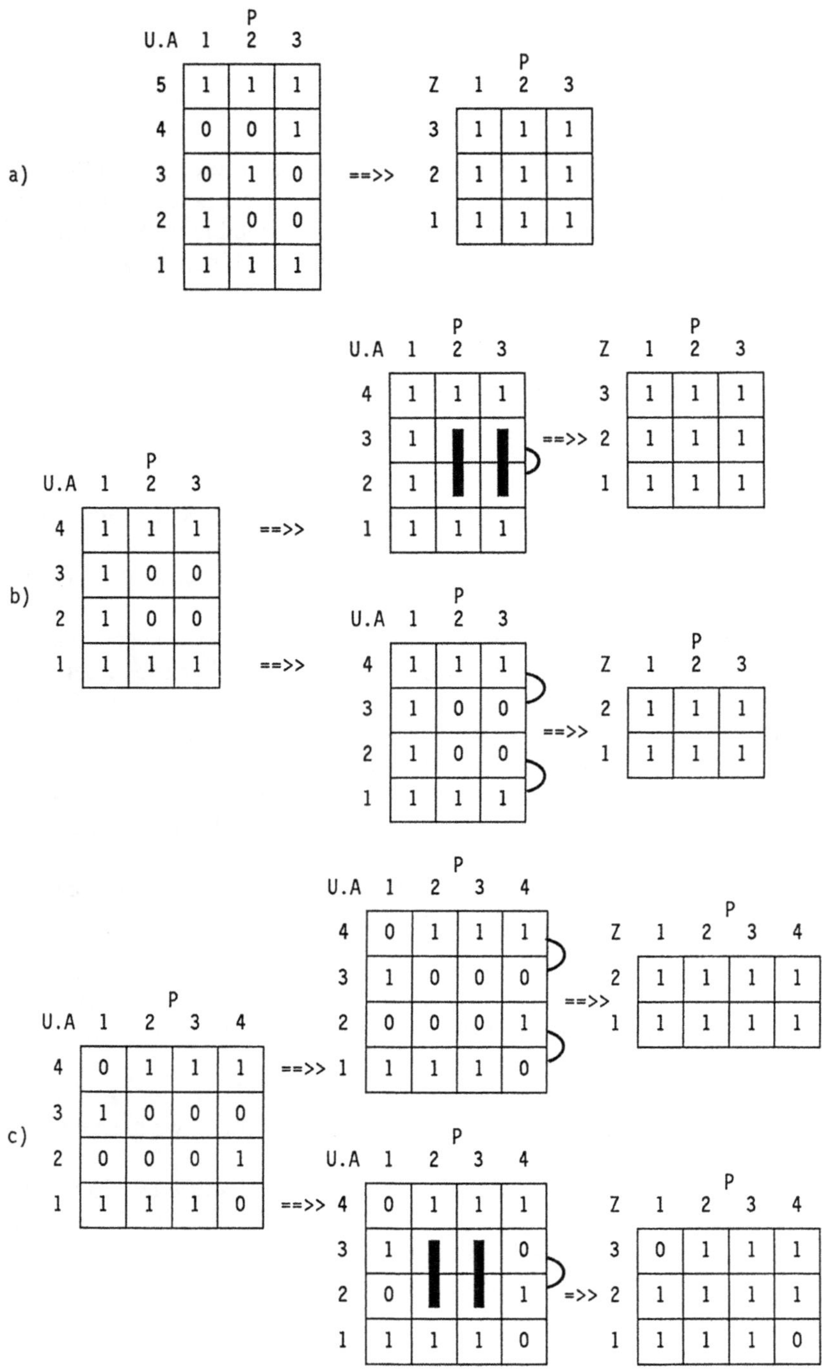

Fig. 5.8 a–c. Empirical union of unitary assocations: three typical situations. *P*, sections. *U.A.*, unitary associations. *Z*, zones. 1's, identified unitary associations. *Black rectangles*, stricly identified unions of unitary associations. *Black rectangles*, strictly identified unions of unitary associations

5.6.2.3 *Example*

The reproducibility matrix and the correlation tables in Fig. 5.6 show that: (1) U. A.2, 4 and 6 are respectively identifiable in only one location; they have no chronologic value. (2) The unions of U. A.1 and 2; U. A.4 and 5; and U. A.6 and 7 are strictly identified in sections 5, 6 and 10.

In this concrete case, the empirical solution consists of defining four biochronozones, denoted *A* to *D* in Fig. 5.6.

5.6.3 Computing the Unions of U. A.: the Graph G'_k

5.6.3.1 *Method*

We can calculate the unions described above by analyzing directly the information contained in the relevant reproducibility matrix.

By definition each column of this matrix represents the local sequence of characteristic elements of the unitary associations identified in the corresponding section. These local sequences can be transcribed in an oriented graph G'_k whose vertices represent the U. A. and whose arcs represent the superpositional relationships observed locally between the characteristic elements (denoted *c*) of these U. A.

We will define the graph G'_k by its associated matrix $B' = (b'_{ij})$:

$$b'_{ij} = \begin{cases} 1 \text{ if } c_i \text{ is observed below } c_j, \\ 0 \text{ otherwise.} \end{cases}$$

where c_i is a characteristic element of U. A. k_i and c_j is a characteristic element of U. A. k_j.

This graph gives a clear picture of the undetermined superpositional relationships between the different unitary associations.

The idea of the present method is twofold:

1. We consider as a protoreferential the U. A.-species matrix associated to the longest maximal path (L) of G'_k; and
2. We treat the unitary associations of G'_k that do not belong to L as fossil assemblages that must be assigned a position in this referential on the basis of their species content.

In other words, this method is equivalent to reducing G'_k to a unique maximal path by using the procedure for transforming G_k that was described in Section 5.4.3.

5.6.3.2 *Example*

The matrix B' and the graph G'_k constructed from the data in the reproducibility matrix of our standard alveolinid problem (Fig. 5.6) are illustrated in Fig. 5.9.

B'

	1	2	3	4	5	6	7
1	■	0	1	1	0	0	0
2	0	■	1	0	0	0	0
3	0	0	■	0	1	0	0
4	0	0	0	■	0	0	0
5	0	0	0	0	■	0	1
6	0	0	0	0	0	■	0
7	0	0	0	0	0	0	■

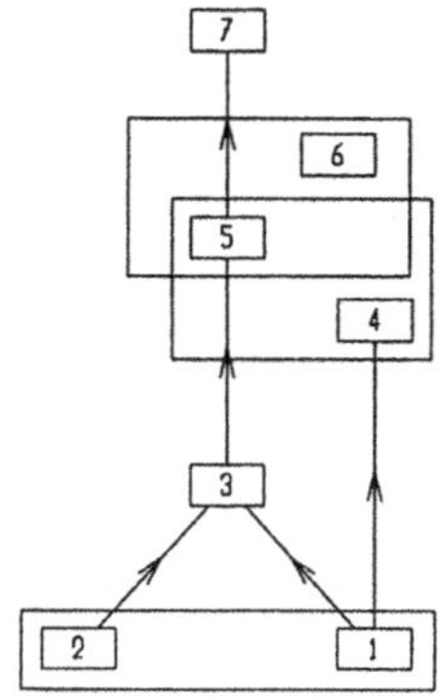

Fig. 5.9. Matrix B' constructed from the data in Fig. 5.6 and the graph G'_k associated to B'

The reduction of this graph to a unique path was studied as an example in Section 5.4.3.2.

We note that the correlations resulting from this procedure are not rigorously identical to those obtained empirically (Sect. 5.6.2): here, U. A.4 and U. A.6 are merged with U. A.5, while in the empirical solution U. A.4 is merged with U. A.5 and U. A.6 is merged with U. A.7.

This is because the procedure described here gives more weight to the global resemblance (similarity) of the species content of the unitary associations than to their reproducibility.

In the absence of external information (i.e., information that are not purely biostratigraphic, such as lithologic criteria), it is not possible to decide which solution gives the better chronology. That decision is necessarily left up to the user of the method.

5.6.4 Contemporaneous Unitary Biofacies and Graph G'_k

The reproducibility of unitary associations is usually very poor when the biogeographic distribution of the species is strongly controlled by ecological or taphonomic factors (e.g., latitude, depth, temperature, facies dependence of the species, differential sorting related to gravitational settling etc.).

In such cases, the reproducibility matrix and its associated graph G'_k display very useful information about the lateral juxtapositions of synchronous "unitary biofacies" that are characterized by ecology-dependent fossil assemblages

Two examples of ecological control over the distribution of such assemblages are given in Fig. 5.10. The symbols A_i, B_i, and C_i denote both the biofacies and their characteristic assemblages.

Figure 5.10.a illustrates an imaginary shallow progradational phase (biofacies $B_1 - B_3$) framed by deeper and more evenly distributed biofacies (A and C). In this case, the graph G'_k contains three maximal paths generated by biofacies B_1 to B_3.

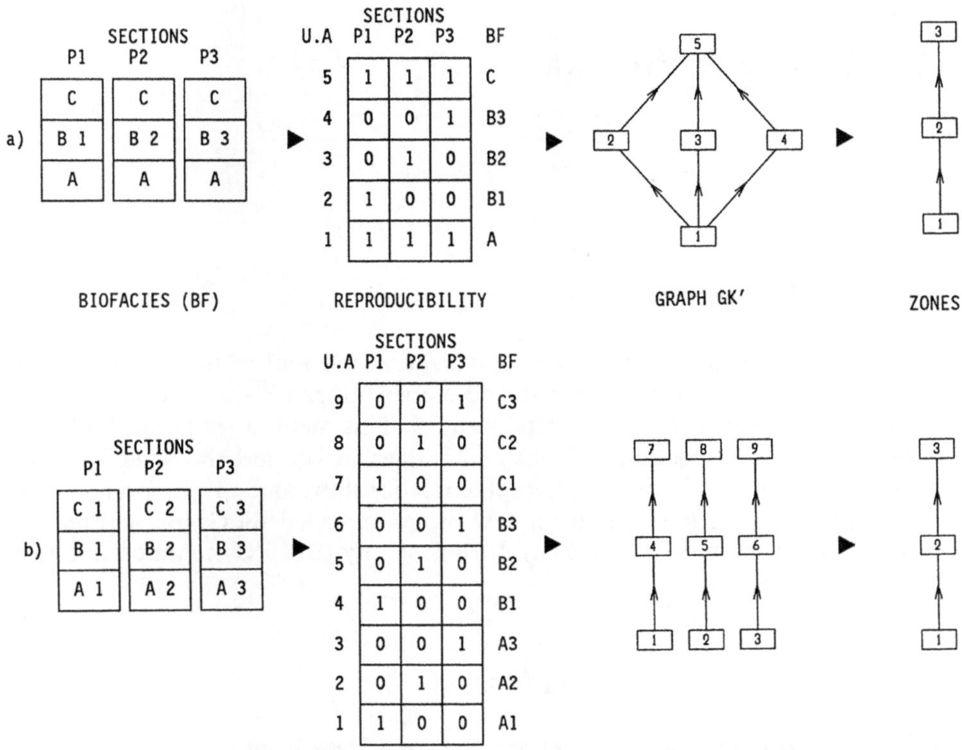

Fig. 5.10. Relationships between synchronous biofacies, reproducibility matrix, graph G'_k and zones

Figure 5.10.b illustrates a strong depth control over the geographic distribution of nine biofacies (A_1-A_3, B_1-B_3, C_1-C_3). Here, the graph G'_k is made up of three disconnected maximal paths, each of them identified in a single section (P_1-P_3).

Chapter 6
The BioGraph Program

6.1 Introduction

To avoid some of the difficulties of optimization inherent to the combinatorial approach discussed above, we recently developed (Guex 1988) a different method from the one described in Chapters 4 and 5. This method simplifies both the procedure for constructing the maximal cliques of G^* and the detection and treatment of the conflicting stratigraphic relationships among the species. This algorithm served as the basis for a new program, called BioGraph, written by Savary (in Savary and Guex 1991 a). It runs on any IBM PC or compatible and is extremely efficient.

6.2 Conventions and Definitions

6.2.1 Local Distribution of Species and Standard Data Input

We recall that if a fossil species has a discontinuous vertical distribution in a stratigraphic section, it is considered as virtually present in all the horizons that are flanked by its first local appearance and disappearance.

For a given stratigraphic section, the standard BioGraph data input is a triplet of numbers: (1) Species number code; (2) FAD (first appearance "datum") level number code; and (3) LAD (last appearance "datum") level number code.

For example, the alveolinid ranges in Drobne's section 1 are: 2 1 1, 3 4 4, 4 3 3, 6 2 2, 7 3 3, 8 4 4, 12 3 3 (see Fig. 3.21).

The local ranges of all taxa used as case studies in this book are listed in the Appendices' numerical databases.

6.2.2 Residual Maximal Horizons

In a given stratigraphic section, we consider only the levels that register the maximal intersections of local existence intervals of the species. These levels are called maximal horizons (Sect. 2.2.9). Maximal horizons that are strictly distinct from each other (as far as inclusion relationships are concerned) are called "residual" maximal horizons.

6.3 Method

6.3.1 Constructing Maximal Cliques of G^*

To construct the maximal cliques of G^* using residual maximal horizons, the following procedure is used.

Let k_m be the residual horizon with the greatest cardinality, and let $\Gamma(x_i)$ be the set of neighboring vertices of x_i in G^*.

If $k_m - \Gamma(x_i) = \emptyset$, we add x_i to k_m (x_i is compatible with all the elements of k_m). The result is then denoted $k_m^* = \{k_m \cup x_i\}$.

To k_m^* we then add the next element x_j, for which $k_m^* - \Gamma(x_j) = \emptyset$ (k_m^* thus becomes equal to $\{k_m \cup x_i \cup x_j\}$). We thus compare the enlarged k_m^* successively to all the vertices of G^*.

At the end of this operation, we eliminate all the residual horizons that are contained in k_m^*, then we follow the same procedure for the next k_m, etc. The resulting k_m^* are now distinct cliques of G^*.

6.3.2 Stratigraphic Relationships Among the Maximal Cliques

The stratigraphic relationship between two maximal cliques, k_i and k_j, can be deduced from the stratigraphic relationships observed among the species belonging respectively to k_i and to k_j. Three kinds of relationships exist.

1. The clique k_i is above (or below) k_j if there exists at least one species of k_i that is above (or below) a species of k_j (univoque superposition).
2. The relationship between k_i and k_j is undetermined if the species of k_i are not compatible with those of k_j and if their superpositional relationships are undetermined.
3. The relationship between k_i and k_j is conflicting if some species of k_i are above some species of k_j, while other species of k_i are below some species of k_j.

6.3.3 Searching for and Eliminating Contradictions

To transform a biostratigraphic graph into an interval graph by ordering its maximal cliques according to raw biostratigraphic observations, it is first essential to eliminate any contradictory superpositions of species. This can be done in two ways. The first consists of detecting all the forbidden configurations of G^* and replacing certain superpositions by deduced virtual coexistences (Sect. 4.4). The second consists of ignoring certain contradictory and poorly reproducible superpositions between two maximal cliques.

The BioGraph program uses the latter approach. It is based on the following principle: faced with two conflicting superpositions, we are forced to admit (in the absence of nonbiochronologic arguments) that a superposition that is reproducible (i.e., observed over a vast geographic area) has more chronologic "value" than one that is not. In other words, if two distinct cliques of G^* contain species that situate them both above and below each other, we will in each case consider as "true" the superposition that is more reproducible and as "false" (i.e., generated by insufficient documentation or by reworking) the superposition that is less reproducible.

To do this, we proceed as follows: for each pair k_i, k_j of maximal cliques of G^* showing a contradictory stratigraphic relationship, we define two sets of arcs, A and B, where A is the set of arcs that links the elements of k_i to those of k_j (in the direction $k_i \rightarrow k_j$) and where B is the set of arcs (of the opposite orientation) that links the elements of k_j to those of k_i. To each set A and B we attribute a value $V(A)$ [or $V(B)$] equal to the sum of the individual reproducibilities of the arcs belonging to A (or B) added to the number of arcs of A (or B).

If $V(A) > V(B)$ we say that the clique k_j is "located above" k_i (keeping in mind that this is an abuse of language and that the superposition is conflicting).

If $V(A) = V(B)$ we say that the stratigraphic relationship between k_i and k_j is undetermined.

This procedure is equivalent to making a global (and no longer individual) search for the forbidden configurations of G^* (see Fig. 6.2.c).

6.3.4 Matrix *M* and Graph G_k

Once the stratigraphic relationships of the cliques have been established, they are entered in a "maximal clique-maximal clique" matrix M.

The relationships represented by M are denoted as follows:

$k_i \ll k_j = k_i$ above k_j without contradiction
$k_i \gg k_j = k_i$ below k_j without contradiction
$k_i \leftarrow k_j = k_i$ above k_j with contradiction
$k_i \rightarrow k_j = k_i$ below k_j with contradiction
$k_i\ ??\ k_j =$ undetermined relationship .

Using this matrix, we next contruct the oriented graph G_k that is associated to it.

6.3.5 Strongly Connected Components of G_k

6.3.5.1 Detection and Destruction

Poorly documented biostratigraphic data (polluted by undetected reworking or by false taxonomic identifications) often are responsible for the presence of strongly connected components (cycles) in G_k. These components must of course be eliminated.

The algorithm used to detect them is similar to the one described by Carre (1979, p. 47): we will not discuss the latter further here.

In our program, the connected components are destroyed according to the following rule.

Consider, in a given component of G_k, the set of pairs (k_i, k_j) of cliques whose deduced stratigraphic relationship is conflicting (i.e., $\rightarrow$ or $\leftarrow$ in the matrix M). We denote A the set of arcs that link k_i to k_j and B the set of arcs of opposite orientation. To each set A and B we attribute a value $V(A)$ [or $V(B)$] equal to the sum of the individual reproducibilities of the arcs belonging to A (or B) added to the number of arcs of A (or B); (see above). To each arc $k_i \rightarrow k_j$ we now attribute a value

$$C_{ij} = \min[V(A);\ V(B)]/\max[V(A);\ V(B)]\,.$$

The arc $k_i \rightarrow k_j$ for which the value C_{ij} is the greatest is destroyed (the relationship k_i, k_j is considered to be undetermined). If this undetermination is not sufficient to destroy the connected component, we proceed in the same way for the next couple (k_i, k_j), etc.

6.3.5.2 Technical Remark

It is essential to stress the fact that in the case of particularly poor data, the destruction of the connected components takes place in a quasi-arbitrary fashion, and depends on the order in which G_k is examined. Such situations can be detected by permuting the order of the stratigraphic sections in the input (which has the effect of modifying the indexing of the maximal horizons and consequently that of the vertices of G_k). If the solutions thus obtained differ from one output to the next, it is then necessary to generate virtual coexistences artificially by grouping the unitary associations whose characteristic elements were permuted. The different positions that were assigned to them in the different solutions tested result from the fact that their real chronologic relationships are impossible to determine.

6.3.6 Maximal Paths of G_k

When the graph obtained at the end of these operations consists of several maximal paths, we reduce it to a unique path using the technique described in Section 5.4.3. The BioGraph program adds an additional constraint to this technique: a vertex that does not belong to the longest maximal path of $G_k(L)$ can be united only with a successor of its immediate predecessor in L (or predecessor of its immediate successor in L; when such elements exist).

6.3.7 Matrix TGH and Unitary Associations

The last step of the method consists of transcribing the species content of the cliques of the reduced graph L in the maximal cliques – species incidence matrix associated to it.

This matrix, denoted TGH in the output of BioGraph, records the discontinuities in the stratigraphic distribution of the species (see Sects. 6.4.6, 7.3.1). To give it the consecutive 1's property, it is sufficient to replace with 1's the 0's that are flanked by 1's in the columns.

In the end only the rows of the resulting matrix that correspond to maximal cliques are conserved (the others are eliminated): each row of this final compacted matrix corresponds to a unitary association and the graph associated with it is an interval graph.

6.3.8 Residual Virtual Edges

The procedure described above does not take into account the virtual edges in the initial step of the calculations. This is because (by definition) the pairs of species that correspond to these edges are not observed in any maximal horizon. In general these edges are generated artificially during the transformation of G^* into an interval graph. When this is not the case, we use a detecting procedure that enables us to recover them during the last step of the calculations.

6.3.9 Residual Arcs

Certain conflicting stratigraphic superpositions destroyed during the transformation of G^* can be restored during the last step of the procedure. This is done as follows.

In each unitary association k_i of the first protoreferential calculated by the program (i.e., constructed immediately after compacting the matrix associated

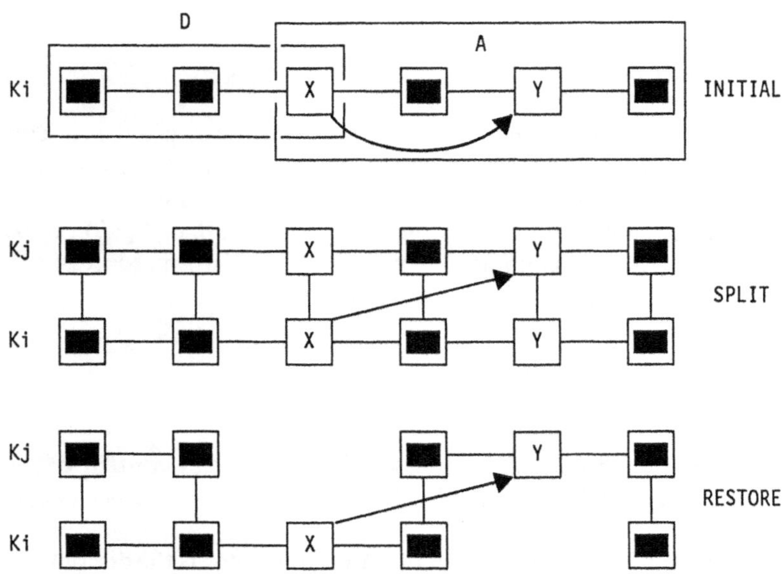

Fig. 6.1. Restoring residual are x → y (see text)

with L), we define two sets of species, A and D, where A is the set of species that appear in k_i and where D is the set of species that disappear in k_i.

If there exists an arc $x \rightarrow y$ between an element x of D and an element y of A, we turn k_i into two unitary associations, k_i and k_j (initially $k_i = k_j$). Next we remove species x from k_j and species y from k_i (Fig. 6.1).

If there exist several distinct arcs of the type $x \rightarrow y$, we repeat the operation for each arc, verifying at each step that the removal of x from k_j (or y from k_i) does not lead to the destruction of a real edge [e.g., (x, z) or (y, z)]. If it does, the species is not removed.

When this operation has been completed for all the unitary associations, the test of maximality is applied once more and sets that are not maximal are eliminated.

6.3.10 Identification of Unitary Associations and Correlations

The last operation performed by our program consists of constructing correlation tables based on the unitary associations. This is done as follows:

We denote X_{ij} the specific content of the ith level of the jth section.

We denote $K = \{k_m, m = 1 \text{ to } p\}$ the set of unitary associations obtained.

For each level of each section we calculate

$$X_{ij} - k_m = I.$$

If $|I| = 0$, the level is assigned to the corresponding unitary association (or unitary associations).

6.4 Example of Application of BioGraph

6.4.1 Introduction

The simplest way to describe the different stages of the program is to give a simple example of its application. To do this we will consider once more our standard alveolinid problem.

6.4.2 Maximal Horizons and the Graph G^*

The first step of the calculation consists of constructing the residual maximal horizons; this first output is given in Table 6.1. The residual horizons are denoted k_{ij}, and the indices i and j signify "j^{th} horizon of the ith section."

Using the original database (Fig. 3.21), we establish the biostratigraphic graph G^*. This graph takes the form of two separate lists, the first giving the set of its edges, the second giving the set of its arcs with their respective reproducibilities [Table 6.2; example: the arc (3, 1) has a reproducibility equal to 1].

Table 6.1. Residual maximal horizons (output of the BioGraph program)

$k_{2,3}$	= {1,6,7,9,11,12}
$k_{4,2}$	= {1,8,9,11,12}
$k_{4,4}$	= {9,10,14,15}
$k_{4,3}$	= {4,9,10,14}
$k_{7,4}$	= {7,13,15}
$k_{1,3}$	= {4,7,12}
$k_{3,2}$	= {3,6,7}
$k_{2,2}$	= {6,7,13}
$k_{2,1}$	= {3,5,7}
$k_{9,2}$	= {6,10,12}
$k_{1,4}$	= {3,8}
$k_{3,5}$	= {11,15}
$k_{8,1}$	= {2,5}
$k_{8,5}$	= {8,14}
$k_{3,4}$	= {12,15}

Table 6.2. Graph G^* (output of the BioGraph program)

Edges	*Arcs*
1 : 6,7,8,9,11,12	1 : [1]<-3,[2]->4,[1]<-5,[3]->10, [2]<-13,[3]->14,[1]->15
2 : 5	2 : [3]->3,[1]->4,[3]->6,[4]->7,[2]->8, [1]->9,[2]->10,[1]->11,[2]->12, [3]->13,[2]->14,[2]->15
3 : 5,6,7,8,12	3 : [1]<-4,[2]->9,[1]->10,[2]->11, [3]->13,[2]->14,[2]->15
4 : 7,8,9,10,12,14	4 : [1]<-5,[1]<-6,[1]<-11,[1]<-13, [1]->15
5 : 7	5 : [2]->6,[1]->8,[1]->9,[1]->10,[1]->11, [1]->12,[3]->13,[2]->14,[1]->15
6 : 7,9,10,11,12,13,15	6 : [3]->8,[3]->14
7 : 9,11,12,13,15	7 : [3]->8,[4]->10,[5]->14
8 : 9,10,11,12,14	8 : [1]<-13,[1]->15
9 : 10,11,12,14,15	9 : [3]<-13
10: 12,14,15	10: [3]<-11,[3]<-13
11: 12,15	11: [2]<-13,[3]->14
12: 15	12: [2]<-13,[3]->14
13: 15	13: [4]->14
14: 15	

6.4.3 Construction of Maximal Cliques

The second calculation consists of constructing the maximal cliques from the residual horizons, and eliminating horizons that are not maximal at the end of this operation.

Residual horizons whose initial content has been increased are denoted $k^*_{i,j}$ in the following figures.

In the example studied here, we note that the horizon $k_{7,4}$ (= {7, 13, 15}) is included in $\Gamma(6)$ (see Table 6.2). This set thus becomes $k^*_{7,4}$ = {6, 7, 13, 15}. Similarly, we see that $k_{2,2}$ (= {6, 7, 13}) is included in $\Gamma(15)$. This set thus becomes $k^*_{2,2}$ = {6, 7, 13, 15}. Since $k^*_{2,2} = k^*_{7,4}$ we eliminate $k^*_{2,2}$ from the list of maximal cliques (Table 6.3). We also see that $k_{4,3}$ (= {4, 9, 10, 14}) is included in $\Gamma(8)$, so it becomes $k^*_{4,3}$ = {4, 8, 9, 10, 14}. Similarly, $k_{8,5}$ (= {8, 14}) is included in $\Gamma(4)$, $\Gamma(9)$ and $\Gamma(10)$ (with 4, 9 and 10 all mutually compatible). $K_{8,5}$ thus becomes

Table 6.3. Maximal cliques of G^* (output of the BioGraph program)

$k_{3,4*}$	[6]	= {6,7,9,11,12,15}
$k_{2,3}$	[6]	= {1,6,7,9,11,12}
$k_{8,5*}$	[5]	= {4,8,9,10,14}
$k_{4,2}$	[5]	= {1,8,9,11,12}
$k_{9,2*}$	[5]	= {6,9,10,12,15}
$k_{4,4}$	[4]	= {9,10,14,15}
$k_{7,4*}$	[4]	= {6,7,13,15}
$k_{1,3*}$	[4]	= {4,7,9,12}
$k_{3,2*}$	[4]	= {3,6,7,12}
$k_{1,4*}$	[3]	= {3,8,12}
$k_{2,1}$	[3]	= {3,5,7}
$k_{8,1}$	[2]	= {2,5}

$k^*_{8,5} = \{4, 8, 9, 10, 14\}$. In other words, $k^*_{8,5}$ is identical to $k^*_{4,3}$, which is then eliminated.

6.4.4 Constructing the Matrix M and the Graph G_k

The next step consists of establishing the relationships between the cliques. These relationships are given in the matrix M of the Table 6.4. The definitions of the symbols used in this table were given above (Sect. 6.3.4).

Example: the relationship between $k^*_{7,4}$ and $k_{2,3}$ is conflicting [$V(A) = 9 + 4 = 13$ and $V(B) = 1 + 1 = 2$]; (Fig. 6.2.a).

The relationship between $k_{2,3}$ and $k^*_{1,4}$ is undetermined [$V(A) = V(B) = 8$]; (Fig. 6.2.b).

We note here that this method of detecting contradictions highlights the forbidden configurations of the subgraph of G^*, which is made up of the elements of the $k_{i,j}$ being compared (Fig. 6.2.c).

Table 6.4. Graph G'_k (output of the Biograph program)

$k_{3.4*}$: $<<k_{2.3}, ->k_{8.5*}, ->k_{4.2}, >>k_{9.2*}, >>k_{4.4}, <<k_{7.4*},$
$->k_{1.3*}, <<k_{3.2*}, <-k_{1.4*}, <<k_{2.1}, <<k_{8.1}$
$k_{2.3}$: $>>k_{8.5*}, >>k_{4.2}, >>k_{9.2*}, >>k_{4.4}, <-k_{7.4*},$
$>>k_{1.3*}, <<k_{3.2*}, ??k_{1.4*}, <<k_{2.1}, <<k_{8.1}$
$k_{8.5*}$: $<<k_{4.2}, <-k_{9.2*}, >>k_{4.4}, <-k_{7.4*}, <<k_{1.3*},$
$<-k_{3.2*}, <-k_{1.4*}, <-k_{2.1}, <<k_{8.1}$
$k_{4.2}$: $->k_{9.2*}, >>k_{4.4}, <-k_{7.4*}, ->k_{1.3*},$
$<<k_{3.2*}, <<k_{1.4*}, <<k_{2.1}, <<k_{8.1}$
$k_{9.2*}$: $>>k_{4.4}, <<k_{7.4*}, <-k_{1.3*}, <<k_{3.2*},$
$<-k_{1.4*}, <<k_{2.1}, <<k_{8.1}$
$k_{4.4}$: $<<k_{7.4*}, <<k_{1.3*}, <<k_{3.2*},$
$<<k_{1.4*}, <<k_{2.1}, <<k_{8.1}$
$k_{7.4*}$: $->k_{1.3*}, <-k_{3.2*}, ->k_{1.4*},$
$<<k_{2.1}, <<k_{8.1}$
$k_{1.3*}$: $<-k_{3.2*}, ->k_{1.4*},$
$<-k_{2.1}, <<k_{8.1}$
$k_{3.2*}$: $>>k_{1.4*}, <<k_{2.1},$
$<<k_{8.1}$
$k_{1.4*}$: $<<k_{2.1},$
$<<k_{8.1}$
$k_{2.1}$: $<<k_{8.1}$

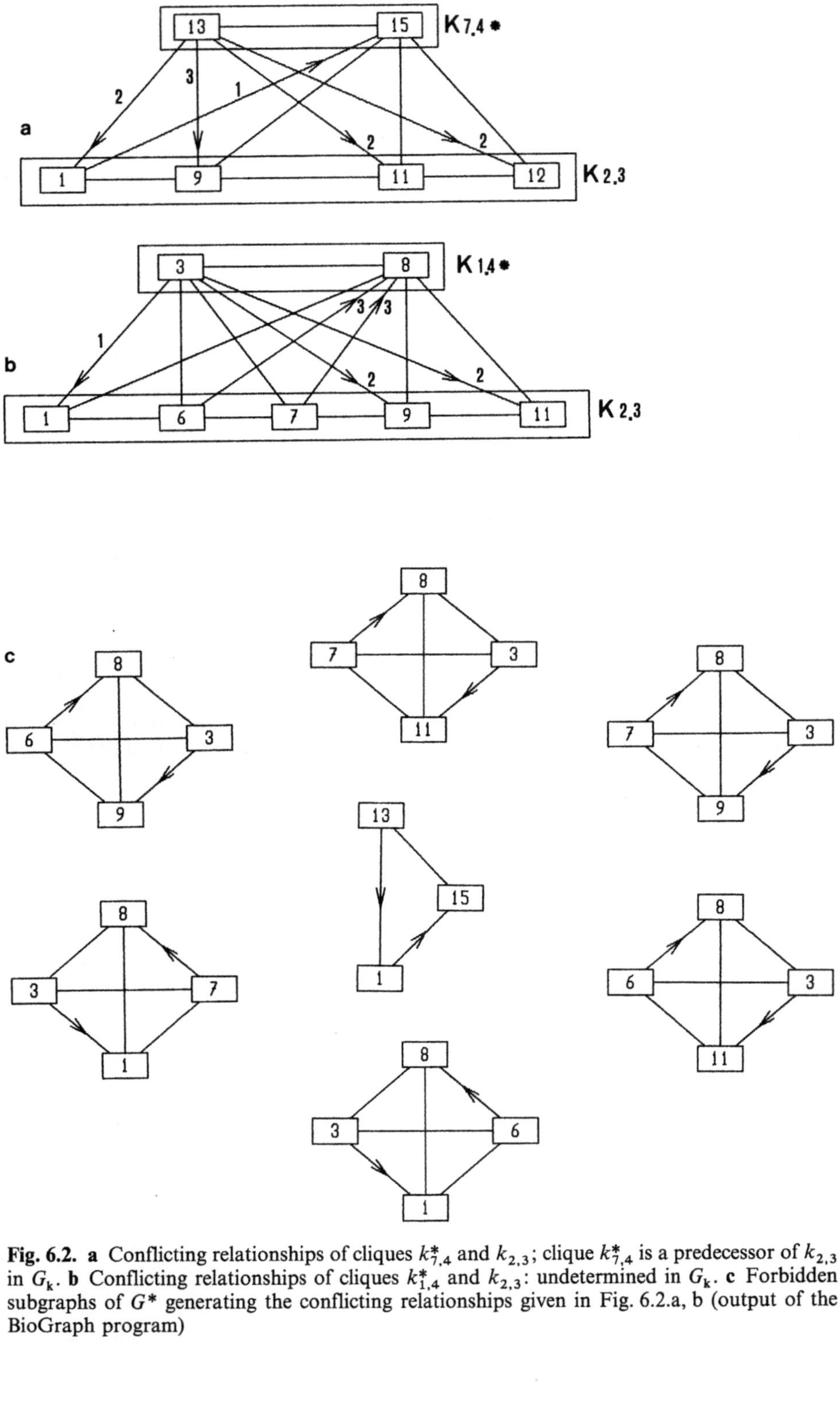

Fig. 6.2. **a** Conflicting relationships of cliques $k^*_{7,4}$ and $k_{2,3}$; clique $k^*_{7,4}$ is a predecessor of $k_{2,3}$ in G_k. **b** Conflicting relationships of cliques $k^*_{1,4}$ and $k_{2,3}$: undetermined in G_k. **c** Forbidden subgraphs of G^* generating the conflicting relationships given in Fig. 6.2.a, b (output of the BioGraph program)

6.4.5 Finding and Destroying the Connected Components of G_k

In our example, the oriented graph G_k associated to the matrix M contains a strongly connected component : $k^*_{3,4}$, $k_{4,2}$, $k^*_{1,3}$, $k^*_{1,4}$ (see Table 6.4). The destruction of the arcs $k^*_{1,4} \rightarrow k^*_{3,4}$ and $k_{4,2} \rightarrow k^*_{1,3}$ results in the destruction of the component.

6.4.6 Maximal Clique-Species Incidence Matrix, Residual Arcs and Unitary Associations

The maximal clique-species incidence matrix associated with the longest reduced maximal path (see Sect.5.4) of the preceding graph is given in Fig. 6.3.a. Figure 6.3.b shows the same matrix with the consecutive 1's in the columns, and Fig. 6.3.c shows the unitary associations after eliminating the nonmaximal cliques.

	SPECIES														
K_i	1	2	3	4	5	6	7	8	9	10	11	12	13	14	15
12	0	0	0	0	0	0	0	0	1	1	0	0	0	1	1
11	0	0	0	1	0	0	0	1	1	1	0	0	0	1	0
10	0	0	0	0	0	1	0	0	1	1	0	1	0	0	1
9	1	0	0	0	0	0	0	1	1	0	1	1	0	0	0
8	0	0	1	0	0	0	0	1	0	0	0	1	0	0	0
7	0	0	0	1	0	0	1	0	1	0	0	1	0	0	0
6	0	0	0	0	0	1	1	0	1	0	1	1	0	0	1
5	1	0	0	0	0	1	1	0	1	0	1	1	0	0	0
4	0	0	0	0	0	1	1	0	0	0	0	0	1	0	1
3	0	0	1	0	0	1	1	0	0	0	0	1	0	0	0
2	0	0	1	0	1	0	1	0	0	0	0	0	0	0	0
1	0	1	0	0	1	0	0	0	0	0	0	0	0	0	0

a)

	SPECIES														
K_i	1	2	3	4	5	6	7	8	9	10	11	12	13	14	15
12	0	0	0	0	0	0	0	0	1	1	0	0	0	1	1
11	0	0	0	1	0	0	0	1	1	1	0	0	0	1	1
10	0	0	0	1	0	1	0	1	1	1	0	1	0	0	1
9	1	0	0	1	0	1	0	1	1	0	1	1	0	0	1
8	1	0	1	1	0	1	0	1	1	0	1	1	0	0	1
7	1	0	1	1	0	1	1	0	1	0	1	1	0	0	1
6	1	0	1	0	0	1	1	0	1	0	1	1	0	0	1
5	1	0	1	0	0	1	1	0	1	0	1	1	0	0	1
4	0	0	1	0	0	1	1	0	0	0	0	1	1	0	1
3	0	0	1	0	0	1	1	0	0	0	0	1	0	0	0
2	0	0	1	0	1	0	1	0	0	0	0	0	0	0	0
1	0	1	0	0	1	0	0	0	0	0	0	0	0	0	0

b)

	SPECIES														
U.A	1	2	3	4	5	6	7	8	9	10	11	12	13	14	15
7	0	0	0	1	0	0	0	1	1	1	0	0	0	1	1
6	0	0	0	1	0	1	0	1	1	1	0	1	0	0	1
5	1	0	1	1	0	1	0	1	1	0	1	1	0	0	1
4	1	0	1	1	0	1	1	0	1	0	1	1	0	0	1
3	0	0	1	0	0	1	1	0	0	0	0	1	1	0	1
2	0	0	1	0	1	0	1	0	0	0	0	0	0	0	0
1	0	1	0	0	1	0	0	0	0	0	0	0	0	0	0

c)

Fig. 6.3. a Initial TGH matrix. **b** TGH with the consecutive 1's. **c** Compacted TGH (output of the BioGraph program)

A residual arc (13 → 12) is restored in the U. A.3 of the present solution. The details of this restoration are given in Fig. 6.4, and the final table of unitary associations is given in Fig. 6.5.

6.4.7 Technical Remark

When constructing virtual edges using the method described here, the greatest weight is given to the reproducibility of the arcs, while the technique described earlier (Sect. 4.4.2) gives most weight to their frequency of implication in forbidden configurations. This explains why the unitary association matrix of Fig. 6.5 differs slightly from that given in Fig. 5.4: in BioGraph, the conflict between the arcs 7 → 8 on the one hand and 3 → 9, 3 → 11 on the other is resolved in favor of the first arc, while in DV-86 it is resolved in favor of the two others (less repro-

SPLIT	1	2	3	4	5	6	7	8	9	10	11	12	13	14	15
U.A.3 (j)	0	0	1	0	0	1	1	0	0	0	0	1	1	0	1
U.A.3 (i)	0	0	1	0	0	1	1	0	0	0	0	1	1	0	1
RESTORE	**1**	**2**	**3**	**4**	**5**	**6**	**7**	**8**	**9**	**10**	**11**	**12**	**13**	**14**	**15**
U.A.3 (j)	0	0	1	0	0	1	1	0	0	0	0	1	0	0	1
U.A.3 (i)	0	0	1	0	0	1	1	0	0	0	0	0	1	0	1
REDUCE	**1**	**2**	**3**	**4**	**5**	**6**	**7**	**8**	**9**	**10**	**11**	**12**	**13**	**14**	**15**
U.A.4 = ↓	1	0	1	1	0	1	1	0	1	0	1	1	0	0	1
U.A.3 (j)	0	0	1	0	0	1	1	0	0	0	0	1	0	0	1
U.A.3	0	0	1	0	0	1	1	0	0	0	0	0	1	0	1

Fig. 6.4. Restoring residual arc 13 → 11

RESULT	1	2	3	4	5	6	7	8	9	10	11	12	13	14	15
U.A. 7	0	0	0	1	0	0	0	1	1	1	0	0	0	1	1
U.A. 6	0	0	0	1	0	1	0	1	1	1	0	1	0	0	1
U.A. 5	1	0	1	1	0	1	0	1	1	0	1	1	0	0	1
U.A. 4	1	0	1	1	0	1	1	0	1	0	1	1	0	0	1
U.A. 3	0	0	1	0	0	1	1	0	0	0	0	0	1	0	1
U.A. 2	0	0	1	0	1	0	1	0	0	0	0	0	0	0	0
U.A. 1	0	1	0	0	1	0	0	0	0	0	0	0	0	0	0

Fig. 6.5. Unitary associations table (output of the BioGraph program)

ducible than 7 → 8). This minor difference does not, however, change the correlations.

6.4.8 Correlations

The last stage of the procedure (identifying the U. A. and constructing the reproducibility matrix) is identical to that described in Section 5.5.

The correlations table and reproducibility matrix produced by BioGraph for the problem under study are illustrated in Fig. 6.6.

```
L  SEC.1    L  SEC.2    L  SEC.3    L  SEC.4    L  SEC.5
4: 5-  5    4: 7-  7    7: 7-  7    4: 7-  7    2: 4-  7
3: 4-  4    3: 4-  4    6: 7-  7    3: 7-  7    1: 1-  2
2: 3-  6    2: 3-  3    5: 4-  5    2: 5-  5
1: 1-  1    1: 2-  2    4: 4-  6    1: 4-  4
                        3: 3-  3
                        2: 3-  4
                        1: 1-  1

L  SEC.6    L  SEC.7    L  SEC.8    L  SEC.9    L  SEC.10
4: 7-  7    4: 3-  3    6: 7-  7    2: 6-  6    3: 5-  7
3: 4-  7    3: 2-  4    5: 7-  7    1: 6-  7    2: 6-  7
2: 4-  4    2: 1-  2    4: 3-  3                1: 4-  5
1: 3-  3    1: 1-  1    3: 3-  6
                        2: 2-  4                L  SEC.11
                        1: 1-  1                1: 7-  7
```

U.A	1	2	3	4	5	6	7	8	9	10	11	Z
7	0	1	1	1	█	1	0	1	0	█	1	D
6	0	0	0	0	█	0	0	0	1	█	0	D
5	1	0	█	1	█	0	0	0	0	█	0	C
4	1	1	█	1	█	1	0	0	0	█	0	C
3	0	1	1	0	0	1	1	1	0	0	0	B
2	0	1	0	0	█	0	0	0	0	0	0	A
1	1	0	1	0	█	0	1	1	0	0	0	A

Fig. 6.6. Correlations table and reproducibility matrix (*black rectangles*, unions of U.A.'s; *column Z*, zones *A–D*; *column L*, stratigraphic levels) (output of the BioGraph program)

Chapter 7
Detecting Reworking

7.1 Introduction

In biostratigraphy, the term "reworking" applies to a vast range of sedimentary processes. The extremes are:

1. Reorientation and mixing of biogenic debris within a bed as a result of bioturbation (the "vertical mixing" of Berger and Heath 1968);
2. The removal by erosion of an ancient fossil organism and its resedimentation in a younger deposit.

These two types of reworking have very different effects on the structure of biostratigraphic data.

Most of this Chapter will be devoted to identifying reworking of the second category.

7.2 Definitions

7.2.1 Discontinuous Stratigraphic Distribution

Fossil species are obviously never recorded in a continuous manner in a given sequence of sedimentary rock. In other words, their local stratigraphic distribution is always discontinuous.

It is nevertheless important to distinguish between this ordinary kind of discontinuity and discontinuities that give a false picture of the zonal distribution of species: a species can have a discontinuous stratigraphic distribution everywhere and still be present in all the biochronozones that overlap its entire existence interval. Such a species has a continuous biochronologic distribution.

7.2.2 Discontinuous Biochronologic Distribution

Consider a species x and a sequence of three biochronozones, A, B and C. If x exists in zones A and C but is never found in B, we say that x shows a discontinuous biochronologic distribution (i.e., its distribution in time seems to be discontinuous). This discontinuity can be interpreted in three ways: (1) x is absent in B because of gaps in the documentation; (2) x never existed during the time corresponding to B, and its presence in C is due to reworking (or its presence in A is due to downworking); (3) the two occurrences of x actually represent two separate homeomorphic species.

7.2.3 Nondisjunctive Reworking

The local stratigraphic range of a microfossil can be artificially lengthened (vertical transport into a superjacent zone). This can result from in situ mixing of fauna due to bioturbation, or from continuous lateral displacements that are subcontemporaneous with the burial of microorganisms.

This kind of reworking does not create discontinuities in the biochronologic distribution of species. Therefore we call it "nondisjunctive".

Detecting such reworking according to strictly biochronologic criteria is difficult. We will discuss this problem in Chapters 11 and 12.

7.2.4 Disjunctive Reworking

Reworking that results in a discontinuity in the biochronologic or temporal distribution of a fossil species is called "disjunctive".

The current Chapter is devoted to detecting anomalies of biochronologic distribution of fossil species that are due to such reworking. This can be difficult in the absence of sedimentological criteria (e.g., obvious resedimentation, physical deterioration of fossils, etc.) or any taphonomic criterion that allows one to identify it with certainty.

Four methods can be used to detect discontinuities in the biochronologic distribution of species. These methods can be applied only after meaningful biochronologic zones have been established.

7.3 Methods

7.3.1 Direct Approach

7.3.1.1 Technical Description

In the intermediate stage of the calculations, the BioGraph program constructs the TGH matrix, which shows the discontinuities in the stratigraphic distribution of some species (Sect. 6.3.7).

To distinguish minor discontinuities from those with an amplitude greater than a zone, the following procedure is used:

First, the TGH matrix is completed by adding the residual arcs and virtual edges calculated during the last steps of the procedure (Sect. 6.3.8 and 6.3.9). Next, the chronologically significant zones (established earlier) are transcribed in this matrix.

The resulting matrix TGH′ enables us to identify the desired discontinuities: all species that are present in two nonadjacent zones, *A* and *C*, and that are absent from zone *B* between *A* and *C* have, by definition, a discontinuous biochronologic distribution.

7.3.1.2 Example

The alveolinid zones *A*–*D* established above (Sect. 6.4.8) are entered in the column *Z* of the matrix TGH′ of Fig. 7.1. This matrix shows that species 3 is absent from zone *B*, which is identified in sections 2, 3, 6, 7 and 8 (Fig. 6.6). To interpret the discontinuous biochronologic distribution of this species, it is necessary to refer to the biostratigraphic database (Fig. 3.21) and to the correlations table (Fig. 6.6). Doing this, we note that species 3 is present in the levels that belong to zone *A* in profiles 2, 3 and 7, and also in level 4 of section 1, which is assigned to zone *C*.

This leads us to suspect that this level contains a reworked fauna, a conclusion that is confirmed by Drobne's sedimentological observations (loc.cit. p. 82), which showed that the alveolinids from this horizon were mechanically rounded and deteriorated.

7.3.2 Detection by Removing Species

7.3.2.1 Method

The resolution power of our correlation method tends to decrease as the problem becomes more complex (i.e., when G^* has more forbidden subgraphs). The number of virtual edges added to G^* increases as a function of this complexity and, as a result, the number of unitary associations tends to decrease. In extreme cases, this can obscure some potentially meaningful subdivisions.

K_i	Z	1	2	▼ 3	4	5	6	7	8	9	10	11	12	13	14	15
12	D	0	0	0	0	0	0	0	0	1	1	0	0	0	1	1
11	D	0	0	0	1	0	0	0	1	1	1	0	0	0	1	0
10	D	0	0	0	0	0	1	0	0	1	1	0	1	0	0	1
9	C	1	0	0	0	0	0	0	1	1	0	1	1	0	0	0
8	C	0	0	1	0	0	0	0	1	0	0	0	1	0	0	0
7	C	0	0	0	1	0	0	1	0	1	0	0	1	0	0	0
6	C	0	0	0	0	0	1	1	0	1	0	1	1	0	0	1
5	C	1	0	0	0	0	1	1	0	1	0	1	1	0	0	0
4	B	0	0	0	0	0	1	1	0	0	0	0	0	1	0	1
3	A	0	0	1	0	0	1	1	0	0	0	0	0	0	0	0
2	A	0	0	1	0	1	0	1	0	0	0	0	0	0	0	0
1	A	0	1	0	0	1	0	0	0	0	0	0	0	0	0	0

Fig. 7.1. TGH′ matrix. K_i, maximal cliques. *Z*, zones *A*–*D*. Species 3 shows a discontinuous biochronologic distribution (output of the BioGraph program)

In such situations, we must get rid of the species that appear in the greatest number of forbidden subgraphs of G^*. The threshold at which such species are removed depends on the nature and complexity of the problem.

Establishing correlations without these species should enable us to discover which of them have a discontinuous biochronologic distribution: to do this we merely examine the stratigraphic position they occupy in the different biochronozones of the sections under study.

7.3.2.2 Example

Figure 7.2 shows the frequency with which the 15 alveolinid species appear in the forbidden subgraphs of the initial graph G^* (Fig. 4.6).

If we recalculate the biostratigraphic data after removing species 3, 4 and 8, which appear most frequently in these subgraphs, we get the correlations shown in Fig. 7.3.

Examining the positions that the removed species occupy in the different sections (Fig. 3.21) we see that:

1. Species 3 is found in zone *A* of section 2, zone *B* of section 3, below zone *B* in section 7 and above zone *C* in section 1. Its distribution is thus discontinuous.
2. Species 4 is found in zones *C* and/or *D* of sections 1, 4 and 6 and it occupies an undetermined biochronologic position above zone *A* of section 5. Its biochronologic distribution is thus continuous.
3. Species 8 is found in zone *C* and/or *D* of sections 1, 4, 8 and 10; its distribution is thus also continuous (Fig. 7.3.c).

We conclude, as we did above, that only species 3 has a discontinuous biochronologic distribution and its presence in bed 4 of section 1 may be due to reworking.

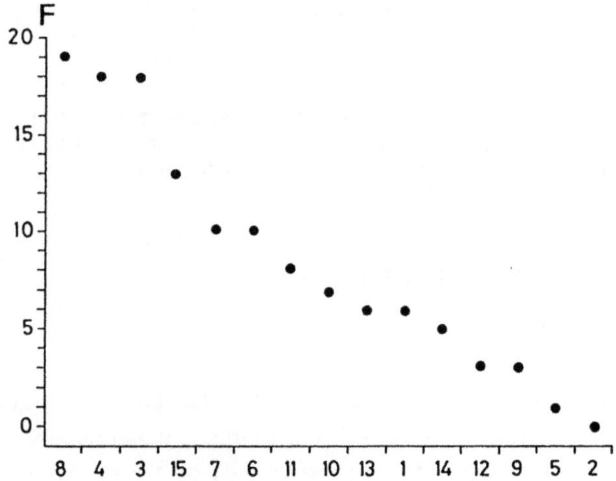

Fig. 7.2. Occurrence frequency (*F*) of alveolinid 1–15 in the forbidden subgraphs of the initial graph G^* (Fig. 4.6)

a) 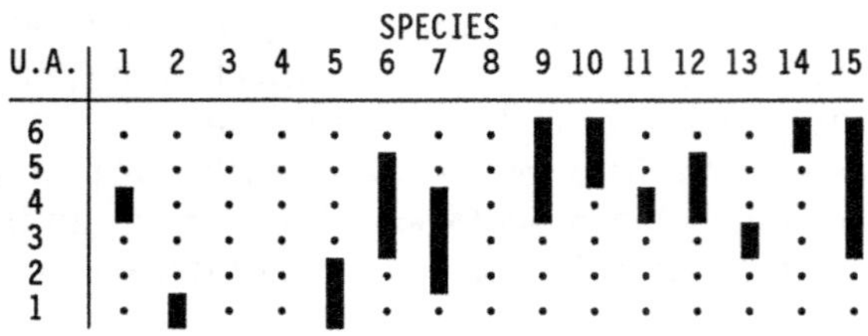

b)

U.A.	1	2	3	4	5	6	7	8	9	10	11	Z
6	0	1	1	1	0	1	0	1	0	■	1	D
5	0	0	0	0	0	0	0	0	1	■	0	D
4	1	1	1	1	0	1	0	0	0	1	0	C
3	0	1	1	0	0	1	1	1	0	0	0	B
2	0	1	0	0	■	0	0	0	0	0	0	A
1	1	0	1	0	■	0	1	1	0	0	0	A

(SECTIONS)

c)

```
L  SEC.1          L  SEC.2        L  SEC.3        L  SEC.4
4: ?-   ? (3,8)   4: D-   D       7: D-   D       4: D-   D
3: C-   C (4)     3: C-   C       6: D-   D       3: D-   D (4)
2: B-   C         2: B-   B       5: C-   C       2: C-   C (8)
1: A-   A         1: A-   A (3)   4: C-   C       1: C-   C
                                  3: B-   B
                                  2: B-   B (3)
                                  1: A-   A

L  SEC.5          L  SEC.6        L  SEC.7        L  SEC.8
2: A-   D (4)     4: D-   D       4: B-   B       6: D-   D
1: A-   A         3: C-   D (4)   3: A-   B (3)   5: D-   D (8)
                  2: C-   C       2: A-   A       4: B-   B
                  1: B-   C       1: A-   A       3: B-   B
                                                  2: A-   B
                                                  1: A-   A

L  SEC.9          L  SEC.10        L  SEC.11
2: D-   D         3: D-   D (8)    1: D-   D
1: D-   D         2: D-   D
                  1: C-   C
```

Fig. 7.3. **a** Protoreferential constructed after removing species 3, 4 and 8. **b** Reproducibility matrix (*column Z*, zones *A–D*). **c** Correlations generated by this protoreferential and biochronologic position of species 3, 4 and 8 in sections 1–11 (output of the BioGraph program)

7.3.3 Interpreting C_3 Cycles

7.3.3.1 Method

In general we can say that all forbidden subgraphs of a biostratigraphic graph may result from species with discontinuous time distribution. Some of these subgraphs can be destroyed easily by adding virtual edges (see Chap. 4), but for others it is more difficult. That is in particular the case for C_n cycles: considered individually, a sequence of species $x \rightarrow y \rightarrow z \rightarrow x$ gives no information about the order in which x, y and z followed each other in time.

The only irreducible cycles that we frequently find in tricky biochronologic problems are those of the C_3 type and their detection is not difficult (Sect.4.3.4). To each arc of a C_3 cycle of G^* we can give a value

$$s = c/r,$$

where c is the number of times that the arc (x, y) appears in a C_3 and where r is the number of times the superposition between x and y is observed in the stratigraphic sections.

The greater the value of s of an arc (x, y), the more likely this arc is to represent an abnormal stratigraphic relationship caused by a species with discontinuous biochronologic distribution. Therefore, to destroy the C_3 cycles of G^*, we simply remove the arcs with the highest values of s. We do this as follows: if it appears that a superposition $x \rightarrow y$ is abnormal, we ignore the presence of y in the beds where it is observed in sequence with species x.

7.3.3.2 Example

The alveolinid biostratigraphic graph (Fig. 3.22) contains three imbricated C_3 cycles. The value of s of each of the arcs of these cycles is given in Fig. 7.4.

It is evident that the coefficient s of the arc (4, 3) has a much higher value than the others, so we can assume that this arc represents an abnormal stratigraphic superposition. This purely technical deduction allows us to detect immediately the only stratigraphic relationship here that is likely to result from reworking.

7.3.4 Interpreting Semioriented Virtual S_3 Circuits

7.3.4.1 Definition

Remember that two species x and y are virtually associated when x is observed below y in a given location and the inverse stratigraphic relationship is observed in another location (Sect. 2.2.1). Most initial virtual coexistences are due to the

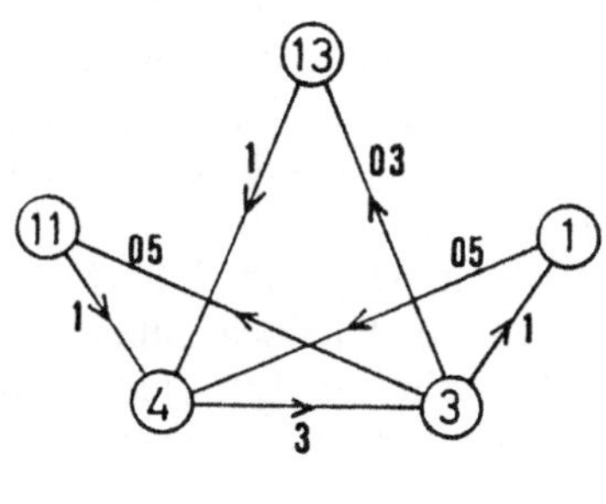

Fig. 7.4. Three imbricated C_3 cycles: the values 0.3, 0.5, and 1 are those of the coefficient s of the different arcs (Guex 1982)

fragmentary nature of the fossil record. However, they can also result from reworking or contamination.

Examined individually, these virtual associations give few clues as to their origin (insufficient documentation or reworking). This is not the case when they give rise to semioriented S_3 circuits in the biostratigraphic graph. Such circuits are called virtual S_3.

Two different kinds of stratigraphic situations can generate such circuits.

7.3.4.2 Virtual S_3 of Type 1

An S_3 generated by the relationships of three species, x, y and z, is a virtual S_3 of type 1 (denoted S_3') if: (1) a sequence $x \to y \to z$ is observed in more than one location; and (2) the superposition $z \to x$ (generating the virtual edge of the S_3) is observed in an isolated location.

In this case, it is reasonable to assume that the relationship $z \to x$ is caused by a discontinuous biochronologic distribution of species x (in other words, by possible reworking).

7.3.4.3 Virtual S_3 of Type 2

An S_3 resulting from the relationships of three species, x, y and z, is a virtual S_3 of type 2 (denoted S_3'') if the superpositions $x \to y$, $y \to z$, $x \to z$ and $z \to x$ are observed in different locations.

These S_3'' do not allow us to detect discontinuities in the temporal distribution of species.

7.3.4.4 Example

Four virtual coexistences are represented by edges in our alveolinid graph of Fig. 3.22. They are the pairs: {4, 8}, {3, 12}, {6, 15}, and {8, 10}.

One of these virtual edges appears in a virtual S_3 of type 1: the pair {3, 12}, which forms a semioriented circuit with species 13. The local superpositions observed among these three species are given in Fig. 7.5.

We note that the sequence $3 \to 13 \to 12$ is observed in sections 2 and 3, the sequence $3 \to 13$ in section 7 and the superposition $12 \to 3$ only in section 1. (It is this superposition that gives rise to the virtual edge {3, 12}).

In this case, it is clearly the superposition $12 \to 3$ that results from a discontinuous biochronologic distribution of species 3. The detection of this S_3' allows us once more to assume that bed 4 of section 1 contains a species that was probably reworked.

Last, we note that the alveolinid graph also contains two imbricated S_3'' (Fig. 7.6). The local superpositions of the four species jointly implicated in these subgraphs are illustrated in Fig. 7.6. This figure shows in a simple way that in this case we cannot interpret these subgraphs as discontinuous biochronologic distri-

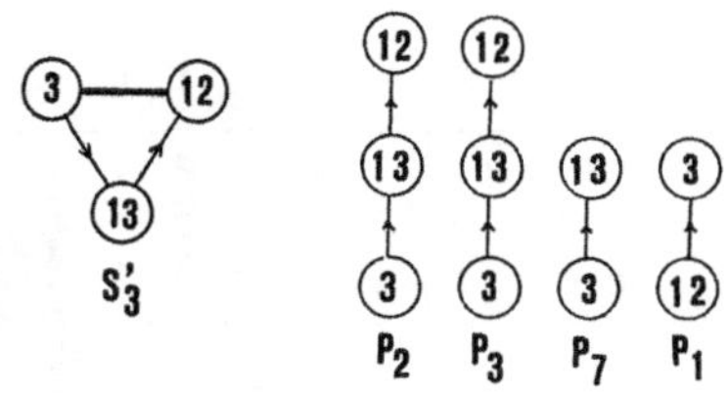

Fig. 7.5. Stratigraphic position of species 3, 12 and 13 in sections P_1, P_2, P_3 and P_7 (Fig. 3.21). The semioriented S'_3 circuit represents these relationships

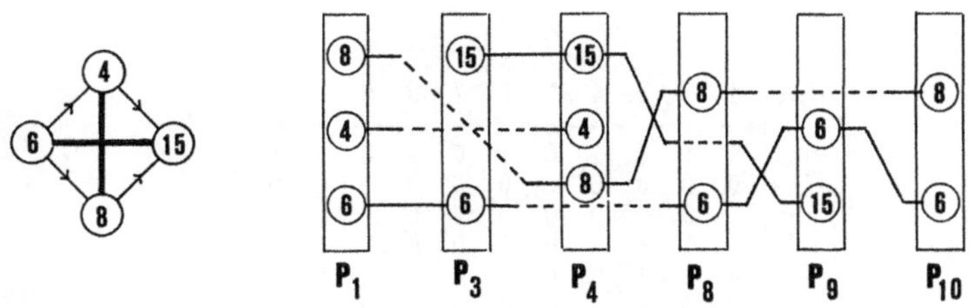

Fig. 7.6. Two imbricated semioriented S'_3 circuits representing the stratigraphic relationships of the species and their relative positions in sections P_1-P_{10}

butions. It is clear that the extremely contradictory nature of the local stratigraphic relationships among the alveolinids of Drobne's sections has three possible causes: strong variations in the quality of the local fossil record, strong ecological control on the distribution of the species (see Fontaine and Rey 1984), or insufficient sampling of some localities.

7.4 Conclusion to the Alveolinid Problem

If we treat Drobne's data after eliminating bed 4 of section 1 (Fig. 3.21), which contains a reworked fauna, we get an initial sequence of eight unitary associations (Fig. 7.7.a). The reproducibility matrix of that sequence and the graph G'_k that represents the superpositions among the characteristic elements of these unitary associations are illustrated in Figs. 7.7.b and 7.7.c. The reduction of G'_k to a unique path results in five biochronozones (Fig. 7.7.d).

The correlations corresponding to this solution are given in Fig. 7.8. They are very similar to those established empirically by Drobne (loc.cit.): zones 1–5 of the referential in Fig. 7.7.d correspond closely to Drobne's informal horizons 2–6.

7.5 Conclusion to the Reworking Problem

The above procedures for detecting reworking cannot be used to make sense of conflicting stratigraphic relationships among fossil species collected in turbidites, cuttings, karstic fillings, or other collections of refuse. Such collections can be dated only indirectly (and approximately), using biochronologic scales already constructed in other areas.

a)

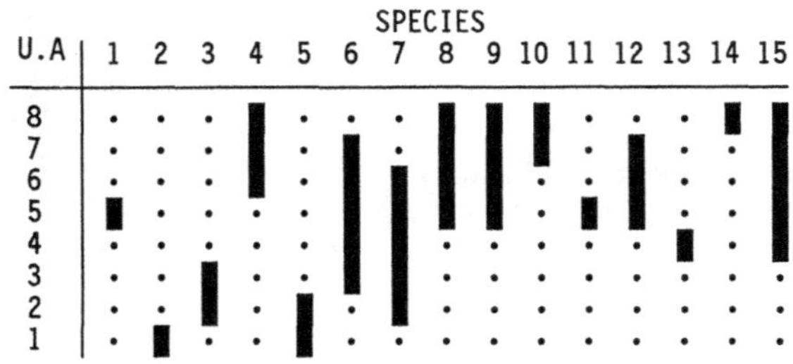

b)

SECTIONS

U.A.	1	2	3	4	5	6	7	8	9	10	11
8	0	1	1	1	0	1	0	1	0	0	1
7	0	0	0	0	0	0	0	0	1	0	0
6	1	0	0	0	0	0	0	0	0	0	0
5	0	1	1	1	0	1	0	0	0	1	0
4	0	1	1	0	0	1	1	1	0	0	0
3	0	0	1	0	0	0	0	0	0	0	0
2	0	1	0	0	0	0	0	0	0	0	0
1	1	0	1	0	0	0	1	1	0	0	0

c)

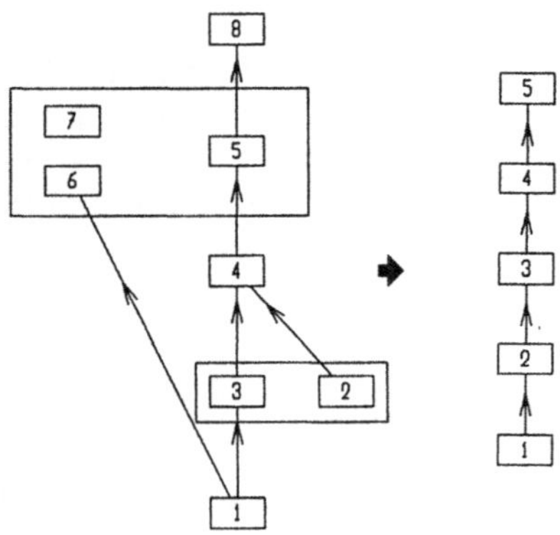

d)

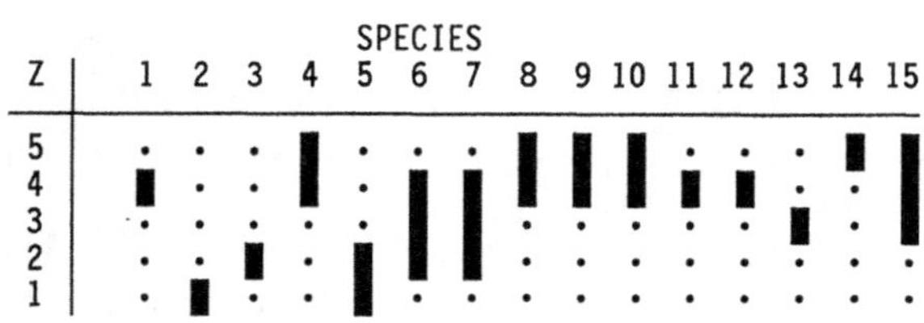

e)

SECTIONS

Z	1	2	3	4	5	6	7	8	9	10	11
5	0	1	1	1	■	1	0	1	0	0	1
4	1	1	1	1	■	1	0	0	1	1	0
3	0	1	1	0	0	1	1	1	0	0	0
2	0	1	1	0	■	0	1	0	0	0	0
1	1	0	1	0	■	0	1	1	0	0	0

Fig. 7.7. a Initial protoreferential. *Solid bars*, stratigraphic distribution of species 1–15. **b** Initial reproducibility matrix. **c** Graph G'_k reduced to a unique path L'. **d** Zonal referential. *Solid bars*, stratigraphic distribution of species 1–15. **e** Reproducibility of zones 1–5. *Black rectangle*, strictly identified unions of zones

```
L SEC.1    L SEC.2    L SEC.3    L SEC.4    L SEC.5
3 4 - 4    4 5 - 5    7 5 - 5    4 5 - 5    2 4 - 5
2 2 - 4    3 4 - 4    6 5 - 5    3 5 - 5    1 1 - 2
1 1 - 1    2 3 - 3    5 4 - 4    2 4 - 4
           1 2 - 2    4 4 - 4    1 4 - 4
                      3 3 - 3
                      2 2 - 2
                      1 1 - 1

L SEC.6    L SEC.7    L SEC.8    L SEC.9    L SEC.10
4 5 - 5    4 3 - 3    6 5 - 5    2 4 - 4    3 4 - 5
3 4 - 5    3 2 - 2    5 5 - 5    1 4 - 5    2 4 - 5
2 4 - 4    2 1 - 2    4 3 - 3               1 4 - 4
1 3 - 3    1 1 - 1    3 2 - 4
                      2 2 - 4               L SEC.11
                      1 1 - 1               1 5 - 5
```

Fig. 7.8. Correlation of Drobne's (1977) 11 sections based on the zonal referential of Fig. 7.7. d. *L*, stratigraphic levels (output of the BioGraph program)

Chapter 8
Quality of the Fossil Record

8.1 Introduction

Three main criteria can be used to evaluate the quality of a fossil species' vertical and lateral record:

1. The biochronologic dispersion of its datums;
2. Its actual lateral range (as opposed to the ideal);
3. The number of forbidden subgraphs resulting from conflicting superpositions.

We define below several coefficients that assign a simple numerical value to these variables. The functional relationships among these factors are sketched in Section 8.4 and these relationships will be studied in detail in Chapters 10–12.

8.2 Definitions

8.2.1 Datum, FAD and LAD

The term "datum" is used here to mean the first local appearance or disappearance of a species, without preconceived phylogenetic connotations.

We will systematically use the standard abbreviations FAD ("first appearance datum") and LAD ("last appearance datum").

8.2.2 Diachronism of Datums

Paleontologists are well aware that the FADs and LADs of fossil species are never really synchronous in locations that are remote from each other. If a datum is found in different biochronozones in different locations, we say it is "diachronous" (Fig. 8.1). In the following Chapters we are interested in the detection of diachronisms whose amplitude is greater than a biochronozone.

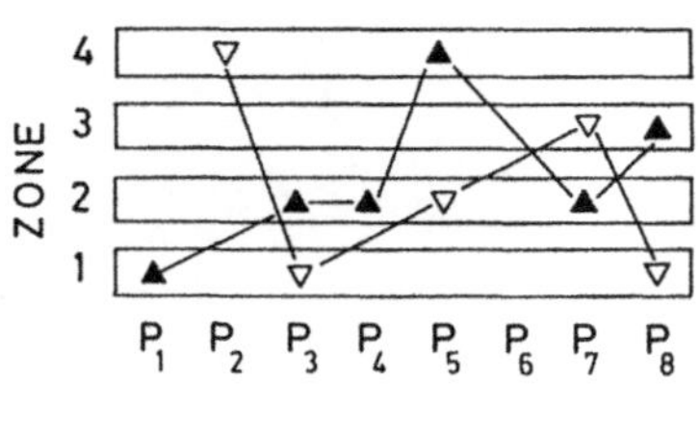

Fig. 8.1. Diachronous FADs (*solid triangles*) and LADs (*white triangles*). P_1–P_8, stratigraphic sections

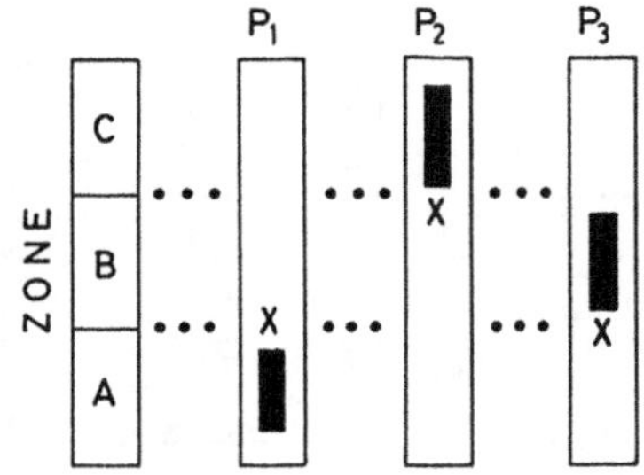

Fig. 8.2. Species x that is diachronous on a regional scale. $P_1 - P_3$, stratigraphic sections. *Black bars*, stratigraphic distribution of species x

8.2.3 Diachronous Species

If, in a given stratigraphic section, a species' FAD is located in a more recent biochronozone than its LAD in another section, the species is diachronous on a regional scale (Fig. 8.2).

We should note that such a species does not necessarily have a discontinuous biochronologic distribution (Sect. 7.2.2).

8.2.4 Biochronologic Dispersion

The "biochronologic dispersion" of a datum corresponds to the amplitude of its diachronism. A numerical characterization of this dispersion is given in Section 8.3.

8.2.5 Apparent Longevity of a Species

The "apparent longevity" $L_a(x)$ of a species x (i.e., its apparent existence interval) equals the number of biochronozones in which it exists (actually or virtually).

8.2.6 Actual Geofrequency and Ideal Geofrequency

The "actual geofrequency" of a species is the number of sections in which the species is found. Its "ideal geofrequency" is the number of sections in which it would be found if it were preserved (and discovered) in all the sections where at least one of the biochronozones of its apparent existence interval ($L_a(x)$) is recorded.

An example is given in Fig. 8.3. It shows a sequence of four zones (*A*, *B*, *C*, *D*) generated by the superpositions of ten species in four sections, $P_1 - P_4$. We will assume that these sections contain respectively zones *A* and *B*, *B* and *C*, *C* and *D*, and *A*, *B*, *D*.

In this imaginary case the ideal geofrequencies of species 1–10 of the zonal referential of Fig. 8.3 are 4, 4, 3, 2, 4, 4, 3, 3, 2, 2 respectively (for example: if its

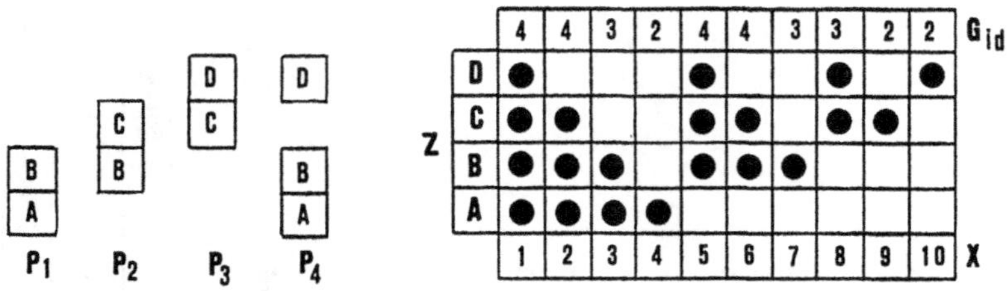

Fig. 8.3. $P_1 - P_4$, four imaginary sections. G_{id}, ideal geofrequency. X, 10 species (1–10). Z, zones $A-D$

lateral record were perfect, species 4 should be found in sections 1 and 4; similarly, species 1, 2, 5 and 6, should be found in all four sections, etc.).

8.3 Coefficients

8.3.1 Biochronologic Dispersion

8.3.1.1 Coefficients $D_a(x)$, $D_d(x)$ and $D_t(x)$

The amplitude of the biochronologic dispersion of a FAD (or LAD) is equal to n ($n = j - i$) when this FAD (or LAD) is found in the ith zone of one section and in the jth zone of another (with $j > i$ and $i = 1, 2, \ldots, m$).

$D_a(x)$ denotes the amplitude of the biochronologic dispersion of the FAD of a species x and $D_d(x)$ denotes that of its LAD.

The total diachronism of the FAD and LAD of a species [denoted $D_t(x)$] equals the sum of the amplitudes $D_a(x)$ and $D_d(x)$ of its FAD and its LAD:

$$D_t(x) = D_a(x) + D_d(x).$$

We will use the simplified notation D_a and D_d when referring to the biochronologic dispersion of the local appearances and disappearances of species in general.

8.3.1.2 Example

The values of the coefficients $D_a(x)$ and $D_d(x)$ of the 15 species of alveolinids of our real example are given in Table 8.1. These values are based on the zonal correlations given in Section 7.4.

8.3.1.3 Technical Remark

If the lower (or upper) limit of a biochronozone is outside the stratigraphic interval recorded in a given section, we must ignore the first local appearances (or disappearances) that occur at the base (or top) of this section when establishing the values of D_a and D_d.

Table 8.1. Biochronologic dispersion

X	D_a	D_d	D_t
1	0	0	0
2	0	0	0
3	0	0	0
4	1	1	2
5	1	1	2
6	2	0	2
7	2	0	2
8	1	1	2
9	1	0	1
10	1	1	2
11	0	0	0
12	0	0	0
13	0	0	0
14	0	0	0
15	2	2	4

Figure 8.4 shows the local ranges of three species in three sections, P_1, P_2, and P_3, where three zones (*A*, *B*, and *C*) are partially recorded.

Species 1 exists in zones *B* and *C*. It appears in zone *B* in sections 1 and 3 and in zone *C* in section 2. The amplitude of its dispersion D_a equals 1. It disappears in zone *C* of sections 1 and 2 and its range is truncated at the top of zone *B* in section 3. This truncation must be ignored and its dispersion D_d thus equals zero.

Species 2 exists in zones *B* and *C*. It appears in zone *B* in sections 1 and 2, and its range is truncated at the top of zone *B* in section 3: its dispersion D_d equals zero.

Species 3 exists in zones *A*, *B* and *C*. It appears at the base of zone *B* of sections 1 and 2, which is truncated. Its dispersion D_a also equals zero. It disappears in zone *B* of sections 2 and 3 and in zone *C* of P_1: its dispersion D_d equals 1.

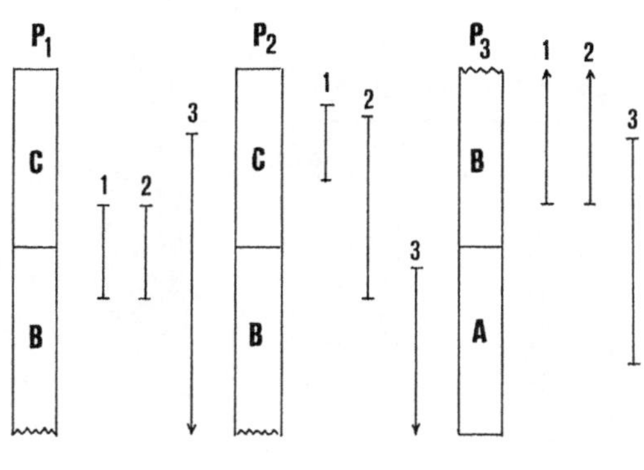

Fig. 8.4. Local range of three species (1–3) in three imaginary sections (P_1–P_3). The base of zone *B* is truncated in sections P_1 and P_2 and its top is truncated in P_3. The biochronologic position of the FAD of species 3 is unknown in P_1 and P_2 and the positions of the LADs of species 1 and 2 are unknown in P_3

8.3.2 Relative Geofrequency

8.3.2.1 *Coefficient $G_r(x)$*

The quality of the lateral record of a species is evaluated with the help of the coefficient

$$G_r(x) = \frac{\text{Actual geofrequency}}{\text{Ideal geofrequency}} .$$

8.3.2.2 *Example*

If species 4 of Fig. 8.3 is found in sections 1 and 4 (P_1 and P_4), its coefficient G_r equals 1, for it exists only in zone A.

If species 5 of the same figure is found only in one section, its G_r equals 0.25, for its apparent existence interval covers zones B–D, which are (partially) recorded in the four sections.

8.3.3 Forbidden Subgraphs

8.3.3.1 *Coefficient $I_t(x)$*

The number of forbidden subgraphs in which species appear reflects in part the quality of their record in the fossil-bearing terrains.

When analyzing the complexity of the problems studied in Chapters 10 and 11 we will use a coefficient $I_t(x)$ equal to the number of times a species appears such conflicting relationships.

The BioGraph program calculates the individual occurrence frequency of taxa in the C_3, S_3, C_4, S_4, Z_4, Z'_4 and Z''_4 subgraphs of a graph G^*.

8.3.3.2 *Example*

Table 8.2 gives the occurrence frequency of the alveolinid species in the forbidden subgraphs of the graph G^* of Fig. 3.22 (N.B.: these subgraphs are shown in Fig. 4.6).

Table 8.2. Coefficient I_t

X	I_t	X	I_t
1	6	9	3
2	0	10	7
3	18	11	8
4	18	12	3
5	1	13	6
6	10	14	5
7	10	15	13
8	19		

8.4 Relationships Between the Coefficients

8.4.1 Graphical Representation

Some of the relationships between the coefficients defined here can be shown graphically in double-entry tables like the one in Fig. 8.5. These tables essentially record how many species belong in the various sectors of the table.

Consider, for example, 100 species, each with two coefficients, C_1 varying from 1 to 5 and C_2 varying from 1 to 4.

If C_1 and C_2 equal 1 for 40 species, we enter the value 40 in the entry (1, 1) of the table (Fig. 8.5). If C_1 equals 3 and C_2 equals 4 for 20 species, we write 20 in the entry (3, 4) of the table, and so on.

8.4.2 Quality of the Vertical Record of Species

8.4.2.1 Relationship Between Longevity and Biochronologic Dispersion

Two relationships are obvious.

1. The greater the apparent longevity ($L_a(x)$) of a species, the more likely it is to have a large biochronologic dispersion.
2. Species with an apparent longevity of 1 have (by definition) a total dispersion equal to zero $[D_a(x) = D_d(x) = 0]$.

The statistical correlation between apparent longevity and biochronologic dispersion is necessarily positive: if we plot the relationship between $L_a(x)$ and $D_a(x)$ [or $D_d(x)$ or $D_t(x)$] and if that plot is a straight line, the slope of this line will always be equal to or greater than zero. In the best case, $D_a(x) = D_d(x) = 0$, and in the worst case, $D_a(x) = D_d(x) = L_a(x) - 1$.

The functions $D_a(x) = f[L_a(x)]$ and $D_d(x) = f[L_a(x)]$ thus give a partial description of the quality of the vertical record of the group under study.

In Chapters 10–12, the relationships between the biochronologic dispersion of the datums and the longevity of species will be represented by tables constructed according to the rule described above. These tables will also contain information about the percentage of species that occupy their various sectors: $D_a(x) = L_a(x) - 1$, $D_a(x) = L_a(x) - 2$, $D_a(x) = L_a(x) - 3$, etc. or $D_t(x) = L_a(x) + 1$, $D_t(x) = L_a(x)$, $D_t(x) = L_a(x) - 1$, etc.

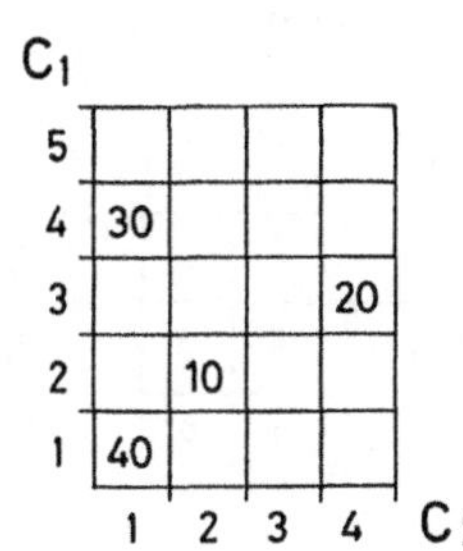

Fig. 8.5. Standard representation of the relationships between any two coefficients C_1 and C_2. For example, $C_1 = 4$ and $C_2 = 1$ for 30 species, $C_1 = C_2 = 1$ for 40 species

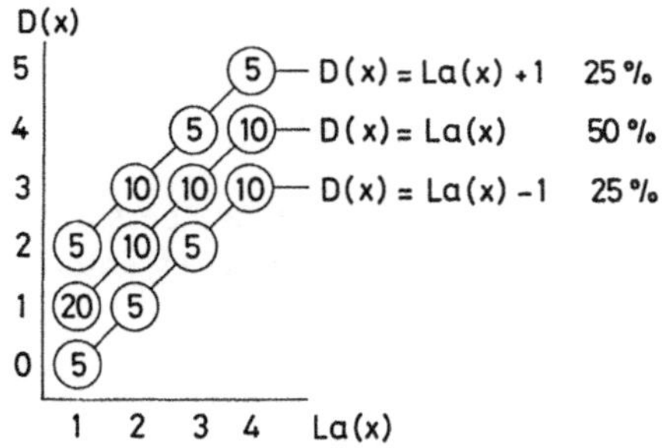

Fig. 8.6. Relationship between $D(x)$ and $L_a(x)$ in a set of 100 species

Figure 8.6 shows that, in an imaginary set of 100 species, 25% have a dispersion $D_a(x)$ equal to $L_a(x) - 1$, 50% have a dispersion $D_a(x)$ equal to $L_a(x)$, etc.

8.4.2.2 Relationship Between $D_a(x)$ and $D_d(x)$

In addition to being of purely paleontological interest, knowing whether the FADs of a given group are more diachronous than their LADs is of direct practical interest when using Interval Zones defined by these datums (Sect.16.4). We can determine this with the help of a double-entry table (like the one in Fig. 8.5), showing the frequency of the different combinations of the values D_a and D_d within a set of species.

The table of Fig. 8.7 shows the frequency of these combinations for the 15 alveolinid species of our standard example (Table 8.1).

The values in the entries along the diagonal of this table correspond to the number of species for which D_a equals D_d. The values in the entries above (or below) this diagonal correspond to the number of species for which $D_a < D_d$ (respectively $D_a > D_d$).

In this case we see that: (1) fewer than half the species have a biochronologic dispersion equal to zero ($D_a = D_d = 0$); and (2) the FADs of these species are clearly more diachronous than their LADs.

This is probably due to the fact that phenotypical variations of the alveolinid are significantly controlled by changes in the environment (see Fontaine and Rey 1984).

8.4.2.3 Variations in the Ratio $D_t(x)/L_a(x)$

Variations in the ratio $D_t(x)/L_a(x)$ (total dispersion over apparent longevity) give immediate and interesting information about the global quality of the vertical stratigraphic record of species.

Dd \ Da	0	1	2
2	0	0	1
1	0	4	0
0	7	1	2

$D_d(x) > D_a(x)$ (0 species)

$D_a(x) = D_d(x)$ (12 species)

$D_a(x) > D_d(x)$ (4 species)

Fig. 8.7. Relationship between $D_a(x)$ and $D_d(x)$ for Drobne's 15 alveolinid species

The diagram of Fig. 8.8 is divided into five sectors ($A-E$), which show the limits within which this ratio can fluctuate.

1. Sector A records all species with a total biochronologic dispersion of zero.
2. Sectors B and C record species with a moderate (B) or strong (C) dispersion. The local FADs of these species are anterior everywhere to their local LADs.
3. Sector D records species whose distribution is diachronous on a regional scale (i.e., species found in different biochronozones in different places).
4. Sector E is always empty since the total dispersion of a species cannot exceed $2[L_a(x) - 1]$.

Consider, for example, a sequence of eight zones (1–8) and three species, with apparent longevities of 8, 6, and 3 respectively. The local biochronologic (and stratigraphic) distribution of these species in three imaginary sections is given in Fig. 8.9.

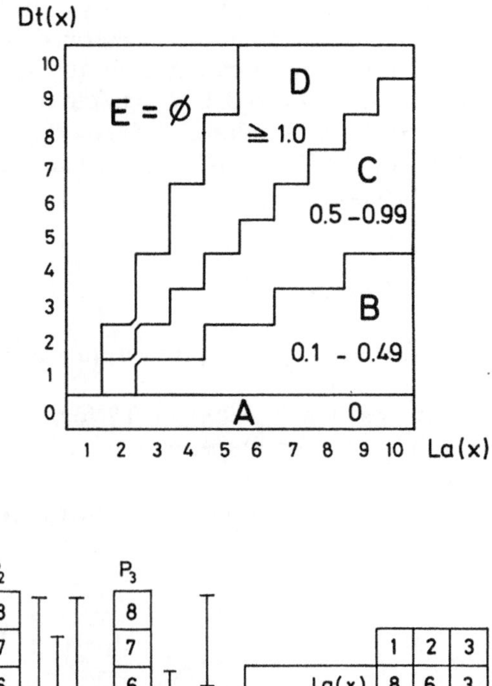

Fig. 8.8. Ratio ($D_t(x)/L_a(x)$: standard sectors $A-E$

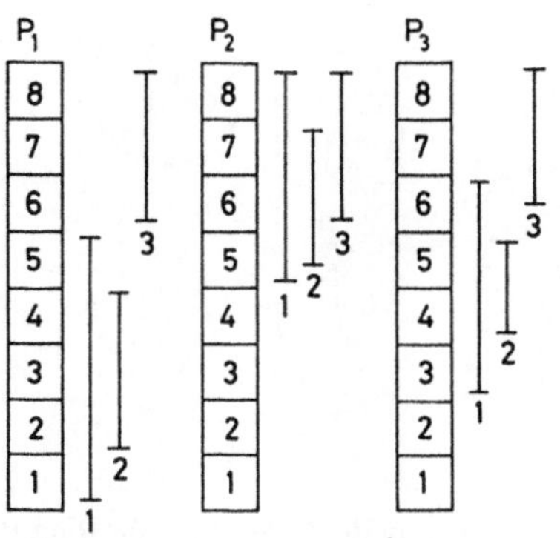

	1	2	3
La(x)	8	6	3
Da(x)	4	3	0
Dd(x)	3	3	0
Dt(x)/La(x)	0.88	1	0

Fig. 8.9. Values of coefficients $L_a(x)$, $D_a(x)$ and $D_d(x)$ and of the ratio $D_t(x)/L_a(x)$ of species 1–3 in zones 1–8 of sections P_1-P_3

The ratio $D_t(x)/L_a(x)$ of species 1 equals 0.88, that of species 2 equals 1, and that of species 3 equals 0. These values are obtained as follows:

Species 1 is found in zones 1 to 5 of section 1, in zones 5 to 8 of section 2 and in zones 3 to 6 of section 3. Its dispersion D_a equals 4 and its dispersion D_d equals 3: $D_t(x) = 7$ and $L_a(x) = 8$, so $D_t(x)/L_a(x) = 0.88$.

Species 2 is found in zones 2 to 4 of section 1, in zones 5 to 7 of section 2 and in zones 4 to 5 of section 3: $D_a = 3$, $D_d = 3$ and $L_a = 6$, so $D_t(x)/L_a(x) = 1$.

The total dispersion of species 3 equals zero: that species is found everywhere in zones 6 to 8.

8.4.3 Quality of the Lateral Record of Species

8.4.3.1 Relationship Between $G_r(x)$, $L_a(x)$, and $D_t(x)$

To see whether a species' apparent longevity [$L_a(x)$] and total biochronologic dispersion [$D_t(x)$] affect the quality of its lateral record [$G_r(x)$], we use diagrams that are different from those described above.

First we rank species by their quality, according to the value of their coefficient G_r: $G_r = 0.1$ to 0.2, $G_r = 0.3$ to $0.4, \ldots, G_r = 0.9$ to 1.0.

Next, for each value of $L_a(x)$ or $D_t(x)$ we construct a histogram showing the frequency distribution of the species as a function of the quality of their respective lateral record.

8.4.3.2 Example

Consider an imaginary problem in which a group of fossils have an apparent longevity of 1.

Suppose that $G_r = 0.1$ or 0.2 for 10 species, $G_r = 0.3$ or 0.4 for 20 species, $G_r = 0.5$ or 0.6 for 30 species, $G_r = 0.7$ or 0.8 for 20 species, and $G_r = 0.9$ or 1 for 5 species.

The histogram showing how many species have the different values of G_r is given in Fig. 8.10.

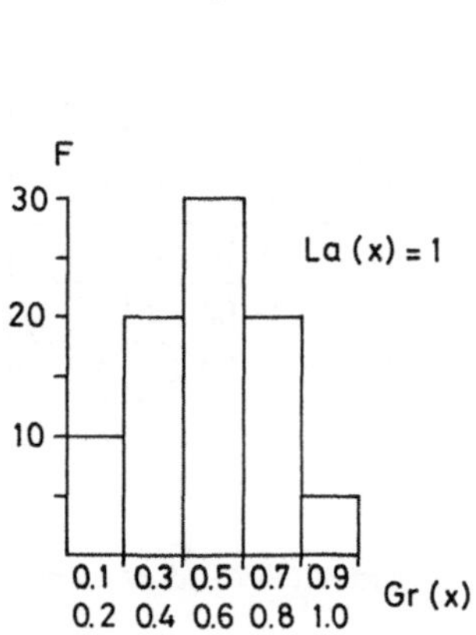

Fig. 8.10. Histogram showing the variation in the frequency (*F*) of the values of G_r in an imaginary set of 85 species with apparent longevity of 1

Chapter 9
Contradictory Biochronologic Correlations

9.1 The Lower Triassic of the Salt Ranges

No one disputes the fact that the boundaries of biochronologic zones based on different taxonomic groups generally do not coincide. It is clear, after all, that the geometry of the existence domains of different fossil groups usually generates biochronologic zones with different boundaries.

The problem studied here illustrates this. It can be summarized as follows: how can a careful taxonomic study of the Lower Triassic conodonts in the Pakistani Salt Ranges and a correct synthesis of their local superpositions (Sweet 1970) produce correlations that strongly contradict both lithostratigraphic correlations and the known ammonite zones in these same sections?

The data in this example will be analyzed semiempirically. This will enable us to describe and apply some general criteria for simplifying biochronologic problems.

9.2 Lithostratigraphic Framework and Ammonite Zones

The Lower Triassic of the Salt Ranges is classically divided (from bottom to top) into seven lithologic units:

1. The Kathwai Member (K. M.);
2. The Lower Ceratite Limestones (L. C. L.)
3. The Ceratite Marls (C. M.)
4. The Ceratite Sandstones (C. S.)
5. The Upper Ceratite Limestones (U. C. L.);
6. The Intermediate Beds (I. B.);
7. The Topmost Limestones (T. L.).

The ammonites collected in these sediments (Waagen 1895; Schindewolf 1953; Kummel 1966; Guex 1978 a) enable us to assign them without great difficulty to the Triassic standard zones established by Tozer (1967).

The correlations based on ammonites suggest that the lithologic units listed above are approximately synchronous in all the locations where they are found (Guex 1978 a; Nakasawa and Dickins 1985):

1. K. M.: *Ophiceras* zone (Upper Griesbachian)
2. L. C. L. and C. M.: *Proptychites* zone (Lower Nammalian)
3. C. S. and U. C. L.: *Flemingites* and *Anasibirites* zones (Upper Nammalian)

4. I. B.: *Columbites* zone (Lower Spathian)
5. T. L.: *Subcolumbites* zone (Middle Spathian)

9.3 Stratigraphic Distribution of Conodonts

In the early 1960s, the Lower Triassic of the Pakistani Salt Ranges was sampled by Kummel and Teichert in an attempt to elucidate certain problems concerning the Permian-Triassic boundary. Their collections were submitted to various specialists, who published their results in a collective work (Kummel and Teichert 1970).

Sweet (1970) gave complete and precise information on the content of the conodont-bearing samples given to him. We will use some of his data here to compare correlations based on conodonts with those based on ammonites.

We will discuss only the distribution of conodonts from the middle of the Ceratite Marls to the Topmost Limestones, for that interval contains the most interesting contradictory correlations.

The conodont-bearing samples come mainly from five sections: (1) Landu; (2) Narmia; (3) Zaluch; (4) Nammal; and (5) Chhidru (Fig. 9.1.c). They include 20 species, which we will identify by the numbers used by Sweet (1970 p. 271; see Appendix C.1).

Figure 9.1.a shows the stratigraphic sections in the localities mentioned above, the position of the conodont-bearing samples in these sections and their species content.

9.4 Biochronologic Interpretation

9.4.1 Simplifying the Data

When working without a computer on combinatorial problems, it is always much easier if the sets under consideration contain the smallest possible number of elements.

Three simple criteria can reduce the size of a biochronologic problem.

1. Eliminating species that are found in only one section.
2. Eliminating species that coexist with all the others. In the framework of a restricted problem, it is clear that these omniassociated species are unlikely to generate biochronologic subdivisions that are useful regionally.
3. Searching for equivalent species. Species with identical neighborhoods generally play an equivalent role in the construction of unitary associations. There-

Fig. 9.1. a Lithostratigraphic sections (1–5). **b** Distribution of conodonts in sections 1–5 (X, virtual presence of species in the samples; *L*, stratigraphic levels; *U.A.*, unitary associations; *black bars*, stratigraphic distribution of species). For names of species, see Appendix C.1. **c** Geographic location of sections 1–5. K, Karachi; Q, Quetta; R, Rawalpindi; M, Mianwali

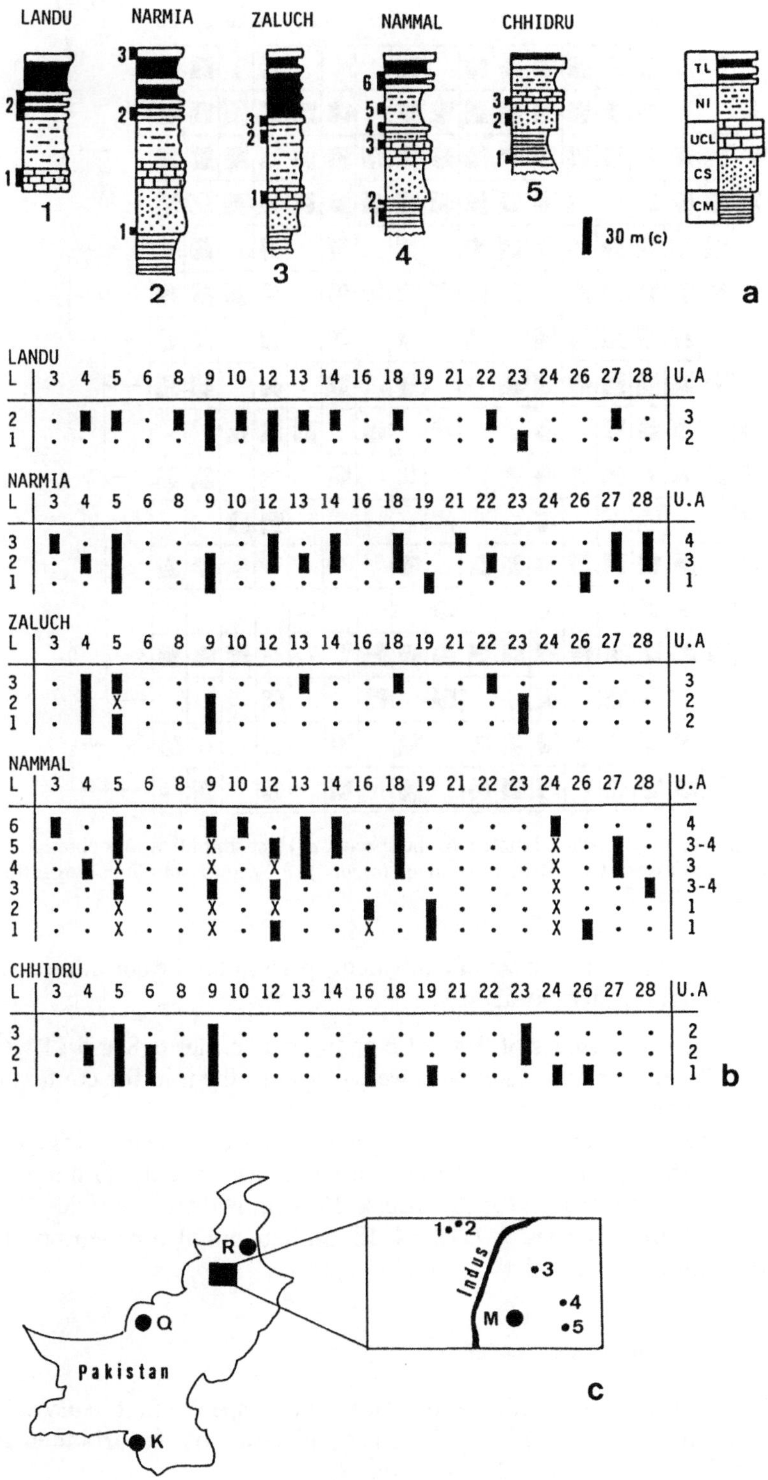
LANDU
NARMIA
ZALUCH
NAMMAL
CHHIDRU
TL
NI
UCL
CS
CM
30 m (c)
a
L
U.A
b
Indus
Pakistan
R
Q
K
M
c

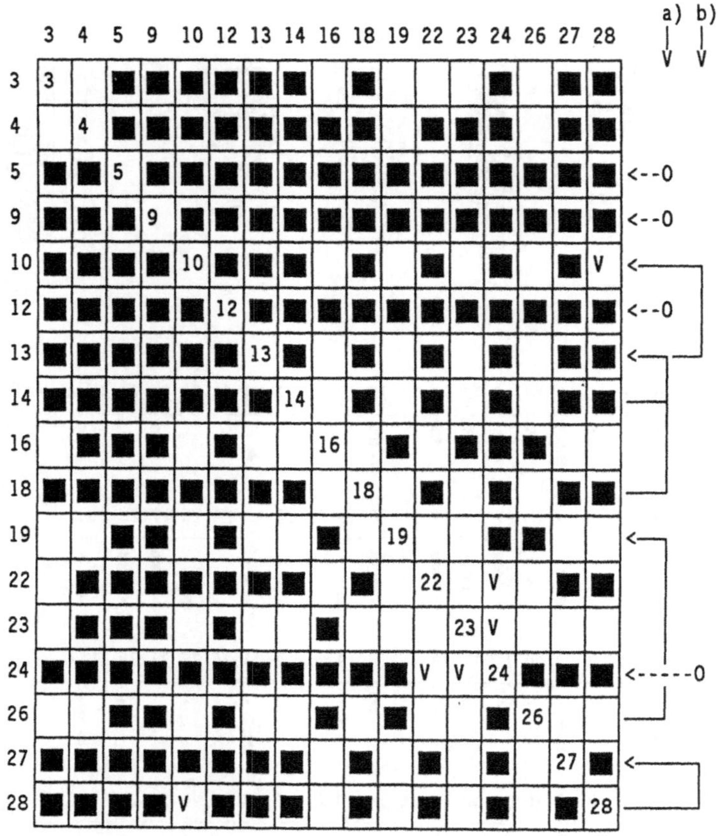

Fig. 9.2. Neighborhood matrix of the species. *a* First simplification: elimination of species that are omniassociated (0) or equivalent (*arrows*). *b* Second simplification, after adding virtual edges (V)

fore, only one in a set of equivalent species need be considered. The others can be ignored temporarily.

The range charts of Fig. 9.1.b show that species 6, 8 and 21 were each found in only one section. Therefore we will ignore them in the context of the present discussion.

The coexistences of the remaining species are listed in Fig. 9.2. This matrix shows that: (1) species 5, 9 and 12 are omniassociated; (2) species 14 and 18 are equivalent to species 13; (3) species 26 is equivalent to species 19.

Therefore species 5, 9, 12, 14, 18, and 26 will also be temporarily ignored in the following analysis.

9.4.2 Virtual Coexistences

The stratigraphic relationships among the 11 species that "survived" our simplification are shown in Fig. 9.3. This graph contains six forbidden generated subgraphs: one Z_4', two Z_4'' and three S_3.

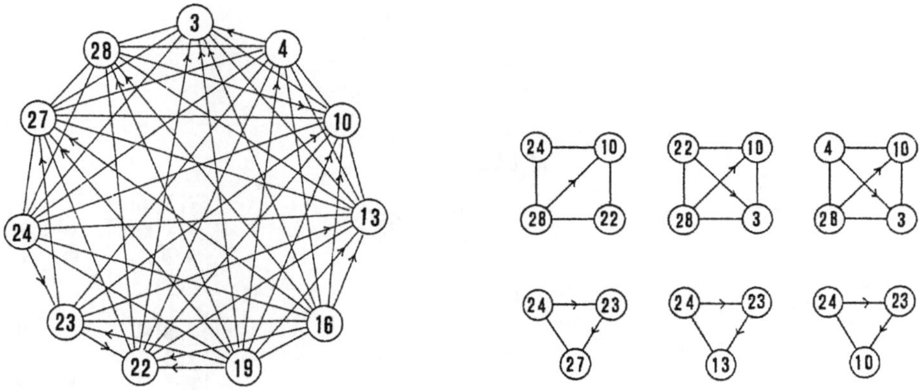

Fig. 9.3. Biostratigraphic graph representing the relationships among the remaining 11 species, and the graph's forbidden subgraphs

We can eliminate these subgraphs by following the rules for adding virtual edges established in Section 4.4:

1. Z_4'. If the superposition of species 28 and species 10 (arc (28, 10) in Fig. 9.3.b) corresponds to a true chronologic relationship, then species 22 and 24 virtually coexisted. Therefore we draw a virtual edge to join the vertices representing those two species.
2. Z_4''. We saw in Section 4.4 that when we destroy Z_4'' circuits we should preserve as much information as possible about the superpositions observed among the species. Here, replacing arc (28, 10) by a virtual edge eliminates both imbricated Z_4'' circuits (Fig. 9.3.b). Therefore we replace the arc by a virtual edge.
3. S_3. It is equally clear that replacing arc (24, 23) by a virtual edge is an optimal solution, since it preserves the superpositions observed among the pairs of species (23, 27), (23, 13), and (23, 10).

We complete the neighborhood matrix in Fig. 9.2 by adding the virtual coexistences that were just established: species 24 now becomes virtually omniassociated, species 28 becomes virtually equivalent to species 27, and species 13 becomes virtually equivalent to species 10.

We can now proceed to a second simplification by ignoring species 10, 24, and 28 when establishing unitary associations. Figure 9.4 illustrates both the neighborhood matrix that results from these successive simplifications, and the graph G' associated to it. We now use the coexistences (actual and virtual) shown in this graph to construct the unitary associations.

9.4.3 Unitary Associations and Protoreferential

Looking at Fig. 9.4, we see that our graph G' is made up of four unitary associations:

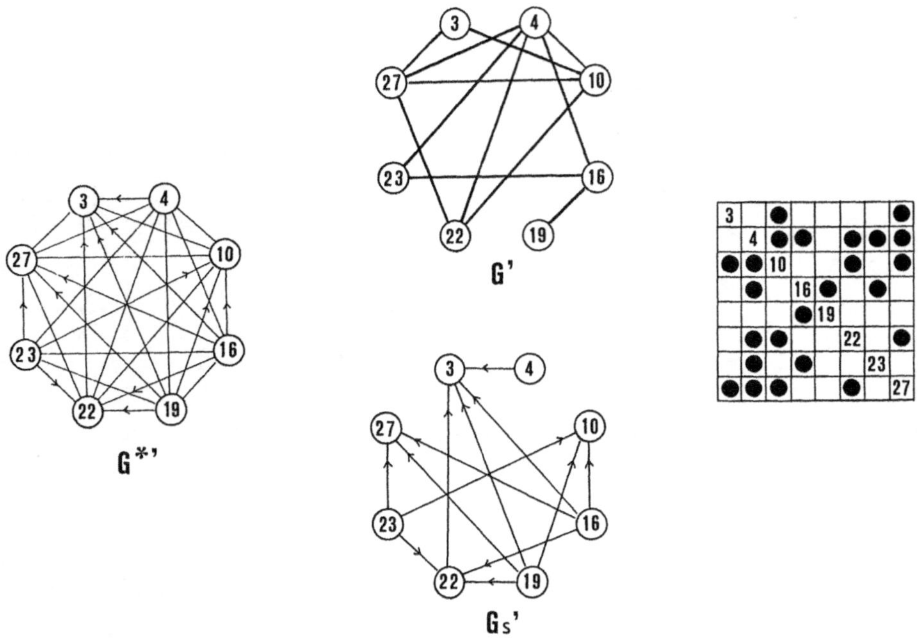

Fig. 9.4. (Reduced) graphs $G^{*\prime}$, G', and G'_s and adjacency matrix of the reduced graph G' (*black circles*, coexistences of species)

1. U. A.4 = {3, 10, 27}
2. U. A.3 = {4, 10, 22, 27}
3. U. A.2 = {4, 16, 23}
4. U. A.1 = {16, 19}

We can immediately establish the superpositional order of these unitary associations by using the graph G'_s shown in Fig. 9.4:

U. A.4 → U. A.3 → U. A.2 → U. A.1

Introducing species 13, 14, 18 (equivalent to species 10), species 26 (equivalent to 19), and species 28 (equivalent to 27) into these restricted unitary associations gives the protoreferential shown in Fig. 9.5.

This solution is almost identical to that established by Sweet (loc.cit., p. 213): his zone 6 corresponds to U. A.1; his zone 7 corresponds to U. A.2; his lower zone 8 corresponds to U. A.3; his upper zone 8 corresponds to U. A.4.

The positions of the characteristic elements of these unitary associations in the stratigraphic sections are given in column *U. A.* of Fig. 9.1.b

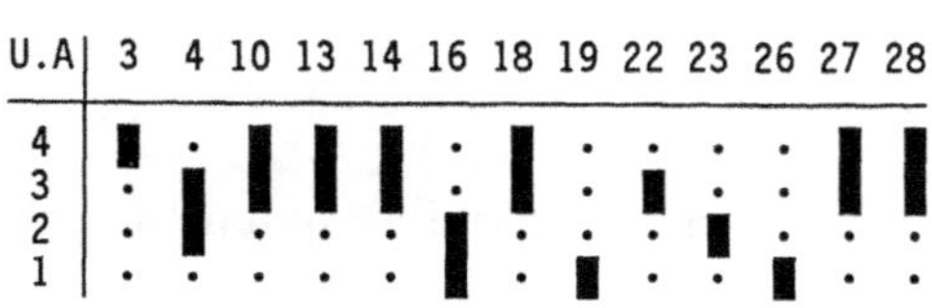

Fig. 9.5. Protoreferential constructed from the data in Fig. 9.4 (see text)

9.5 Discussion

The correlations Sweet established using conodonts are given in Fig. 9.6, where they are compared with the correlations we get using ammonites and lithology.

This figure shows that, using conodonts, sediments of Spathian age at Landu, Narmia, and Zaluch are correlated with beds of Nammalian age at Nammal. This contradiction is easily explained by the fact that the sampling of conodonts available in the 1960s did not allow for more precise correlations than those that could be established with ammonites or using sedimentological criteria. In addition, Sweet's scale (1970, p. 213, Fig. 4) is actually a protoreferential, and most of its subdivisions are not chronogically significant (see Guex 1978 b).

In the case of the Salt Range and adjacent regions, Waagen's purely lithologic correlations (expressed in terms of Ceratite Sandstones, Ceratite Lst, Topmost Lst, etc.) are actually more precise; they coincide rather well with the zones proposed by Tozer in 1967, which can be identified worldwide.

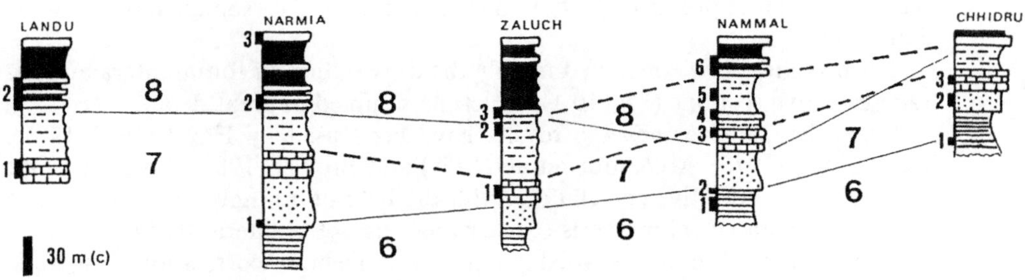

Fig. 9.6. Generalized correlation chart. *Dotted lines*, lithologic correlations + ammonites. *Solid lines*, correlations based on conodonts. *6, 7* and *8*, zones established by Sweet (1970)

Chapter 10
Transgressive-Regressive Cycles

10.1 Paleogene Benthic Foraminifera from Mississippi and Alabama

In the early 1960s, Deboo (1965) began a comprehensive biostratigraphic study of the Paleogene west of Alabama and east of Mississippi. His goal was twofold: first, to solve some problems posed by correlating the lithologic units classically used in this region (Fig. 10.1); and second, to use new biochronologic arguments to determine more precisely the boundary between the Jacksonian and Vicksburgian stages.

His investigations concerned mainly the distribution of foraminifera and ostracods in five sections (Fig. 10.1), which he sampled in great detail.

Since then, his remarkable results have been used by Hazel (1970, 1977), McCammon (1970), Millendorf et al. (1978) and Brower (1985b) in an attempt to establish quantitative correlations with the help of a number of more or less sophisticated statistical methods (cluster analysis, RBV, lateral tracing, etc.).

Their statistical solutions failed to improve on Deboo's correlations, and none of their techniques allowed them to detect major transgressive/regressive cycles (fall and rise of sea level) based on the data they analyzed. (The reasons for this failure will be discussed in Chapter 15.) It is therefore interesting to see what the unitary association method can contribute to these problems.

10.2 Preliminary Remarks

As an introduction, we will briefly examine the imaginary stratigraphic situation sketched in Fig. 10.2. These diagrams show the geometric arrangement of seven biochronologic zones that are identified in three stratigraphic sections aligned along a distal (P_1) → proximal (P_3) transect of a marine sedimentary basin.

The sequence of zones recorded in P_1 is complete, while that of P_3 has many gaps (zones 2, 3, 5 and 6 are absent).

This kind of situation is one of the simplest cases we can encounter when searching for transgressive/regressive cycles.

We will show that studying the relationships of Deboo's benthic foraminifera (Fig. 10.1) enables us to construct a model similar to the one shown in Fig. 10.2.

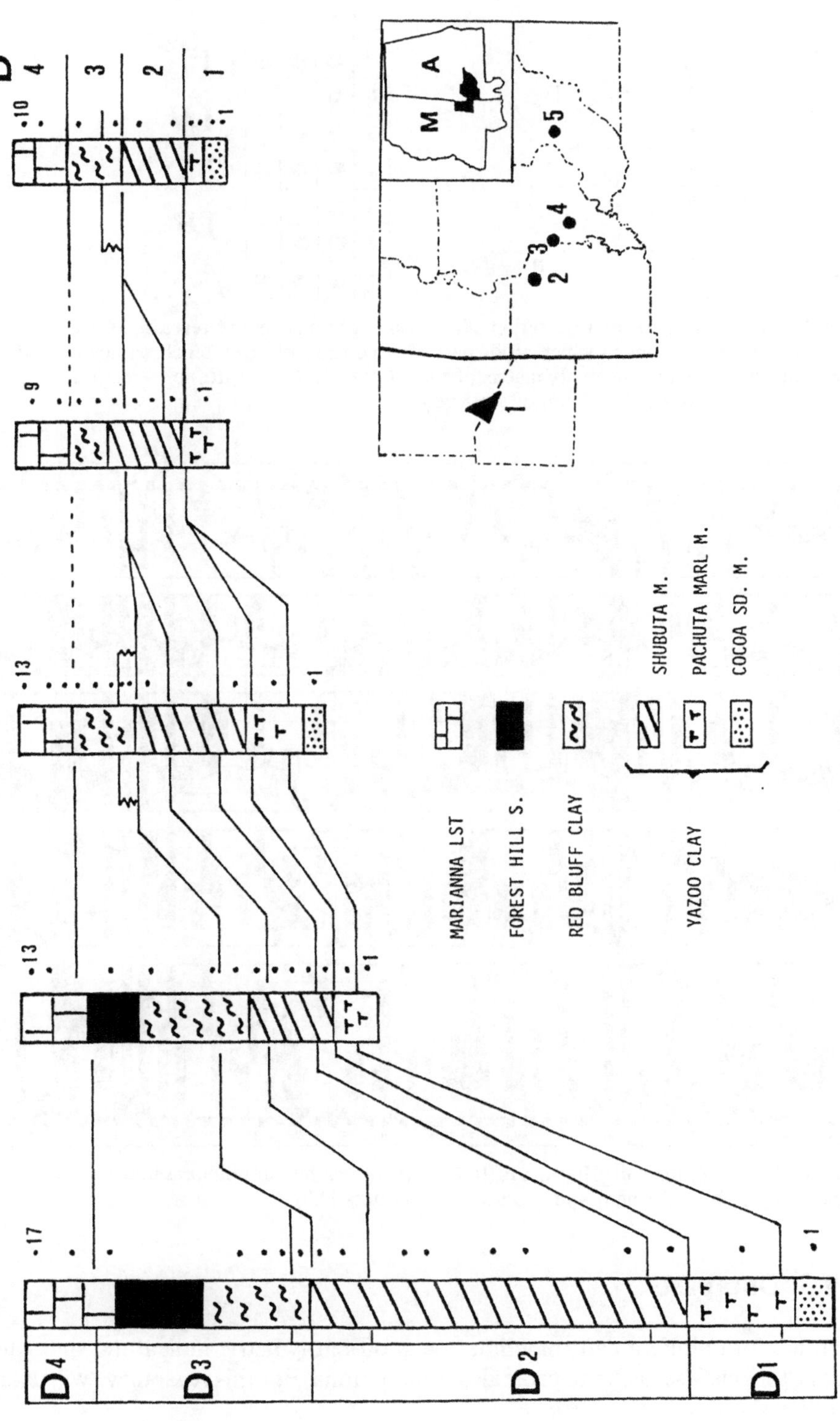

Fig. 10.1. Sections 1–5 (after Deboo 1965, p. 28, Fig. 11; see Appendix D.2). D_1–D_4 are Deboo's (1965) zones 1–4. The correlations drawn in *bold* correspond approximately to the boundaries between these zones. M, Mississippi. A, Alabama

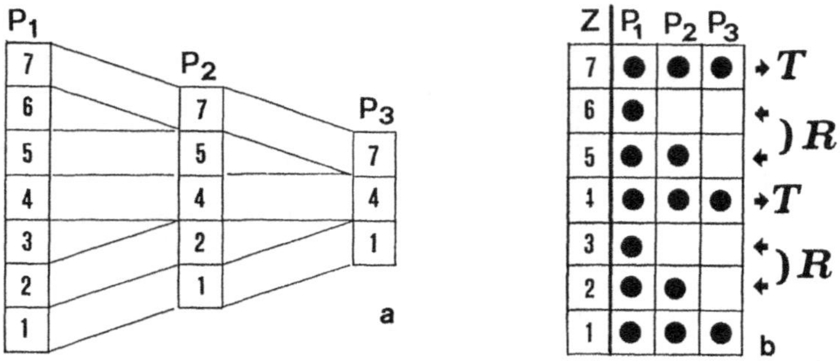

Fig. 10.2 a, b. Schematic representation of a sequence of transgressive/regressive cycles. **a** P_1-P_3, three imaginary sections in which all or part of a sequence of seven biochronologic zones can be recognized. **b** Reproducibility matrix of zones (Z) 1–7. T, transgressions (relative rise of sea level). R, regressions (relative fall of sea level)

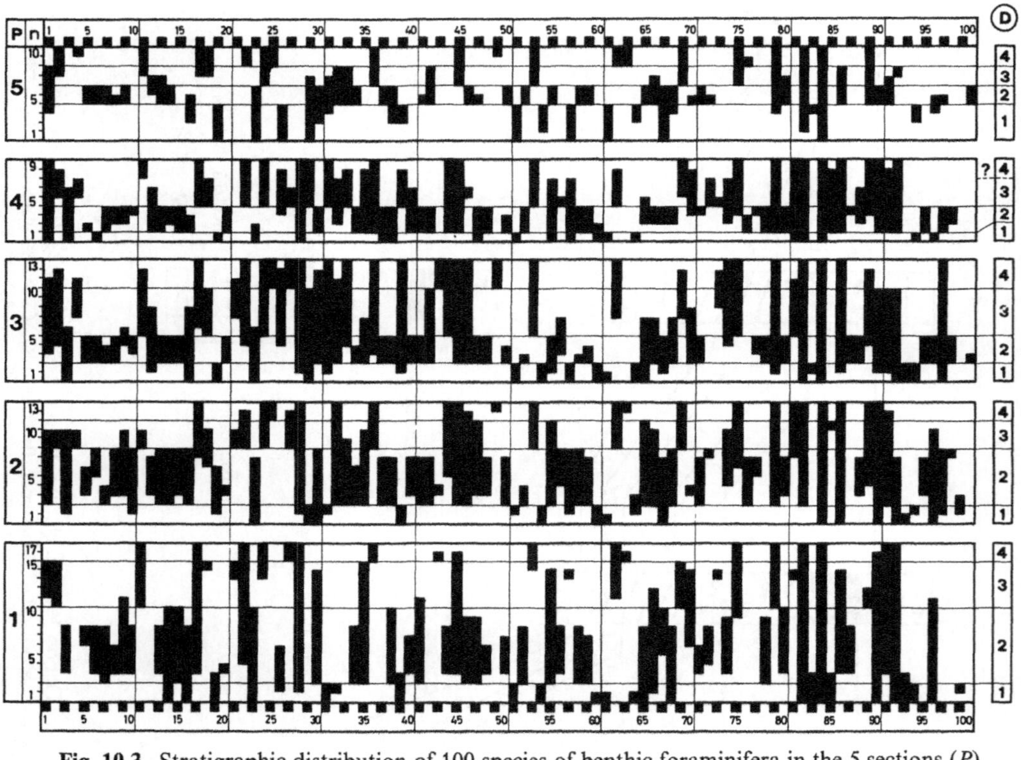

Fig. 10.3. Stratigraphic distribution of 100 species of benthic foraminifera in the 5 sections (P) studied by Deboo (for names of species, see Appendix D.3)

10.3 Database

Benthic foraminifera can sometimes be good bathymetric indicators, since they are notoriously sensitive to ecological fluctuations. For this reason, we will examine this group in detail here.

Of the 147 species Deboo identified, we have chosen 100, according to their geographic distribution and the dissimilarity of their vertical ranges. (In other words, we have ignored some species that are equivalent or quasi-equivalent). The local vertical ranges of these species are given in Fig. 10.3.

10.4 Unitary Associations and Correlations

10.4.1 Processing the Data

Processing the above data results in an initial sequence of 17 unitary associations (Fig. 10.4).

The reproducibility matrix (Fig. 10.4.b) and the correlation chart (Fig. 10.5) that correspond to this protoreferential show very weak superpositional control among most of the unitary associations identified in the five sections. Offhand, this initial result is difficult to interpret.

To extract from it an interesting zonation we begin by analyzing it along with the result we get by using the G_k' reduction procedure described in Section 5.6.3.

We will see that in this case, the initial G_k solution (Sect. 5.4) reveals some useful information that is hidden by the G_k' solution.

10.4.2 Automatic Reduction of G_k'

The graph G_k', constructed from information in the initial reproducibility matrix (Fig. 10.4), and the automatic reduction of this graph to a unique path are illustrated in Fig. 10.6 (output of the DV-86 program). It shows that our reduction procedure produces two particularly drastic mergings of unitary associations. The first (U. A.3 – U. A.6 of the initial sequence) is the new U. A.3, identifiable in all Deboo's sections. The second (U. A.7 – U. A.10 of the initial sequence) is the new U. A.4, identifiable in four sections.

It is easy to see that this simplification of the problem is very useful. In this case, it is in fact difficult to establish biochronozones empirically by grouping together the initially nonreproducible unitary associations.

10.4.3 Correlations

The correlations that result from this reduction are shown by solid lines in Fig. 10.7. Considered in isolation, unitary associations 5 – 7 on the one hand, and 8 and 9 on the other, seem of no particular significance and we have merged them. However, the stratigraphic position of the union of U. A.9 and 10 of the initial solution (Fig. 10.5) has been added to the simplified correlation chart (Fig. 10.7); it is the upper part of U. A.4 of sections 1 – 3, denoted U. A.4′. We will see later that this biochronologic information, hidden by the G_k' reduction procedure, is useful in interpreting relative variations in sea level that can be deduced from these correlations.

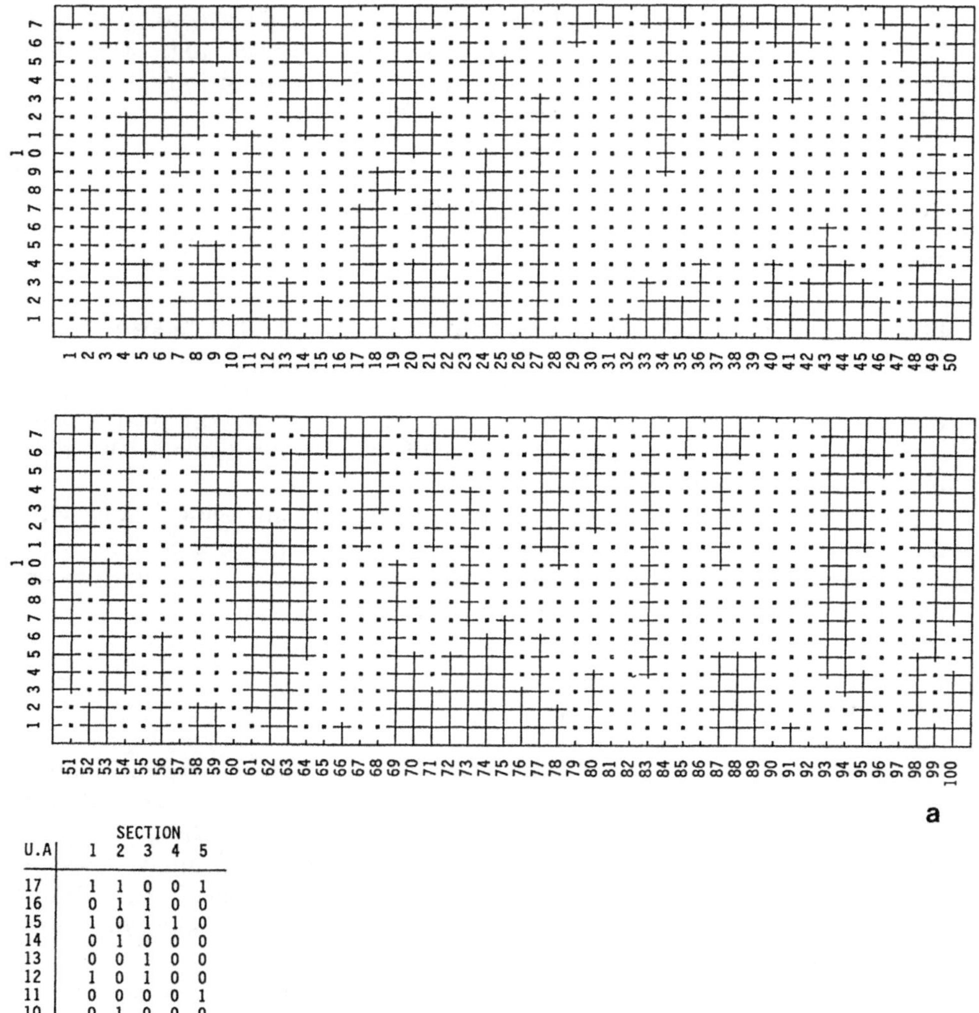

a

U.A	SECTION 1	2	3	4	5
17	1	1	0	0	1
16	0	1	1	0	0
15	1	0	1	1	0
14	0	1	0	0	0
13	0	0	1	0	0
12	1	0	1	0	0
11	0	0	0	0	1
10	0	1	0	0	0
9	0	0	1	0	0
8	1	0	0	1	0
7	0	1	0	0	0
6	0	0	1	0	1
5	0	0	0	1	0
4	0	1	0	0	0
3	1	0	0	0	0
2	1	1	1	0	0
1	1	1	1	1	1

b

Fig. 10.4. a Protoreferential for Deboo's data. **b** Reproducibility matrix (output of DV-86 program)

10.4.4 Comparison with Deboo's Correlations

To correlate his sections with the help of benthic foraminifera, Deboo (1965, p. 27, Fig. 10) made a statistical analysis of the vertical variations in the similarity of adjacent fauna. This enabled him to recognize three major faunal discontinuities delimiting four distinct biochronozones that we will call $D_1 - D_4$, starting

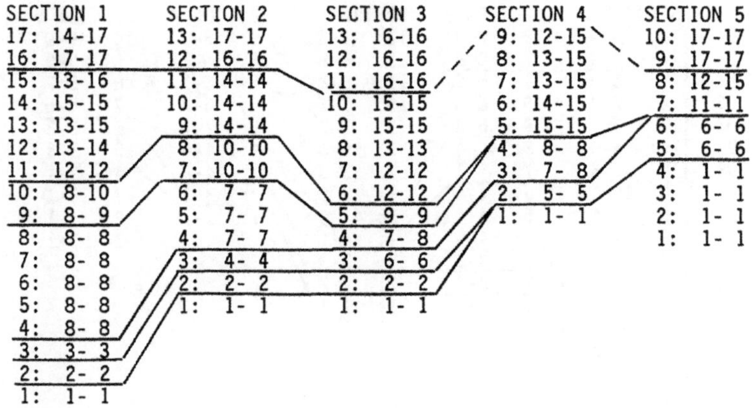

Fig. 10.5. Biochronologic correlations (*solid lines*) generated by the protoreferential of Fig. 10.4. *Dashed lines*, correlation conflicting with Deboo's zonation (output of DV-86 program)

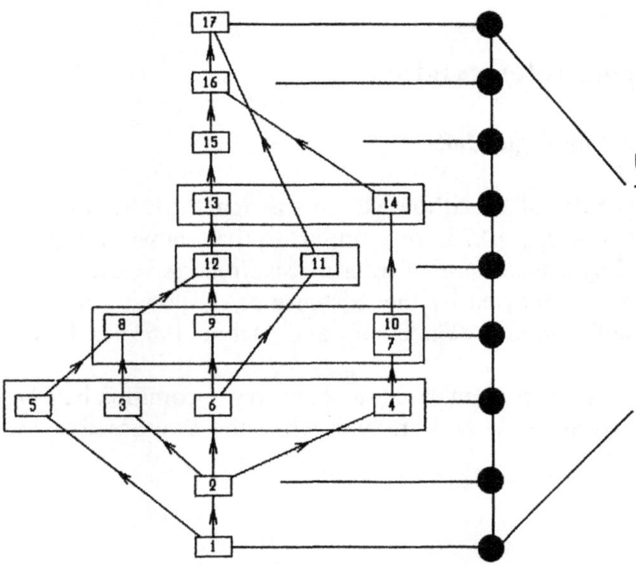

U.A	SECTION 1	2	3	4	5
9	1	1	0	0	1
8	0	1	1	0	0
7	1	0	1	1	0
6	1	1	1	0	0
5	1	0	1	0	1
4	1	1	1	1	0
3	1	1	1	1	1
2	1	1	1	0	0
1	1	1	1	1	1

Fig. 10.6. Graph G'_k reduced to a unique path and the resulting reproducibility matrix (output of DV-86 program)

from the bottom. The correlation of these zones with the sequence of unitary associations discussed above is given in Fig. 10.7.

We should emphasize that our correlations coincide almost exactly with those of Deboo, with one important exception: we assign samples 8 and 9 of section 4 (corresponding to Deboo's original beds 1 and 2) to zone 3, while Deboo assigns them to his zone 4 (see Fig. 10.1).

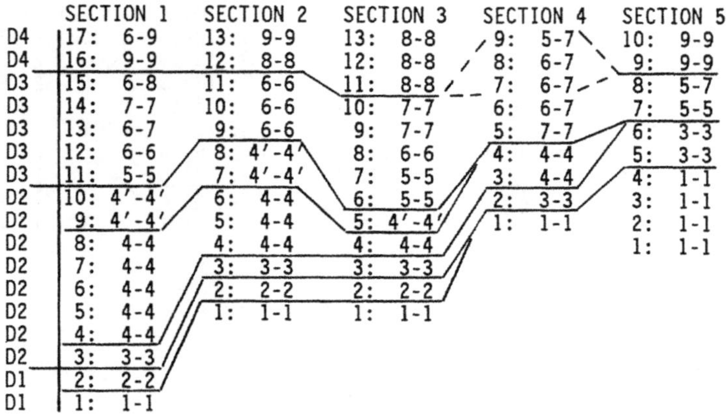

Fig. 10.7. Correlations among Deboo's five sections (option G_k': output of DV-86 program). $D_1 - D_4$, zones recognized by Deboo (*solid lines*). *Dashed lines*, correlation conflicting with Deboo's zonation

10.5 Stratigraphic Interpretation

10.5.1 Geometry of Lithologic Units

The relative arrangement of the lithologic units assigned to the zones discussed above and correlated in Fig. 10.7 is very similar to that shown in Fig. 10.2. It is obviously generated by a succession of transgressive/regressive cycles (Fig. 10.7).

This conclusion is supported by the existence of erosional structures at the base of the Red Bluff Member (Cheetham and Deboo 1963) and by Siesser's (1984) recent work.

We also note that in the present case (and contrary to common belief), it is the regressive phases that are slow and gradual, while the transgressive phases are rapid.

10.5.2 Remarks

Deboo's biostratigraphic observations yield slightly different correlations when we examine the distribution of benthic foraminifera alone than when we examine them together with ostracods (Sect. 10.8.2). Whatever set of species we consider, the arrangement of the unitary biofacies (Sect. 5.6.4) always corresponds to a sequence of two regressive phases located respectively at the base of the Shubuta Member and the base of the Red Bluff Clay Member. In the present example we are interested only in this aspect of the problem.

Siesser (1984) gives a very interesting historical account of research on the stratigraphy of the Paleogene of the eastern coastal plains of the Gulf of Mexico. He notes in particular that the Red Bluff Clay was traditionally considered a shallow deposit and that it was Hazel et al. (1980) who first proved that these deposits are actually deeper than those surrounding them. We see here that the same conclusion can be drawn directly from Deboo's original observations.

10.6 Quality of the Fossil Record of Species

10.6.1 Biochronologic Dispersion of Datums

To evaluate in the simplest possible way the biochronologic dispersion of the datums and the quality of the lateral record of the species studied here, we use the four zones that Deboo established empirically and which cover uniformly the region he studied (Fig. 10.1).

The values of the coefficients $D_a(x)$, $D_d(x)$, $G_r(x)$, $L_a(x)$ and $I_t(x)$ of each species are given in Appendix D.4, and their relationships are given in Figs. 10.8–10.10. These diagrams were constructed according to the procedures described in Chapter 8.

10.6.2 Vertical Record

The variation of $D_a(x)$, $D_d(x)$, and $D_t(x)$ (biochronologic dispersion) as a function of $L_a(x)$ (apparent longevity) is given in Fig. 10.8. These diagrams essentially show that:

1. Forty-four percent of the benthic species analyzed here have a dispersion $D_a(x)$ equal to $L_a(x) - 1$ (the worst possible) and 61% have a dispersion $D_d(x)$ equal to $L_a(x) - 1$ (Fig. 10.8.a, b). This difference between the dispersions D_a and D_d is also clearly emphasized by the diagram recording the number of species with the different possible combinations of the values D_a and D_d (Fig. 10.8.c: 15% have a dispersion D_a higher than D_d and 30% have a dispersion D_d greater than D_a);
2. Thirty-three percent have a total biochronologic dispersion (D_t) equal to zero (Fig. 10.8.c, d);
3. Eighteen percent are diachronous on a regional scale (sector D of Fig. 10.8.e).

10.6.3 Lateral Record of Species

The diagrams representing the relationships between $D_t(x)$ and $G_r(x)$ on the one hand and $L_a(x)$ and $G_r(x)$ on the other (Fig. 10.9) show no obvious correlation between these variables.

They also show that the lateral record of the benthic species is good: almost 60% have an ideal geofrequency ($G_r = 1$).

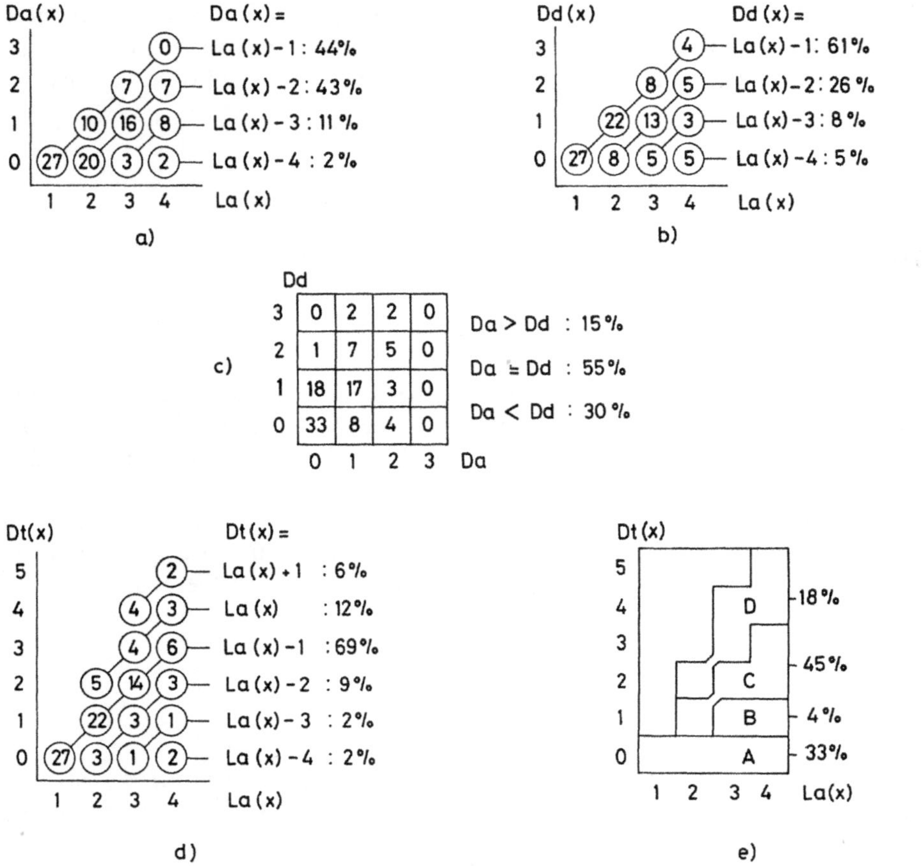

Fig. 10.8 a–e. Relationships among the coefficients $D_a(x)$, $D_d(x)$, $D_t(x)$, $L_a(x)$, and variations in the ratio $D_t(x)/L_a(x)$ (see Appendix D.4)

10.7 Complexity of the Problem

10.7.1 Forbidden Generated Subgraphs

The number of forbidden generated subgraphs in a graph G^* gives an indirect estimate of the complexity of a biochronologic problem, for each such subgraph represents a set of conflicting stratigraphic relationships that must be reconciled. If G^* has no forbidden subgraphs, its nonoriented part (G) is an interval graph and the problem is simple. If G^* has many such subgraphs, the problem will be much more difficult to solve.

It is therefore interesting to try to identify the main factors controlling the number of contradictory stratigraphic relationships between species, for a given problem.

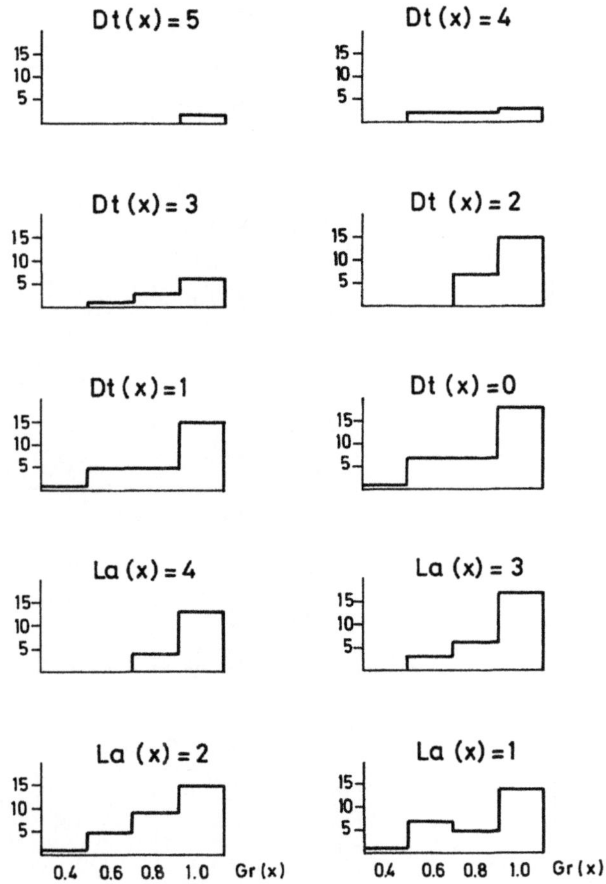

Fig. 10.9. Relationships among $G_r(x)$, $D_t(x)$ and $L_a(x)$

One of these factors was discussed in Chapter 7: disjunctive reworking can result in exponential growth of the number of forbidden subgraphs. The present problem involves no disjunctive reworking; the essential factor determining its complexity is the quality of the lateral record of the species. Figure 10.10.a shows, in fact, a very strong correlation between the coefficients G_r and I_t: the number of forbidden subgraphs in which the species appear (I_t) increases as their relative geofrequency decreases.

This is not surprising; we showed above in a simple example (Sect. 3.4.3.3), that constructing composite sections before processing the data had the effect of suppressing all conflicting stratigraphic relationships among the species in the merged sections. Clearly, such an operation is the same as artificially increasing the value of the coefficient G_r of the species in question.

Finally, we note that the average value of the coefficient $I_t(x)$ varies in a relatively erratic way with the total biochronologic dispersion of the species (Fig. 10.10.b).

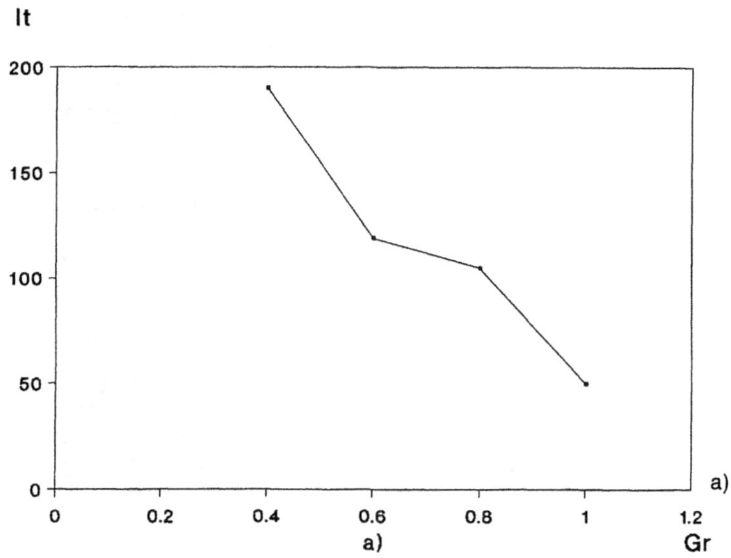

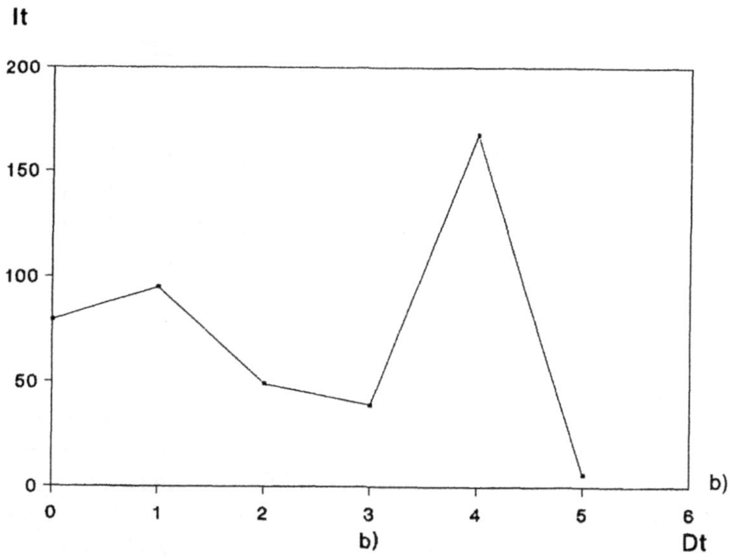

Fig. 10.10 a, b. Variation of I_t (average) as a function of $G_r(x)$ and $D_t(x)$

10.7.2 Remark

The graphs of Fig. 10.10 were constructed by calculating the average value of $I_t(x)$ for each distinct value of $G_r(x)$ and $D_t(x)$; (Appendix D.4).

For example: species 26, 29, 32, 43, 57, 76, and 92 have a dispersion D_t of 4 and an I_t of 162, 281, 141, 315, 31, 215, and 26 respectively (total = 1171). The average value of I_t for the 7 species with D_t of 4 is thus 167.

10.8 An Example of Integrated Biochronology

10.8.1 Benthic Foraminifera and Ostracods

As an example of integrated biochronology, we have completed Deboo's data studied in the preceding sections by adding his observations on the local ranges of 55 ostracod species (see Appendix D.5: numerical database). That augmented data set of 155 species is analyzed by means of the BioGraph program (Chap. 6) which can process much larger data sets than DV-86.

10.8.2 Processing the Data

The BioGraph program produces an initial protoreferential made up of 31 unitary associations (Fig. 10.11). Their individual reproducibility is very poor: 28 of the 31 are identified in only one location (Fig. 10.12).

To extract a zonation from that raw result, it is once more necessary to study the relationships between the reproducibility matrix and the graph G_k' (Fig. 10.13).

This graph gives a clear picture of lateral imbrication of contemporaneous biofacies, as briefly discussed in Section 5.6.4. It allows us to establish eight zones (A–H: Fig. 10.14) whose geometric arrangement is similar to that of the zones based only on foraminifera.

10.8.3 Conclusion

In the present case, integrating the ostracod and benthic foraminifera data results in slightly different but more precise correlations, compared to those obtained by processing the foraminifera alone.

These new correlations confirm that the last two beds of section 4 (samples 8 and 9 at Jackson) are older than Deboo's zone D_4.

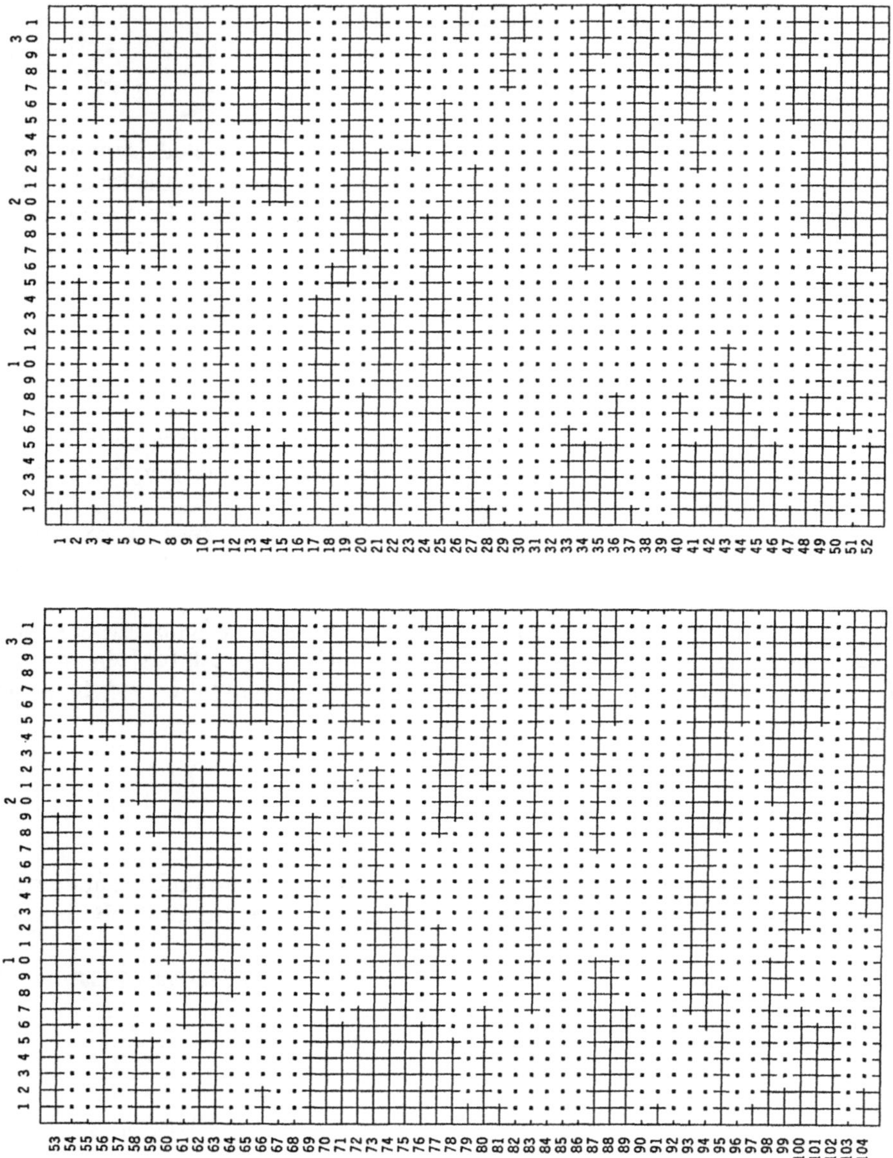

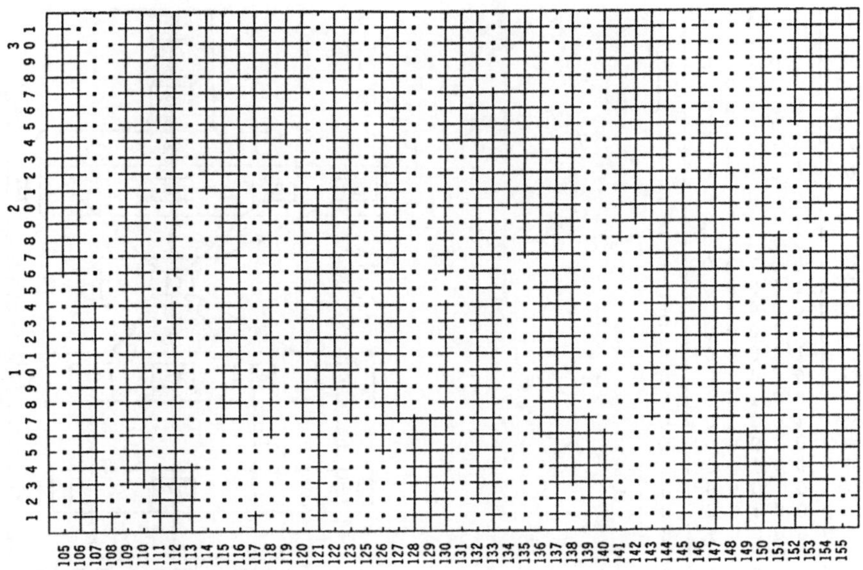

Fig. 10.11. Protoreferential constructed by the BioGraph program for Deboo's 155 species of benthic foraminifera and ostracods (see numerical database in Appendix D.5)

U.A	1	2	3	4	5	Z
31	0	0	1	0	0	H
30	0	0	0	0	1	H
29	0	0	1	0	0	H
28	1	0	0	0	0	G
27	0	1	0	0	0	G
26	0	0	1	0	0	G
25	0	0	0	1	0	G
24	1	1	1	1	0	F
23	0	0	0	1	0	E
22	0	0	1	0	0	E
21	1	0	0	0	0	E
20	0	0	0	0	1	E
19	0	1	0	0	0	E
18	1	0	0	0	0	D
17	0	1	0	0	0	D
16	0	0	1	0	0	D
15	1	0	0	1	0	C
14	0	1	0	0	0	C
13	0	0	1	0	0	C
12	0	1	0	0	0	B
11	0	0	1	0	0	B
10	0	0	0	0	1	B
9	0	0	0	1	0	B
8	0	0	0	0	1	B
7	0	1	0	0	0	B
6	1	0	0	0	0	B
5	0	0	0	0	1	A
4	0	1	0	0	0	A
3	1	0	0	0	0	A
2	0	0	1	1	1	A
1	1	0	0	0	0	A

Fig. 10.12. Reproducibility matrix. *Column Z*, zones *A–H* (output of the BioGraph program)

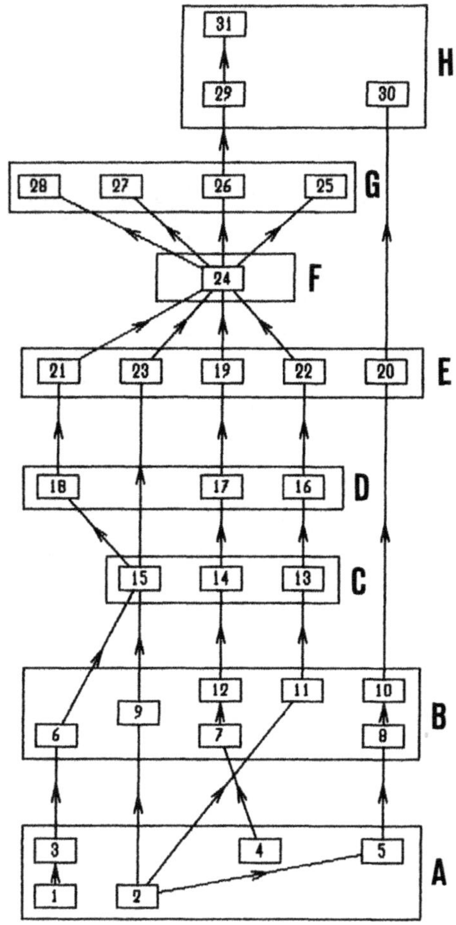

Fig. 10.13. Graph G'_k showing the lateral and vertical distribution of contemporaneous unitary biofacies. *A–H*, zones (output of the BioGraph program)

	SECTION 1	SECTION 2	SECTION 3	SECTION 4	SECTION 5
H	17: 28-31	13: 30-31	13: 31-31	9: 25-26	10: 30-31
H	16: 30-31	12: 27-27	12: 29-29	8: 25-25	9: 30-30
G	15: 28-28	11: 23-24	11: 29-29	7: 24-24	8: 23-26
F	14: 24-24	10: 24-24	10: 26-26	6: 24-24	7: 20-20
F	13: 23-24	9: 24-24	9: 24-26	5: 23-23	6: 10-10
E	12: 23-24	8: 19-19	8: 24-24	4: 15-15	5: 8- 8
E	11: 21-21	7: 17-17	7: 22-22	3: 14-15	4: 5- 5
D	10: 18-18	6: 14-14	6: 22-22	2: 9- 9	3: 5- 5
D	9: 18-18	5: 14-14	5: 16-16	1: 2- 2	2: 2- 2
C	8: 15-15	4: 12-12	4: 13-13		1: 1- 2
C	7: 15-15	3: 7- 7	3: 11-11		
C	6: 15-15	2: 4- 4	2: 2- 2		
C	5: 15-15	1: 1- 2	1: 2- 2		
C	4: 15-15				
B	3: 6- 6				
A	2: 3- 3				
A	1: 1- 1				

Fig. 10.14. Correlation table. *Solid lines*, zonal correlations. *Dashed line*, uncertain correlation (output of the BioGraph program)

Chapter 11
Diachronous Datums

11.1 Paleogene Nannoplankton From California

It was, in part, the classic work by Bramlette and Sullivan (1961) and Sullivan (1964, 1965) on the stratigraphic distribution of the Paleogene calcareous nannoplankton in the California coast ranges that inspired Hay (1972) to propose a probabilistic model for detecting sequences of datums that are as invariant as possible from a statistical point of view.

The multitude of contradictory relationships that affect the order in which the various species (morphospecies) of this group appear and disappear in different sections is indeed striking. Bramlette's and Sullivan's work (op.cit.) thus provides excellent material for studying the effect of diachronous datums on biochronologic correlations.

Before discussing this issue we need to establish a chronologic framework of reference that enables us to say what is diachronous, what is not, and what is only a little. We will begin by discussing biochronologic correlations that can be deduced from the local stratigraphic relationships among the species studied by these authors.

11.2 Database

The sections Bramlette and Sullivan studied, as well as their geographic location, are given in Fig. 11.1. The local stratigraphic ranges of 105 species in these sections are given in Fig. 11.2.

A rigorous analysis of these data led Sullivan (1965) to recognize four biochronozones of nannoplankton that can be identified on a regional scale. It is interesting to note that his correlations coincide rather well with those Mallory (1959) established in the same sections, using benthic foraminifera. Both Sullivan's and Mallory's zonations have been completely ignored and implicitly discredited by all those who later exploited their data in an attempt to establish zones based solely on datums.

We will therefore first consider whether Sullivan's (1965) correlations and zones are indeed false and should be rejected.

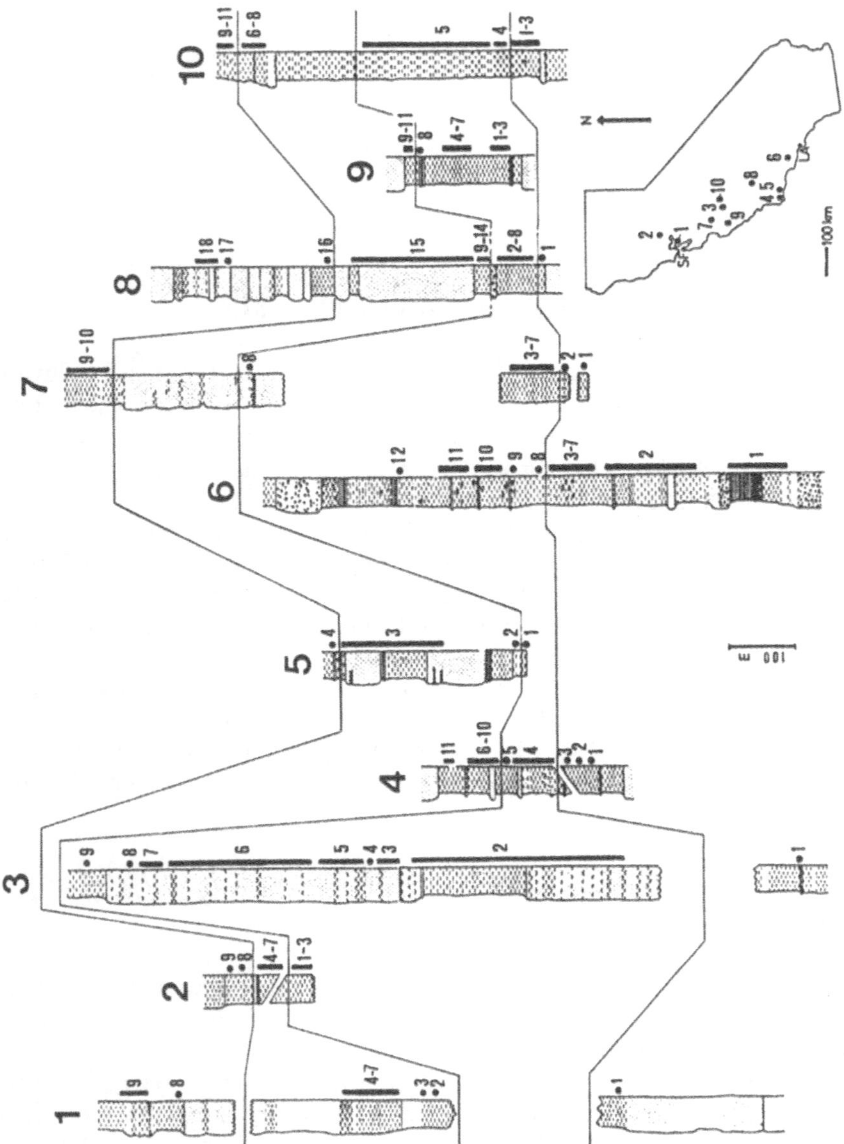

Fig. 11.1. Sullivan's sections 1–10 (for names, see Appendix E.2). Geographic situation of sections 1–10; San Francisco; LA, Los Angeles. $S_1 - S_4$, zones established by Sullivan (1965)

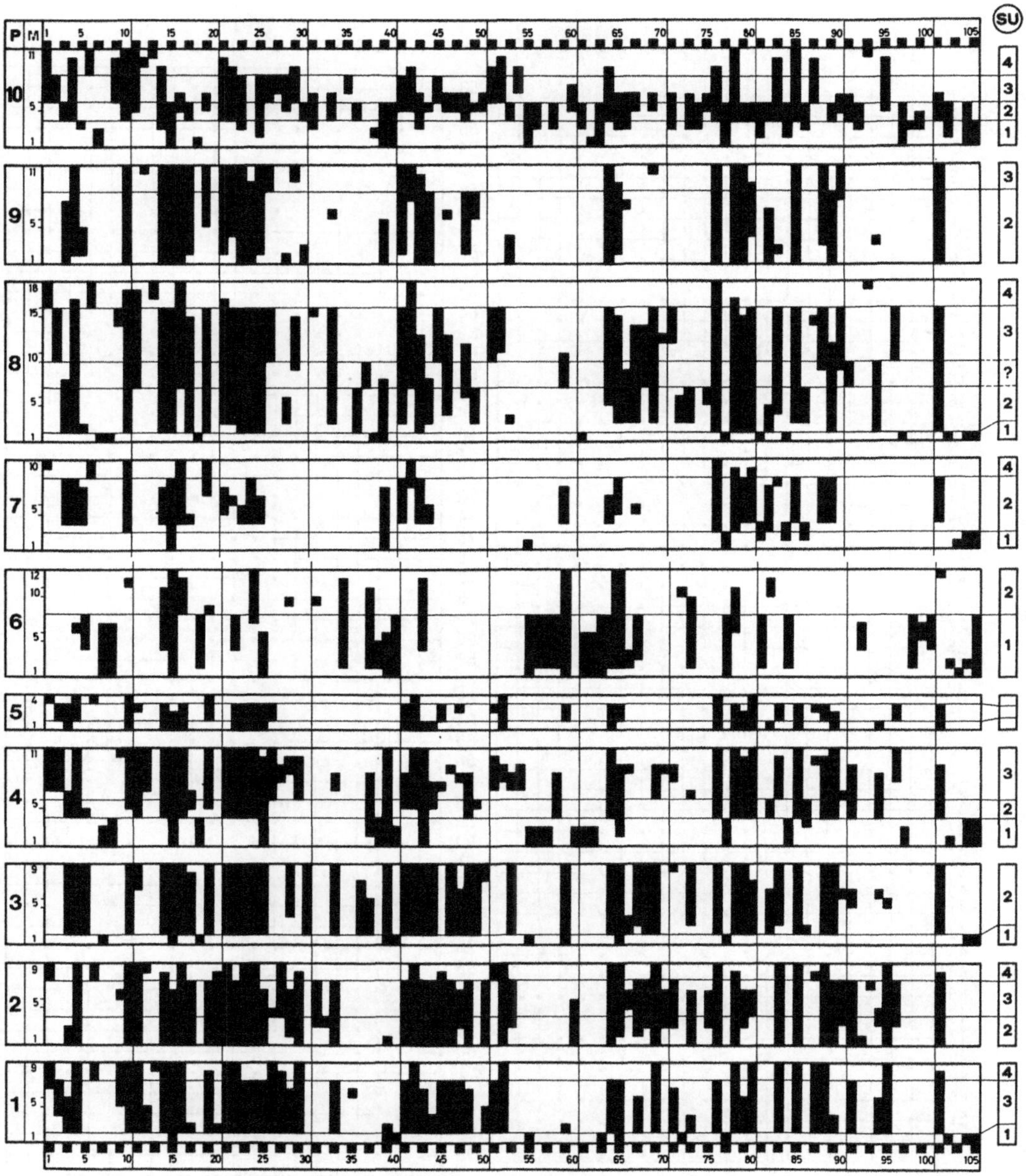

Fig. 11.2. Local stratigraphic ranges of species 1–105 (for names see Appendix E.1). *P*, Sullivan's sections 1–10. *SU*, Sullivan's zones 1–4

11.3 Correlations and Zonation

Processing Sullivan's data with the BioGraph program produces a sequence of 20 unitary associations (Figs. 11.3 and 11.5). To interpret this sequence in terms of biochronozones, we will analyze simultaneously the information given in the

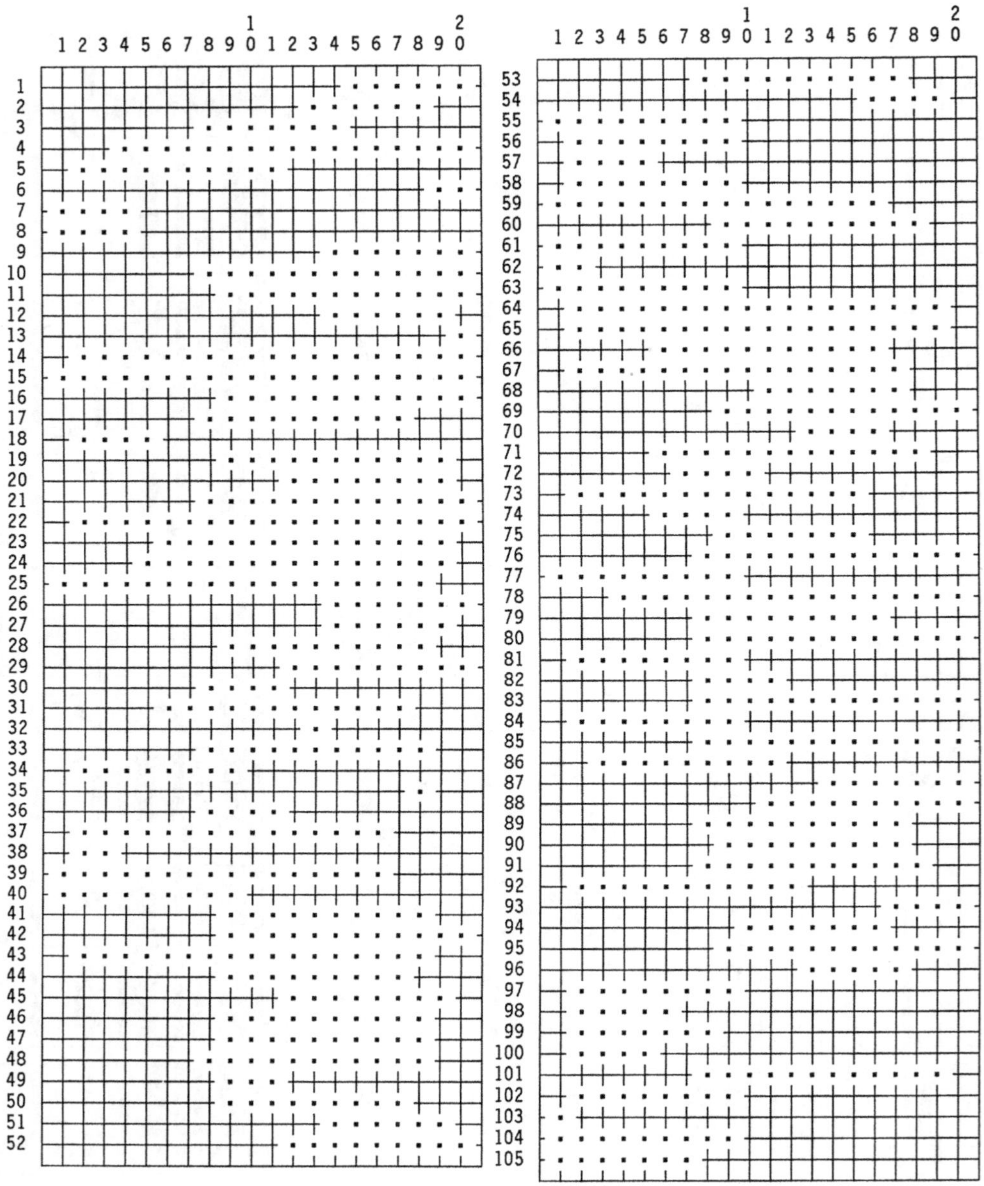

Fig. 11.3. Synthesis of data listed in Fig. 11.2 (output of the BioGraph program)

completed reproducibility matrix (Fig. 11.4.a) and that given in its associated graph G'_k (Fig. 11.4.b).

These diagrams show that we can distinguish three subdivisions which are identifiable in at least half of the sections (union of U.A.10–12 = subzone D; union of U.A.13–15 = subzone E; U.A.19 alone = subzone G) and five others which are identifiable in three or four sections (union of U.A.1 and 2 = subzone

U.A	R	SECTIONS 1	2	3	4	5	6	7	8	9	10	S Z	SUL
20	3	1	0	0	0	0	0	0	1	0	1	H	4
19	5	1	1	0	0	1	0	1	0	0	1	G	4
18	2	1	0	0	0	0	0	0	0	0	1	F	3
17	1	0	1	0	0	0	0	0	0	0	0	F	3
16	1	0	0	0	1	0	0	0	■	■	0	F	3
15	1	0	■	0	0	0	0	0	■	■	1	E	3
14	3	1	■	0	1	1	0	0	■	■	0	E	3
13	1	0	1	0	0	0	0	0	0	0	0	E	3
12	1	0	1	0	0	0	0	0	0	0	0	D	2
11	4	0	0	1	1	■	0	1	0	1	0	D	2
10	1	0	0	0	0	■	0	0	1	0	0	D	2
9	2	0	0	0	0	0	1	■	0	0	1	C	2
8	1	0	0	0	0	0	0	■	0	0	1	C	2
7	1	1	0	0	0	0	0	0	0	0	0	B	1
6	1	0	0	0	0	0	0	0	0	0	1	B	1
5	1	0	0	0	0	0	1	■	0	0	0	B	1
4	1	0	0	■	0	0	1	■	0	0	0	B	1
3	1	0	0	■	1	0	0	■	■	0	0	B	1
2	3	0	0	■	1	0	1	0	■	0	1	A	1
1	2	0	0	0	0	0	1	1	0	0	0	A	1

a)

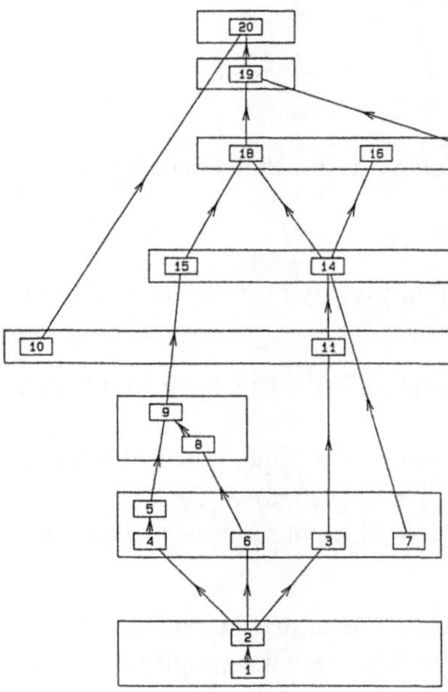

b)

ZONE	SECTIONS 1	2	3	4	5	6	7	8	9	10
SU 4	1	1	0	0	1	0	1	1	0	1
SU 3	1	1	0	1	1	0	0	1	1	1
SU 2	0	1	1	1	1	1	1	1	1	1
SU 1	1	0	1	1	0	1	1	1	0	1

c)

Fig. 11.4. a Reproducibility matrix associated to the protoreferential of Fig. 11.3. The unions of unitary associations (*U.A.*) that are strictly identified in the sections are indicated by *black rectangles*. *R*, reproducibility. *SZ*, subzones *A–H*. *SUL*, Sullivan's zones 1–4. **b** Graph G'_k. **c** Reproducibility of Sullivan's zones 1–4 (*SU*) (output of the BioGraph program)

A; union of U. A.3–7 = subzone B; union of U. A.8 and 9 = subzone C; union of U. A.16–18 = subzone F; U. A.20 alone = subzone H).

The correspondence between these subdivisions (considered here as subzones of Sullivan's zones) and the four biochronozones established empirically by Sullivan (1965) is given in Fig. 11.4.a (column *SUL*). The main boundaries of the two

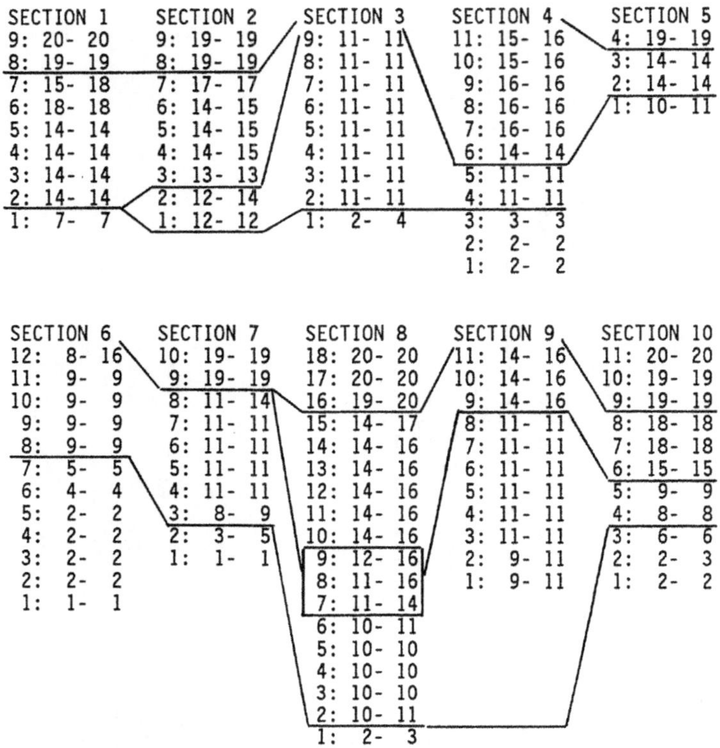

Fig. 11.5. Correlations among Sullivan's sections 1–10 (output of the BioGraph program)

zonations are identical but the present referential is slightly more precise, as seen by the following:

1. Species 62 and 103 are restricted to the lower part of Sullivan's zone 1 (subzone A = U. A.1 and 2);
2. Several pairs of species are restricted to the lower (resp.upper) part of zones 2 and 3: each zone can be subdivided into two parts (subzones C–F);
3. Species 13 is characteristic of the upper part of Sullivan's zone 4 (subzone H = U. A.20).

Sullivan did not refine his zonation by making the minor subzones (A–H) found here. We see, however, that Sullivan's zones have excellent regional reproducibility and very strong superpositional control (Fig. 11.4.c).

11.4 First Conclusions

The above analysis demonstrates unequivocally that Sullivan's (1965) zonation represents a sequence of intervals with almost perfect reproducibility over the entire area he studied.

This fact will be used later (Sect. 16.4) when we discuss the credibility of the so-called "standard" biochronologic scales, composed of "Interval Zones".

Our position is that Sullivan's zonation is an excellent regional biochronologic referential, giving us a solid base for interpreting diachronous datums in the stratigraphic interval under consideration.

We will now proceed with the main object of this chapter, using Sullivan's zonation to describe and interpret the regional diachronism of the datums of the Paleogene nannoplankton of California.

Before doing so, we note that all of Sullivan's species have a continuous biochronologic distribution in the region under study: their local stratigraphic ranges are not perturbed by disjunctive reworking (Sect. 7.2.4).

11.5 Biochronologic Dispersion of Datums

11.5.1 Inventory

The biochronologic dispersion of the datums of each species is given in Appendix E.3.

This quantitative information enables us to see immediately that most of these datums are diachronous. However, it gives no clear information about the origin of the diachronisms.

The question that interests us most is whether the datums' diachronisms are distributed randomly. To answer it, we will begin by showing that a strong statistical correlation exists between the density of the local vertical sampling of the various sections and the variations in the species richness of Sullivan's zones (i.e., the number of species present in each zone).

11.5.2 Species Richness and Sampling Density

Statistics are often the key to a door that stands open. Keeping this in mind, we nevertheless calculated the species richness (S) of the zones in the various sections

Table 11.1. Analysis of species richness (S) of Sullivan's zones 1–4. N, number of samples studied

		Zone 1		Zone 2		Zone 3		Zone 4	
		S	N	S	N	S	N	S	N
S	1	10	2	0	-	58	21	27	5
E	2	0	-	0	-	62	39	29	5
C	3	10	1	55	34	0	-	0	-
T	4	23	11	38	3	59	8	0	-
I	5	0	-	22	1	34	4	9	1
O	6	40	24	18	9	0	-	0	-
N	7	10	4	38	30	0	-	9	4
	8	12	2	43	17	54	21	11	4
	9	0	-	41	15	29	5	0	-
	10	33	9	66	70	44	4	24	3

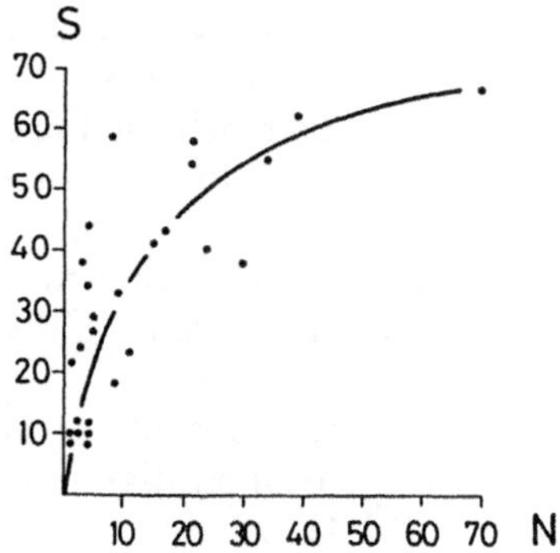

Fig. 11.6. Relationship between the species richness (S) of Sullivan's zones 1–4 and the number (N) of samples collected in each zone (see Table 11.1)

of Bramlette and Sullivan, and analyzed the results in terms of the number of samples (N) from each zone (Table 11.1).

The resulting diagram proves (if proof is necessary) that a very strong correlation exists between sample density and species richness (Fig. 11.6). This naturally leads us to look for a possible relationship between the geographic distribution of the datums' diachronism and the biogeographic distribution of the species.

11.5.3 Relationship with the Geographic Distribution of Species

Consider a set $P = \{p_1, p_2, \ldots, p_n\}$ of stratigraphic sections, and a species x found in set P.

Each species x with dispersion $D_a(x)$ equal to 1 defines two consecutive biochronozones, $A(x)$ and $B(x)$, where:

$A(x)$ is the zone in which x first appears, and

$B(x)$, which immediately follows $A(x)$, is the zone in which it appears last.

Such a species thus divides the set P into two subsets, $S(x)$ and $R(x)$, where:

$S(x)$ is the set of sections in which x appears in zone A (source sections) and

$R(x)$ is the set of sections in which x appears in zone B (receptor sections).

To detect a possible relationship between the geographic distribution of species and the dispersion of their datums, we construct an oriented graph $G'_p = (P, U')$, defined as follows (Fig. 11.7):

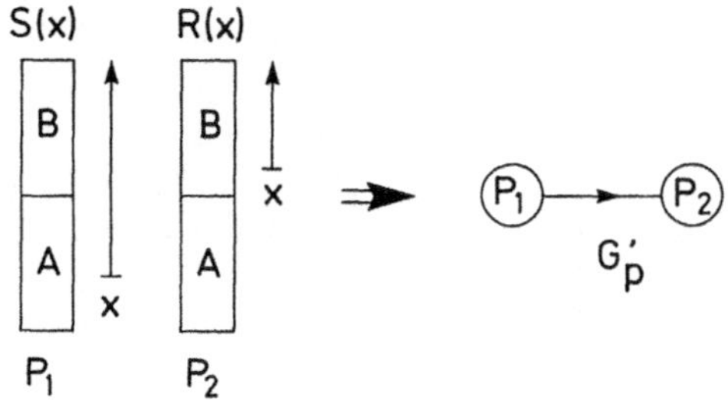

Fig. 11.7. Construction of the graph G'_p. The first appearance of species x occurs in zone A of section P_1 and in zone B of section P_2 : P_1 is the source section and P_2 is the receptor section (Guex 1982)

$$(p_i, p_j) \in U' \leftrightarrow \exists x \text{ such that } p_i \in S(x) \quad \text{and} \quad p_j \in R(x).$$

By definition, two vertices of this graph can be connected by several contradictory arcs (i.e., arcs oriented in opposite directions). If G'_p admits no such contradictory arcs (or only a few) then it necessarily follows that the diachronisms of the datums are not distributed randomly.

Furthermore, if the orientations on G'_p are generated by discontinuities in the local vertical record of the species, then a relationship should exist between the density of the vertical sampling of the different zones and the geographic distribution of the diachronisms.

To see whether this relationship exists, we construct a second graph, $G''_p = (P, U'')$, where:

$P = \{p_i\}$ is the set of sections in which x is found;

n_i is the number of samples studied in zone A of section p_i;

n_j is the number of samples studied in zone A of section p_j;

$(p_i, p_j) \in U'' \leftrightarrow n_i > n_j$.

If the arcs of G''_p are oriented in the same direction as the arcs of G'_p, then the diachronism of the datums is generated exclusively by discontinuities in the regional stratigraphic (i.e., vertical) record of the species.

Figure 11.8 is based on the data listed in Table 11.2. It illustrates the graph G'_p, generated by these diachronous datums (FADs whose dispersion has an amplitude equal to 1) in the biochronologic interval zone 1 → zone 2. It also shows the graph G''_p, which classifies the sections according to the vertical sampling density in zone 1.

The quasi-identity of the two graphs demonstrates clearly that the diachronism of the FADS in this interval is due exclusively to poor sampling in some sections. Applying this test to the intervals zone 2 → zone 3 and zone 3 → zone 4 gives identical results.

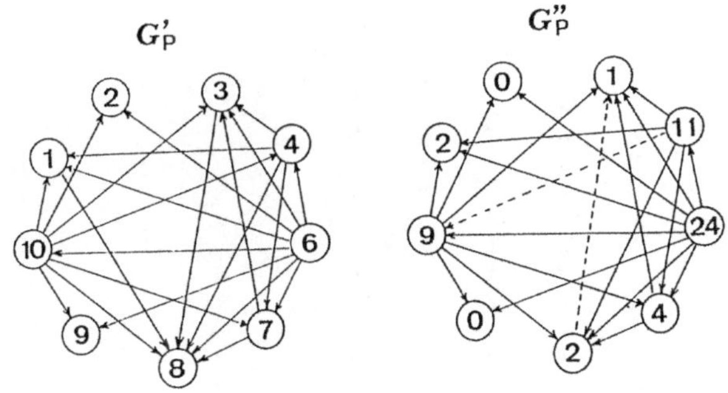

Fig. 11.8. The vertices of the graph G'_p represent the stratigraphic sections. The arcs of this graph are directed in the sense source → receptor (see Table 11.2). The *vertices* of the graph G''_p represent the same sections as those of the graph G'_p but the *numbers* correspond to the number of samples collected in zone 1 of these sections; the *arcs* indicated with *dotted lines* are directed in the opposite sense of those in the graph G'_p

Table 11.2. FADs of the different species (X) in the interval zone 1 → zone 2. $S(x)$, source sections. $R(x)$, receptor sections

X	$S(x)$	$R(x)$
5	6,10	3,7,8,9
14	6,10	1,3,7
15	1,3,4,6,7,9	8
22	6	1,3,4,7,8,9,10
23	10	1,3,4,7,8
24	6	1,3,4,7,8
25	6,10	1,3,7,9
37	4,6	3,8
43	4,6,10	1,3,7,8
64	6	1,3,4,7,8,10
67	6	2,3,7,8
71	10	2,3
72	1	6,8
78	6	1,3,4,7,8,10
86	4,7,10	3,8
92	6,10	3

This conclusion is obvious. It proves, however, that analytical methods that postulate a random distribution of diachronous datums are based on a false assumption (see also Baumgartner 1984a). It also explains the fact that the fluctuations of species richness of the Paleogene Californian nannoplankton in each zone depend strongly on the density of the vertical sampling.

11.5.4 Quality of the Vertical Record of Species

To facilitate our study of the relationship between the datums' dispersion and the apparent longevity of species [$L_a(x)$], we have entered in the first two columns of Table 11.3 all the observed combinations of different values of D_a and D_d listed in Appendix E.3. The number of species with each combination is entered in the columns $L_a(x) = 1$ to $L_a(x) = 4$.

Figures 11.9.a and 11.9.b are constructed from these observations according to the rule described in Section 8.4. They give the occurrence frequency of the

Table 11.3. $L_a(x)$ vs D_a and D_d

		$L_a(x)$			
D_a	D_d	1	2	3	4
0	0	18	6	0	0
0	1	x	27	3	0
0	2	x	x	5	0
1	0	x	5	1	0
1	1	x	12	8	1
1	2	x	x	5	4
1	3	x	x	x	2
2	0	x	x	1	0
2	1	x	x	3	1
2	2	x	x	2	0
2	3	x	x	x	1

species in the different sectors of the graphs $D_a(x) = f[L_a(x)]$ and $D_d(x) = f[L_a(x)]$.

We see in Fig. 11.9.a that the dispersion D_a is $L_a(x) - 1$ (the worst possible) for 39% of the species, $L_a(x) - 2$ for 47%, and $L_a(x) - 3$ for 14%.

For $D_d(x)$ the proportions are different: D_d is $L_a(x) - 1$ for 68% of the species, $L_a(x) - 2$ for 28% and $L_a(x) - 3$ for 4%.

These observations indicate clearly that the LADs of the Californian Paleogene nannoplankton are generally more diachronous than their FADs. The same conclusion can be drawn from Fig. 11.9.c, which gives the occurrence frequency of Sullivan's 105 species for each possible combination of the different values of $D_a(x)$ and $D_d(x)$. This diagram, based on the data in Table 11.3, shows that 10% of the species have FADs that are more diachronous than their LADs, while 45% have LADs more diachronous than their FADs.

We note, incidentally, that the relationship between D_a and D_d is independent of the longevity of species: for each pair of opposite values of D_a and D_d [e.g., (0, 1) and (1, 0); (1, 2) and (2, 1); etc: see Table 11.3], approximately four times as many species have $D_a < D_d$ than the reverse, whatever the value of $L_a(x)$.

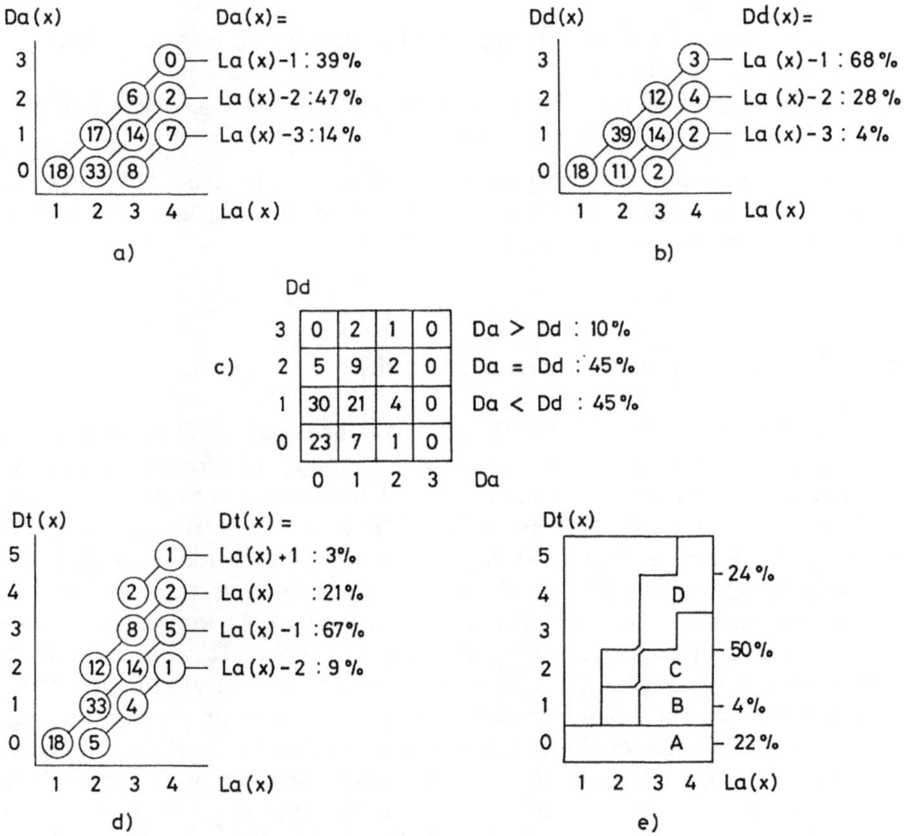

Fig. 11.9 a–e. Relationships among the coefficients $D_a(x)$, $D_d(x)$, $D_t(x)$, $L_a(x)$, and variations in the ratio $D_t(x)/L_a(x)$ (see Appendix E.3)

This implies that if we plot the probability that these species can be found at any point of their actual existence interval, the resulting curves will be similar [in other words, their form is not a function of $L_a(x)$]. The data analyzed here are in any case too limited to permit a reasonable probabilistic model of this particular problem.

Last, it is interesting to examine the relationship between the longevity of species and the total dispersion of their datums ($D_t = D_a + D_d$). This is given in Fig. 11.9.d, which shows that:

1. Almost 90% of the species Sullivan studied are located in the graph sectors described by the equation $D_t(x) = L_a(x) - (0.5 \pm 0.5)$;
2. Long-ranging species [$L_a(x) = 3$ or 4] always have a large total dispersion and none of them occupy the sector $D_t(x) = 0$.

Figure 11.9.e gives a global view of the quality of the vertical record of species; it is based on Fig. 11.9.d and constructed according to the criteria discussed in Section 8.4.2.3. It highlights the following facts:

1. Twenty-four percent of the species (sector *D* of the graph) occur in different zones in different sections; their biochronologic distribution is diachronous on a regional scale.
2. Fifty percent have a very strong total biochronologic dispersion [sector *C*: $D_t(x)/L_a(x) = 0.5$ to 0.9].
3. Twenty-six percent have no or little biochronologic dispersion (sectors *A* and *B*).

These observations, trite in themselves, will serve as the basis of comparisons when we discuss possible causes for the exceptionally high diachronism of the LADs of the group studied here (Chap. 12).

11.5.5 Quality of the Lateral Record of Species

We saw above (Sect. 11.5.2) that the species richness of the different zones depends strongly on the density of the vertical sampling. In addition, it is obvious that only long-ranging species can be recorded in all the sampled sections. Together these two facts imply that there should be a positive statistical correlation between the relative geofrequency (G_r) of species and their longevity [$L_a(x)$].

We note that for Californian Paleogene nannoplankton, the correlation between $L_a(x)$ and $G_r(x)$ is extremely strong (Fig. 11.10.a). All long-ranging species have an ideal or quasi-ideal geofrequency [$G_r(x) = 0.9$ or 1], while most species with a brief existence [$L_a(x) = 1$] have a low value of G_r. The sampling density and the strong biochronologic dispersion affecting this group are thus the major factors controlling the relationship between relative geofrequency and longevity.

These observations, along with those discussed above, imply that there should be a strong positive correlation between the biochronologic dispersion of the datums and the relative geofrequency of the species. This is confirmed by the diagrams of Fig. 11.10.b.

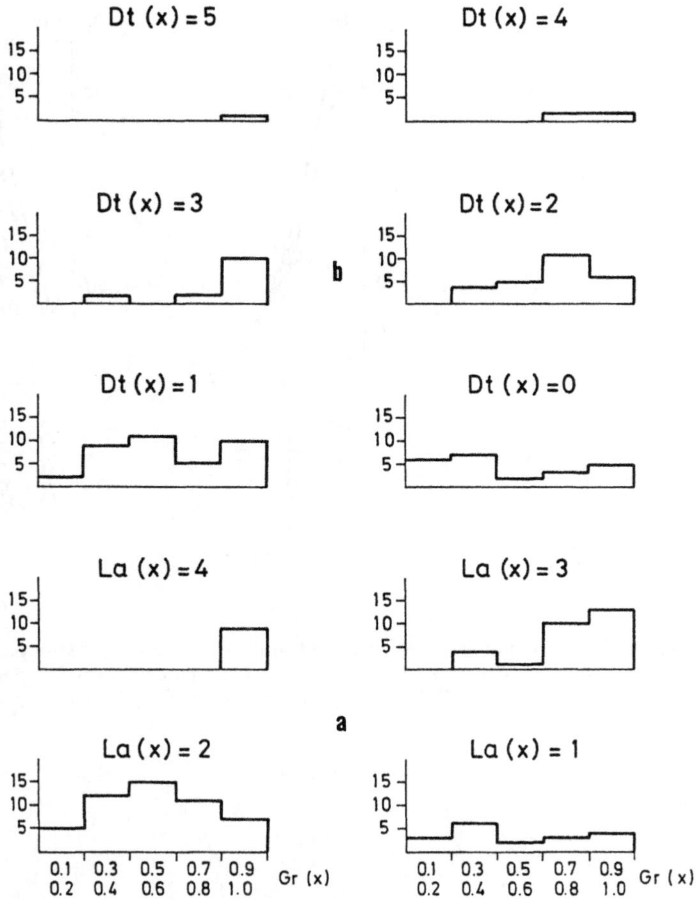

Fig. 11.10 a, b. Relationships among $G_r(x)$, $L_a(x)$ and $D_t(x)$ (see Appendix E.3)

11.6 Comparison with Deboo's Data

11.6.1 Vertical Record

The variation of $D_a(x)$ and $D_d(x)$ as a function of $L_a(x)$ for Deboo's benthics is similar to that observed for Sullivan's nannoplankton (Fig. 11.11.a and 11.11.b).

This is not the case for the relationship between $D_a(x)$ and $D_d(x)$, shown in Fig. 11.11.c. The ratio $(D_a > D_d)$ over $(D_a < D_d)$ is 0.5 for the benthic foraminifera (15 versus 30%) while it is about 0.2 for the Californian nannoplankton (10 versus 45%). In other words, a much greater proportion of Sullivan'S nannoplankton have FADs that are more diachronous than their LADs than is the case for Deboo's benthics.

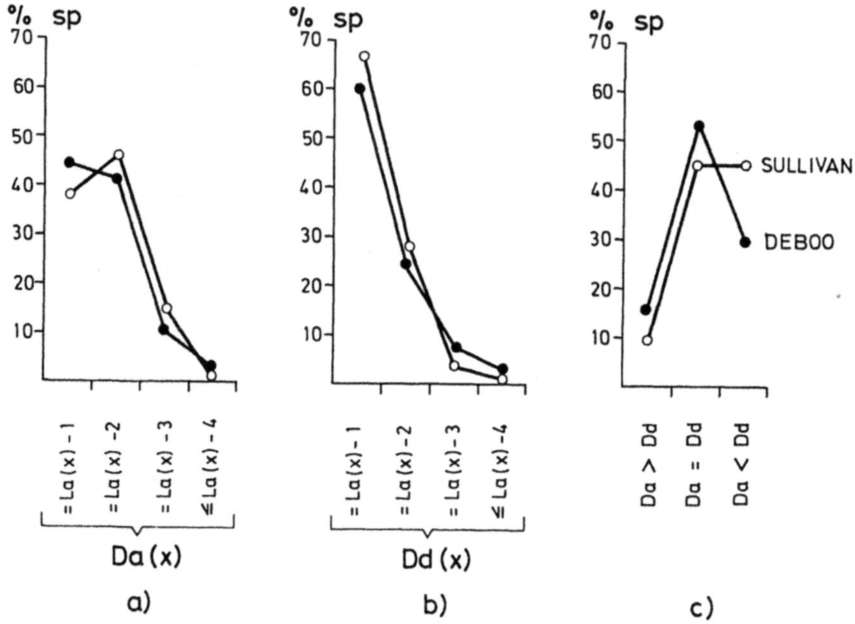

Fig. 11.11 a–c. Comparison of the diachronism of Sullivan's nannoplankton and the diachronism of Deboo's benthics. **a** $D_a(x) = f[(L_a(x)]$: percentage of species. **b** $D_d(x) = f[L_a(x)]$: percentage of species. **c** Relationship between D_a and D_d: percentage of species

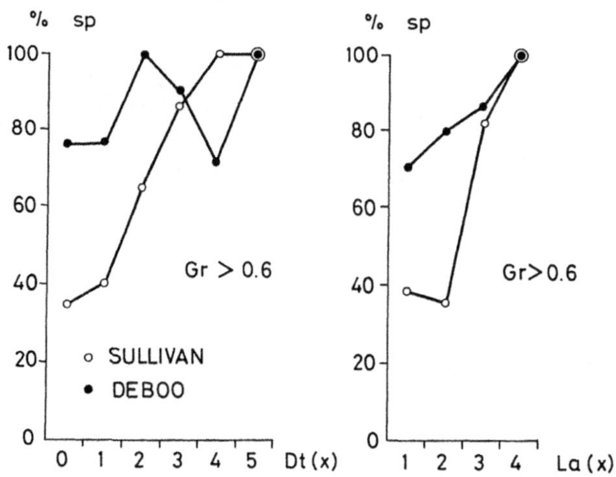

Fig. 11.12. Comparison of Sullivan's and Deboo's data. Percentage of species with a relative geofrequency greater than 0.6: relationship with $D_t(x)$ and $L_a(x)$

11.6.2 Lateral Record

The strong positive correlation observed among the Californian nannoplankton between G_r and D_t on the one hand and G_r and L_a on the other does not exist for Deboo's benthics (Fig. 10.9).

To facilitate a comparison of these two groups we have plotted the percentage of species with a relative geofrequency greater than 0.6 as a function of their total biochronologic dispersion and apparent longevity (Fig. 11.12).

These graphs show that 8 out of 23 species (35%) of Sullivan's nannoplankton with $D_t = 0$ have a relative geofrequency greater than 0.6, compared to 100% with $D_t = 4$ and $D_t = 5$; and 7 out of 18 species of nannoplankton (39%) with $L_a(x) = 1$ have a coefficient $G_r > 0.6$. All the species (nannoplankton and benthics) with an apparent longevity equal to 4 have a relative geofrequency greater than 0.6.

These diagrams emphasize once more the strong positive correlation observed among Sullivan's nannoplankton between $D_t(x)$ and $G_r(x)$ on the one hand and between $L_a(x)$ and $G_r(x)$ on the other.

They also show, however, that among the benthics, $G_r(x)$ and $D_t(x)$ vary independently, and that the correlation between $L_a(x)$ and $G_r(x)$ is weak.

11.7 Complexity of the Problem

11.7.1 Coefficient $I_t(x)$

The complexity of the problem studied here is essentially due to the poor lateral record of almost half of Sullivan's species. Figure 11.13.a shows the relationship between the coefficients $I_t(x)$ (average value: Sect. 11.7.2) and $G_r(x)$. This graph is very similar to the one we obtained in our analysis of the preceding problem (Sect. 10.7, Fig. 10.10). It shows once again the very strong inverse correlation between the relative geofrequency of species and the number of conflicting superpositions in which they appear.

We also note that the value of I_t decreases rapidly as the total biochronologic dispersion (D_t) increases (Fig. 11.13.b). This is a statistical artifice resulting from the strong positive correlation between D_t and G_r (Fig. 11.10.b).

11.7.2 Remark

The graphs of Fig. 11.13 were constructed by calculating the average value of I_t for each distinct value of G_r and D_t (Appendix E.3). For example, species 22, 64, 66, and 67 have dispersions D_t equal to 4, and their values of I_t are 0, 6, 308, and 24 respectively: the average value of I_t for species with D_t equal to 4 is thus 85.

11.8 Constructing a Sequence of Datums

11.8.1 Introduction

Knowing whether a species appears or disappears in the lower, middle or upper part of a biochronozone is a separate problem from that of constructing such a

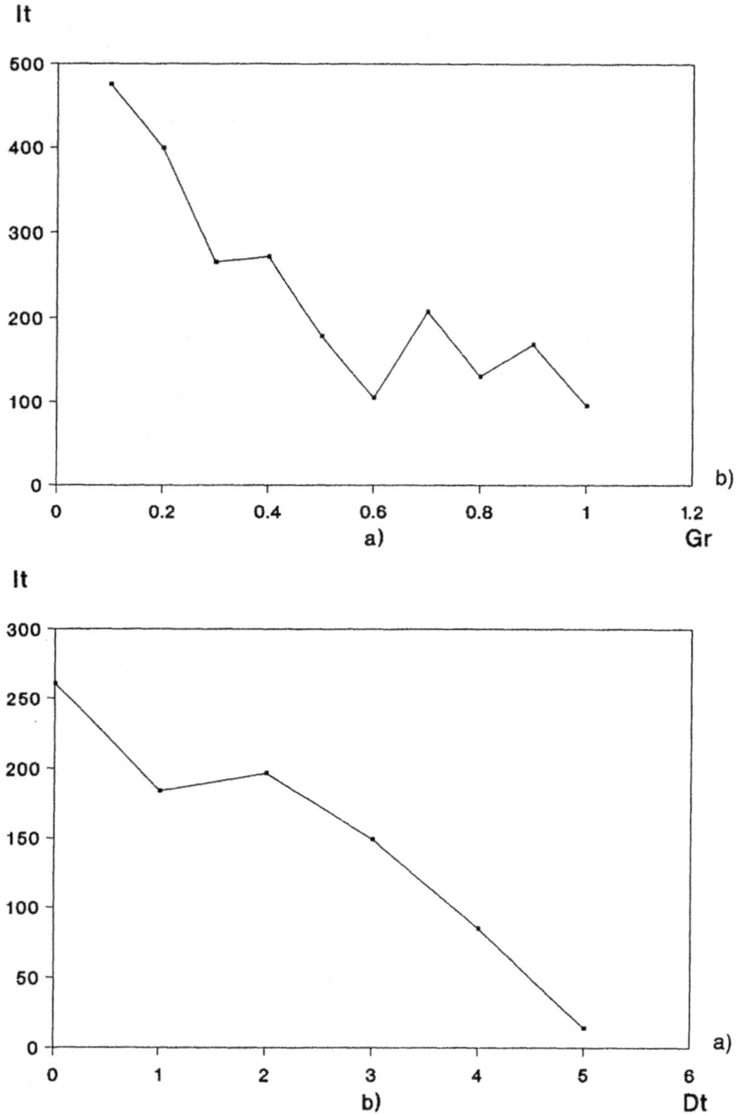

Fig. 11.13 a, b. Variation of I_t (average) as a function of **a** $G_r(x)$ and **b** $D_t(x)$

zone. This is why we have deliberately ignored, up to now, the possibility of increasing the resolution power of unitary associations by taking into account sequences of datums that can sometimes be identified within a zone.

We will see that this problem of ordering the datums is easy compared to that of constructing biochronologically significant zones. The datums can be ordered in two ways. The first is empirical: after delimiting a particular zone in a set of stratigraphic sections, we simply look for species that invariably appear at the base of the zone, a little higher, still a little higher, etc. This can be done visually if the number of species isn't great.

The second method is quantitative. It is also simple: we begin by listing all the species that first appeared in a given zone. We eliminate from this set all species with a dispersion D_a greater than zero (their FADs are a priori unreliable). We enter the local relationships of anteriority and posteriority among the datums into a matrix $E = (e_{ij})$ where

$$e_{ij} = \begin{cases} 1 \text{ if the datum } i \text{ is below the datum } j, \\ 0 \text{ otherwise.} \end{cases}$$

The terms $e_{ij} = e_{ji} = 0$ represent indeterminate relationships between the datums.

The terms $e_{ij} = e_{ji} = 1$ represent contradictory relationships (crossovers): we replace the 1's here by zeros.

Finally, we look for a lower triangular arrangement of the matrix E (i.e., with all the 1's below the main diagonal).

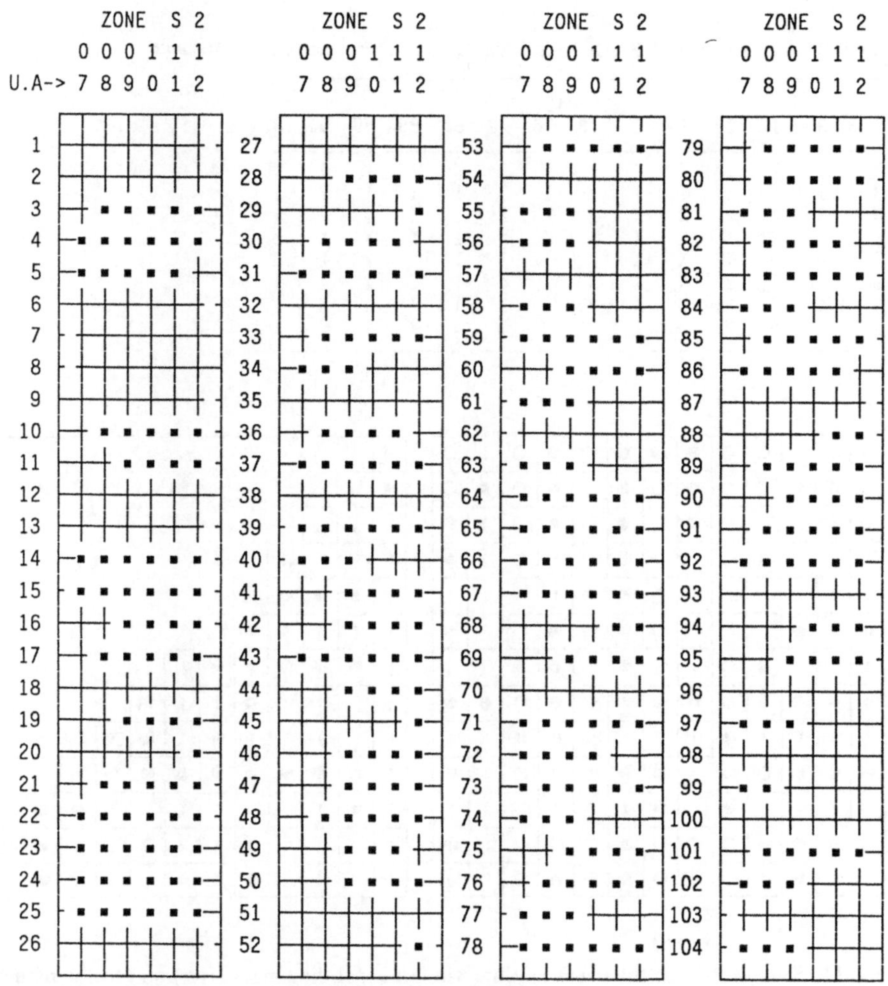

Fig. 11.14. Partial range chart showing the species which first appear in U.A.8 (= base of Sullivan's zone S_2)

In the case of Sullivan's data, we will see that the FADs supposedly characteristic of the base of a biochronozone can very well be located above the species that disappear in the same zone.

11.8.2 Sequence of FADs in Sullivan's Zone 2

As an example we will examine the set of species that appear in Sullivan's zone 2. The base of this zone corresponds to that of U. A.8 of Fig. 11.14.

Sections 1 and 5 are omitted from the list that follows, because zone 2 is absent from the first and its base is truncated in the second.

The species to be studied are: 3, 10, 17, 21, 30, 33, 36, 48, 53, 76, 79, 80, 82, 83, 85, 89, 91, and 101.

Table 11.4. Sequence of FADs in Sullivan's zone 2. *P*, sections. *X*, absence

						SPECIES							
P	3	17	21	30	36	48	76	79	82	85	89	91	101
2	1	1	1	X	X	1	1	2	X	1	1	1	1
3	1	1	1	1	1	1	1	1	1	1	1	2	1
4	3	4	2	X	X	2	2	2	2	1	1	3	3
6	X	X	X	X	X	X	X	X	1	X	X	X	2
7	4	6	7	X	X	X	1	3	1	5	2	X	2
8	1	1	2	X	1	3	1	1	1	2	1	4	1
9	1	2	3	1	X	2	1	1	1	1	1	X	1
10	3	4	2	2	1	2	2	1	1	4	1	5	2

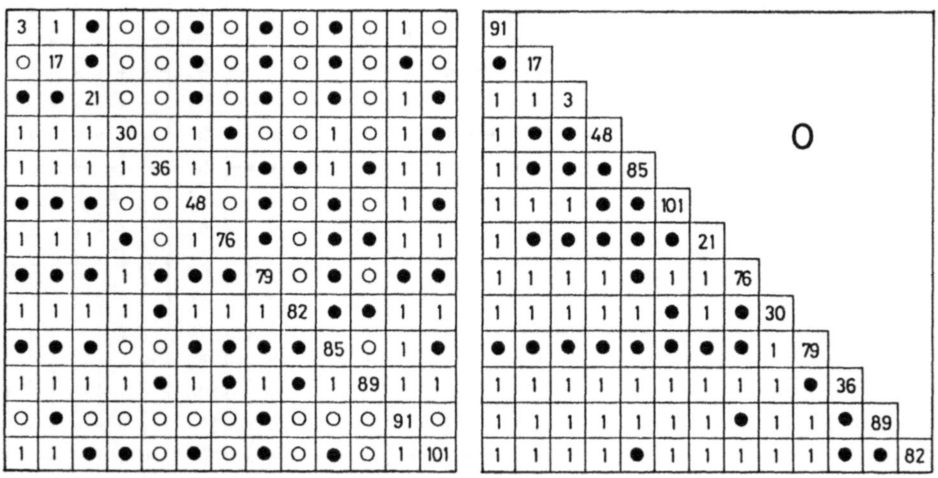

a) b)

Fig. 11.15. **a** Matrix *E*: the *numbers* along the principal diagonal correspond to the names of the species and the *black circles* represent unknown inter-FAD relationships. **b** Lower triangular arrangement of *E*

91	0	3	4	4	3	4	4	2	0	3	4	4
	17	4	0	0	4	0	4	2	0	1	4	4
		3	0	0	2	0	3	1	0	1	3	3
			48	0	0	0	2	1	0	2	4	3
				85	0	0	0	1	0	2	3	0
					101	0	2	0	0	1	2	4
						21	3	1	0	2	5	4
							76	0	0	1	0	1
								30	1	1	1	1
									79	0	3	1
										36	0	0
											89	0
												82

Fig. 11.16. Matrix showing the reproducibility of the inter-FAD superpositions of the 13 species under consideration. The names (*numbers*) of the species are given along the principal diagonal

For the reason given above (Sect. 11.8.1), we eliminate species 10, 33, 53, 80, and 83, which have a dispersion D_a greater than zero (see Appendix E.3).

The order in which the remaining 13 species appear in the different sections is given in Table 11.4, based on the original data of Sullivan (1964, 1965) and Bramlette and Sullivan (1961).

The local sequences of FADs listed in this table are entered in the matrix E of Fig. 11.15.a, and a lower triangular arrangement of E is shown in Fig. 11.15.b. This arrangement represents the desired ordering for the FADs of species 82 to 91 (first FAD of the sequence to last FAD of the sequence).

The reproducibility of the inter-FAD superpositions is given in Fig. 11.16. It shows that 38% of these superpositions are indeterminate ($ij = 0$), and 18% are not reproducible ($ij = 1$).

Most of the indeterminations are generated by crossovers between the first local appearances under consideration. This is not surprising, since 45% of Sullivan's species have a dispersion D_a greater than zero: here we are simply finding again, at the infra-zonal level, diachronisms like those observed globally on a large scale (i.e., diachronisms covering more than one zone).

11.8.3 Selecting "Good" FADs

In our group of 13 species (Table 11.4), we will consider only those with reasonably reproducible FADs, on the assumption that these have the best chance of enabling us to refine Sullivan's zonal correlations.

To classify potentially useful species in a relatively objective way, we will use a coefficient that takes into account simultaneously both the frequency of inter-FAD superpositions and the frequency of indeterminate relationships (crossovers and pure indeterminations).

Let n_1 be the number of indeterminate inter-FAD relationships ($ij = 0$ in Fig. 11.16).

Table 11.5. Coefficient σ characterizing the potential usefulness of the species

X	n_1	n_2	n_3	σ
91	2	0	35	0,06
17	5	1	22	0,23
3	4	2	18	0,33
48	6	1	15	0,46
85	8	1	9	1,00 <=
101	5	1	17	0,35
21	6	1	18	0,39
76	4	2	18	0,33
30	2	8	4	2,50 <=
79	9	2	3	3,66 <=
36	3	5	9	0,89 <=
89	3	1	27	0,15
82	3	3	22	0,27

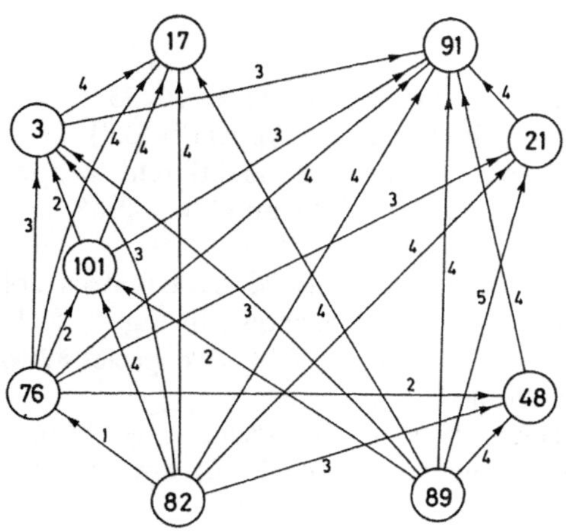

Fig. 11.17. Oriented graph showing the sequences of FADs that are potentially the most useful. The values assigned to the arcs of this graph correspond to the reproducibility of the inter-FAD superpositions

Let n_2 be the number of nonreproducible inter-FAD superpositions ($ij = 1$ in Fig. 11.16).

Let n_3 be the sum of the superpositions (observed in at least two sections) of a given FAD and the other FADs.

Finally, let $\sigma = (n_1 + n_2)/n_3$ be our threshold coefficient: only species with σ less than or equal to 0.5 will be considered.

The values of n_i and of σ for each species are given in Table 11.5, based on the matrix of Fig. 11.16.

We will eliminate species 30, 36, 79, and 85, since these species have σ greater than 0.5 (see Table 11.5). In other words we will consider only the local sequences of FADs of species 3, 17, 21, 48, 76, 82, 89, 91, and 101.

The oriented graph representing these sequences is shown in Fig. 11.17. The values assigned to each arc of this graph correspond to the reproducibility of the observed inter-FAD superpositions.

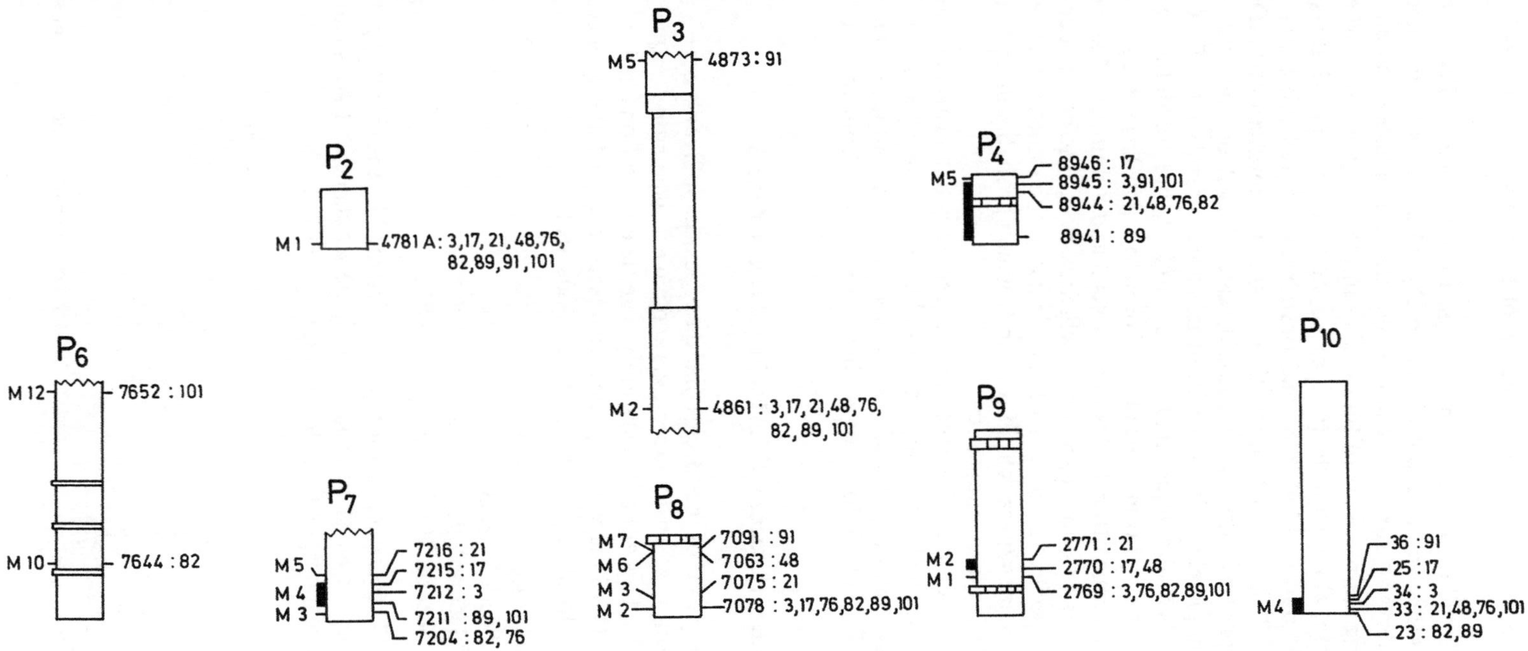

Fig. 11.18. Stratigraphic position of the FADs in sections 2–10 (zone 2). The samples are labeled with Sullivan's original numbers. *M*, maximal horizons

11.8.4 Biochronologic Interest of the Sequence of FADs

We will now consider how to use the sequences of FADs shown in Fig. 11.17. To do this we have reproduced in Fig. 11.18 the stratigraphic intervals assigned to Sullivan's zone 2 in each of the stratigraphic sections under consideration. The original numbers of Sullivan's samples are given next to each stratigraphic column, along with the position of the FADs examined here (see the original tables of Sullivan 1964, 1965 and Bramlette and Sullivan 1961).

This diagram shows that the main sequences of FADs (those with the best reproducibility on a regional scale), based on the matrix of Fig. 11.15.b, fail completely to improve Sullivan's zonal correlations (Sect. 11.3).

We note in fact that: in sections 2, 9, and 10, all the "good" FADs are found in a tiny stratigraphic interval or in the same level; the FAD of species 91 is found above the other FADs in sections 3 and 8, but it is observed at the base of zone 2 in the Lodo section (P_{10}), which is by far the best documented; and the FADs of species 82, 89, and 101 are found in the same level in sections 3, 8 and 9, and they are superposed in section 6, which probably means that the FAD of species 101 is diachronous, etc.

In summary, we see that it is very easy to construct a sequence of datums once a correct zonation has been established. In the present case this exercise is worthless from the point of view of biochronologic applications: the sequence of FADs in Fig. 11.15 does not enable us to improve Sullivan's zonal correlations.

11.8.5 Sequence of LADs in Sullivan's Zone 2

Only four of the species that disappear in Sullivan's zone 2 (species 30, 36, 82, 86) occur in at least two sections and have a dispersion D_d equal to zero (see Appendix E.3). This zone is apparently complete in sections 2, 4, 5, 7, 8, 9 and 10 and the LADs of these four species are observed in partial sequence only in sections 7–10. The local sequences of these LADs are:

1. Section 7: 86 → 82;
2. Section 8: 86 → 82 → 36;
3. Section 9: 30 → 82;
4. Section 10: 36 → 30 → 82 → 86.

These superpositions are represented by the oriented graph of Fig. 11.19. This superb cycle shows that no reliable sequence of LADs can be deduced from the local superpositions listed above.

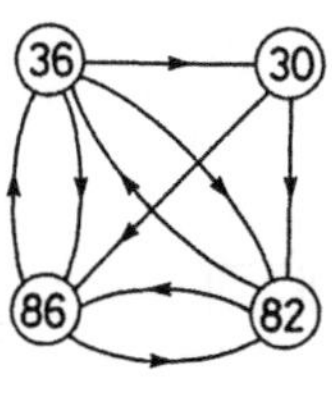

Fig. 11.19. Cyclic sequence of the LADs of species 30, 36, 82, and 86

11.8.6 Comparison with "Good" FADs

Let us remember that a single contradiction in the order in which two species appear (or disappear) is sufficient to prove the diachronism of at least one datum (FAD or LAD). It is therefore interesting to compare the stratigraphic relationships of the potentially useful LADs discussed here and those of the "good" FADs studied earlier (Sect. 11.8.3). To do this we examine the tables giving the species' local stratigraphic ranges (Fig. 11.2). These tables show that:

1. The LAD of species 82 is below the FAD of species 101 in section 6;
2. The LAD of species 86 is below the FADs of species 3, 17, 21, 79, 85, 89, and 101 in section 7
3. The LADs of species 36, 82, and 86 are below the FAD of species 91 in section 8.

These observations prove two things:

1. The FADs and LADs with no biochronologic dispersion ($D_a = 0$ or $D_d = 0$) are still strongly diachronous within the biochronozone under consideration: these datums cannot be used to define Interval Zones (see Sect.16.4).
2. When working with biostratigraphic data similar in structure to the data examined here (e.g., radiolarians, benthic foraminifera, etc.), it is essential to construct zonations based on concurrent ranges of species, following the simple reasoning described in the beginning of this book (Sect. 1.3).

11.8.7 Remark

It may be tempting to tackle the above problem using one of the many probabilistic techniques available in the recent literature and inspired by Hay (1972). We will explain in Chapter 15 why these methods are generally unreliable.

Chapter 12
Diachronism of LADs: Possible Causes

12.1 Introduction

We studied the properties of the diachronism of datums in detail in the preceding chapter. We saw in particular and without surprise that a very strong correlation exists between the density of vertical sampling and the FADs' biochronologic dispersion. That correlation can certainly be explained by the selective dissolution that affects most species in a more or less pronounced way before and during diagenesis.

However, selective dissolution cannot explain why the FADs of Sullivan's nannoplankton are significantly less dispersed than their LADs. If the datums' biochronologic dispersion were exclusively generated by the species' sparse stratigraphic record, the dispersion of FADs and that of LADs should be about equal. However, that is not the case.

We can imagine three possible explanations:

1. Vertical mixing in the sense of Berger and Heath (1968);
2. Relative brevity of the rarity period preceding the maximum biogeographic range of these nannofossils, compared to that preceding their extinction;
3. Continuous lateral displacements generating a local and artificial lengthening of the vertical ranges of some species (i.e., nondisjunctive reworking: see Chapter 7).

The first explanation can be immediately rejected. It is obviously impossible that burrowing organisms could induce vertical (and nondisjuctive) reworking for several million years, through coarse and very thick clastic intercalations (Fig. 11.1).

The two other hypotheses are not mutually exclusive and it is difficult to judge which is better. We will test them indirectly by studying the structure of the diachronisms in a nannofossil group that we have good reason to think was not disturbed by reworking.

12.2 Mediterranean Neogene Nannoplankton

12.2.1 Muller's Data (1978)

We will analyze Muller's (1978) data on the stratigraphic distribution of the Mediterranean Neogene nannoplankton from seven Deep Sea Drilling Project boreholes (leg 42–A; Fig. 12.1).

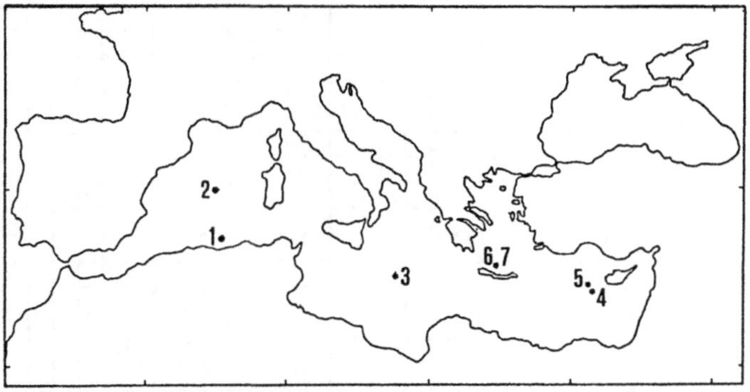

Fig. 12.1. Location of the sites 1–7 studied by Muller (1978) (see Appendix F.2 for DSDP sites' numbers)

Two of the sections studied (372 and 375) reach the Lower Miocene; the other five cover essentially the NN12–NN21 "zones" of the Pliocene and Quaternary.

According to Muller (1978, p. 746) the Pliocene sediments were deposited in an open marine environment with abundant nannofossils, and where virtually no reworking occurred.

It is this great rarity of reworking during the Pliocene interval that interests us. It will allow us to make revealing comparisons between the biochronologic dispersion of datums of the Mediterranean Neogene nannoplankton and that of Sullivan's nannofossils.

12.2.2 Interpreting Muller's Data

Fig. 12.1 shows the location of the seven sections Muller studied; the local ranges of the species identified in these sections are given in Appendix F.5 (numerical database). Processing these data with the BioGraph program results in a sequence of 26 unitary associations (Fig. 12.2). Using that referential we can establish a sequence of seven biochronozones that can be correlated over a large geographic area (Fig. 12.3).

12.2.3 Diachronism of Datums

The sole purpose of the present discussion is to compare the diachronism of the datums of calcareous nannoplankton in two entirely distinct sedimentary environments.

For that reason, we are mainly interested in: (1) the variations of the coefficients D_a, D_d, and D_t as a function of apparent longevity L_a; and (2) the relative frequency of the different possible combinations of the values of D_a and D_d.

The values of these coefficients are based on the zonal correlations shown in Fig. 12.3. They are given in Appendix F.4.

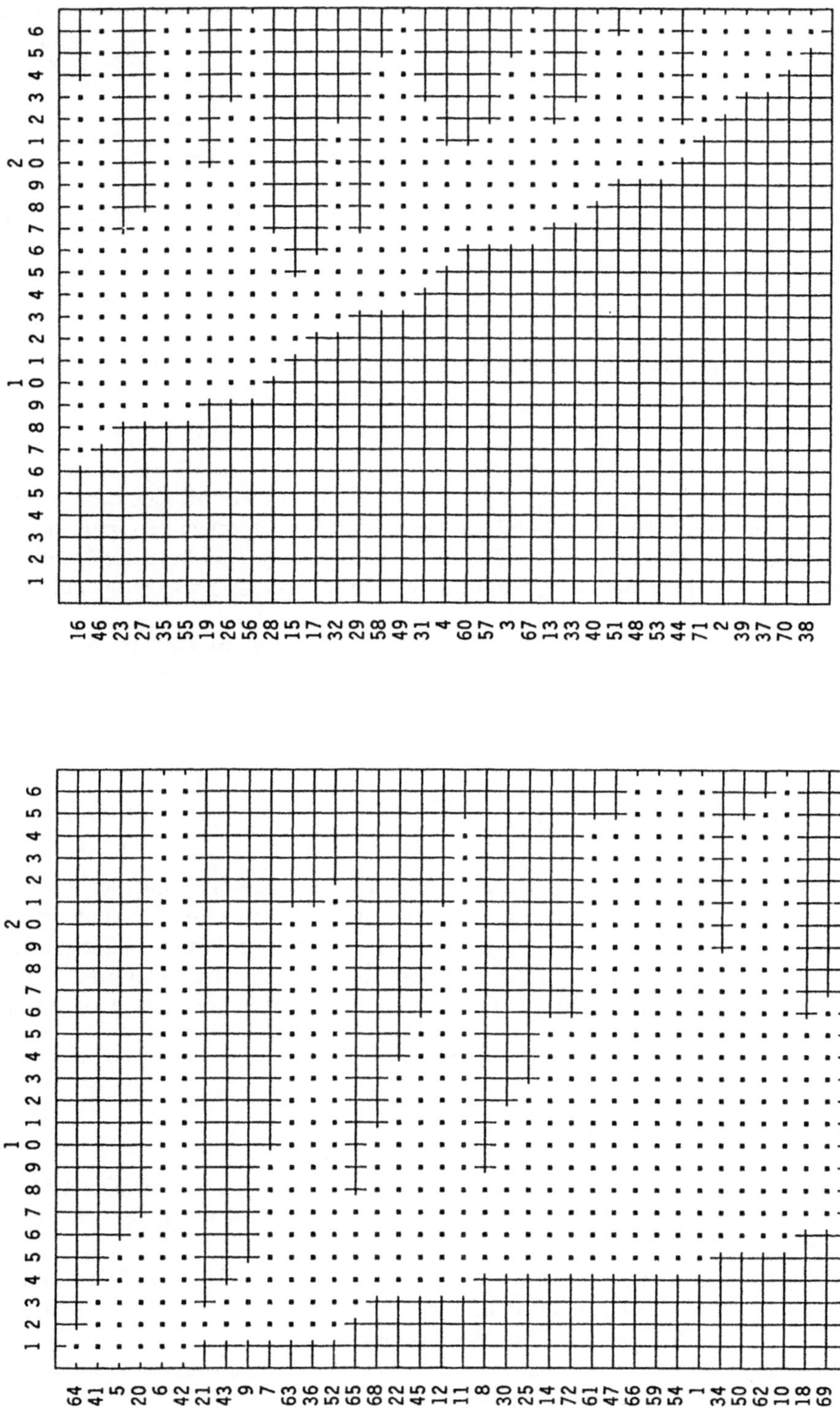

Fig. 12.2. Protoreferential based on the data in Appendix F.5 (output of the BioGraph program)

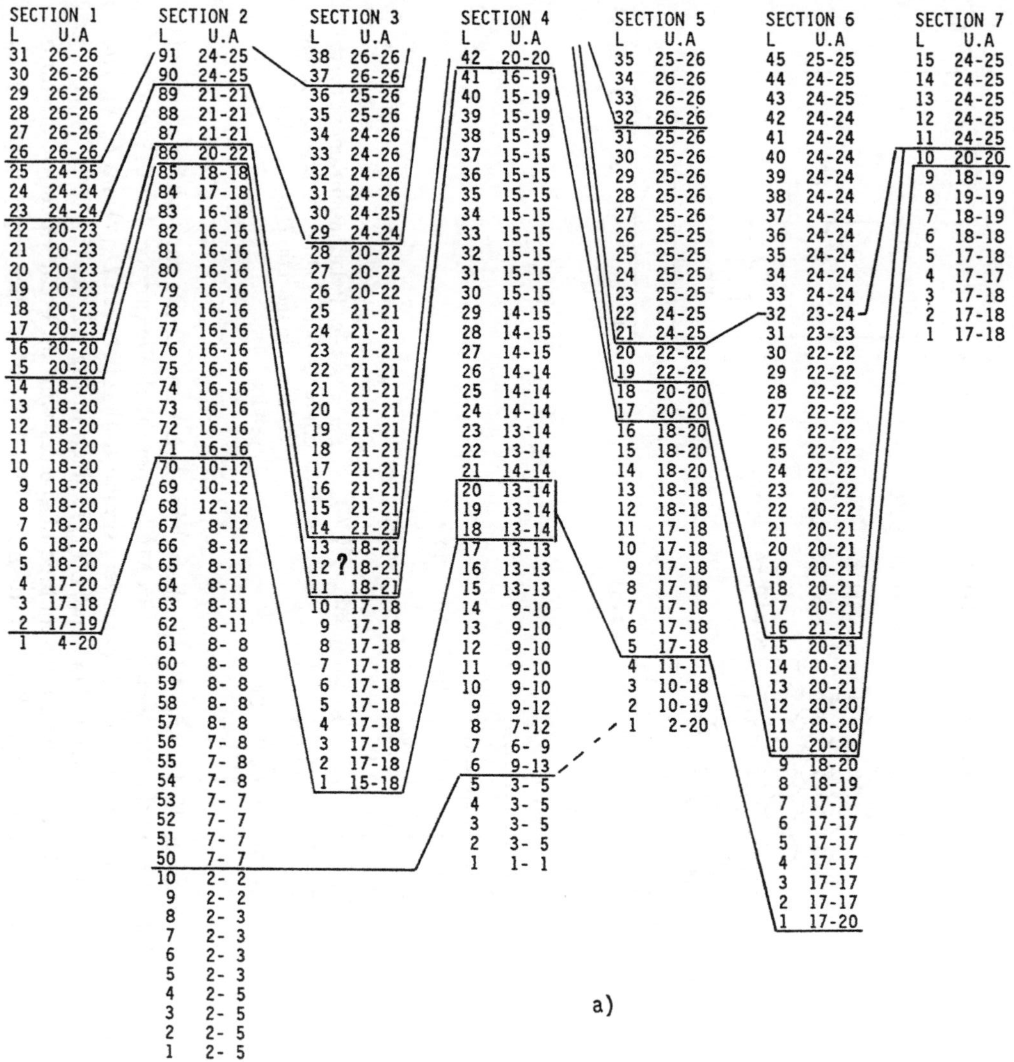

Fig. 12.3. **a** Correlation of Muller's seven sections (see Appendix F.2). **b** Reproducibility matrix associated to the protoreferential in Fig. 12.2. The *black rectangles* represent unions of U.A.'s that are strictly identified in the seven sections. *Column Z*, Zones Z_1-Z_7 (output of the BioGraph program)

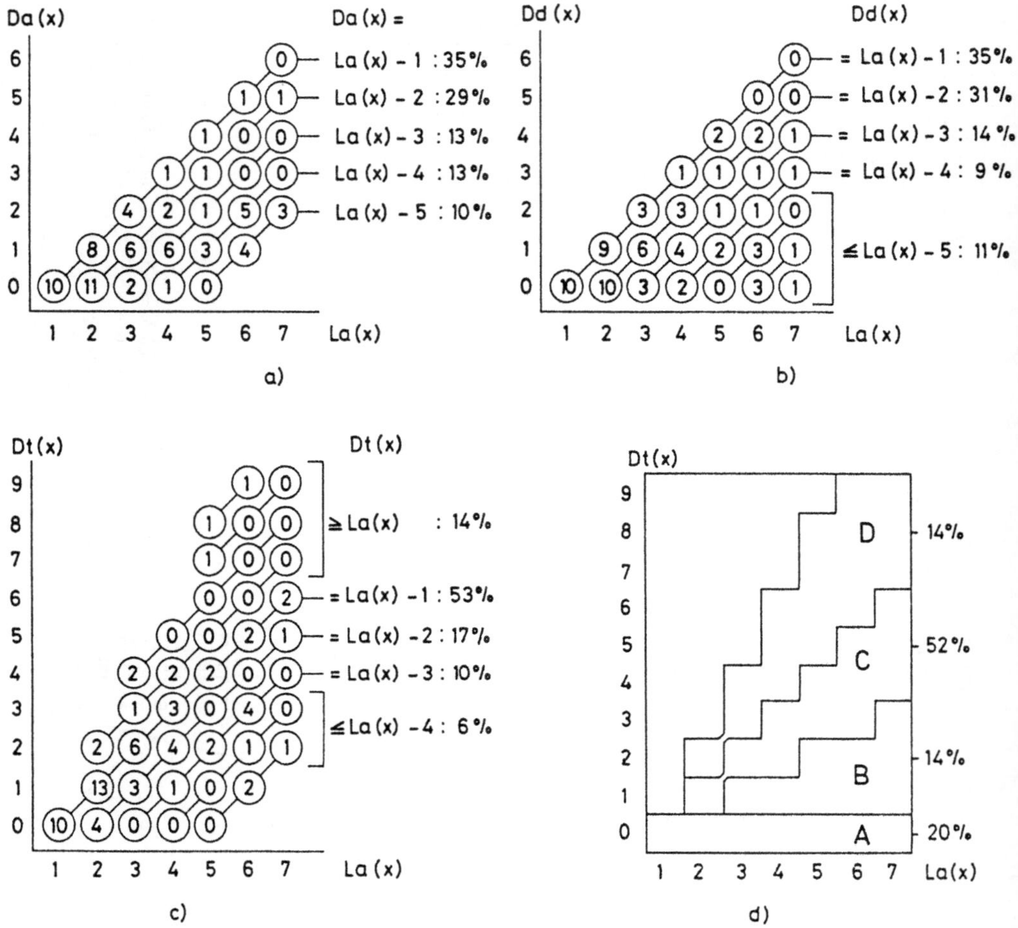

Dd

Dd \ Da	0	1	2	3	4	5
5	0	0	0	0	0	0
4	0	1	1	1	1	1
3	0	2	2	0	0	0
2	3	3	3	0	0	0
1	9	10	5	1	0	1
0	14	10	3	0	0	0

Da > Dd : 30%

Da = Dd : 40%

Da < Dd : 30%

Fig. 12.4 a–e. Relationships among the coefficients $D_a(x)$, $D_d(x)$, $D_t(x)$, $L_a(x)$, and variations in the ratio $D_t(x)/L_a(x)$; (see Appendix F.4)

Figures 12.4.a and 12.4.b show how $D_a(x)$ and $D_d(x)$ vary as functions of $L_a(x)$ for all the Neogene species Muller identified. We can see that the relationship between $D_a(x)$ and $L_a(x)$ is very similar to that between $D_d(x)$ and $L_a(x)$. The relationship between $D_t(x)$ and $L_a(x)$ is given in Fig. 12.4.c.

Fig. 12.4.d gives the percentage of species with different values of the ratio $D_t(x)/L_a(x)$ (see Sect. 8.4.2.3). It shows that 14% of Muller's species [sector D:

$D_t(x)/L_a(x) \geqq 1$] are diachronous at the regional scale (i.e., they are found in different zones in different sections). Fifty-two percent have very high total dispersion [sector C: $D_t(x)/L_a(x) = 0.5$ to 0.9]. Thirty-four percent have no dispersion or very little (sectors A and B).

Fig. 12.4.e studies the relationship between D_a and D_d for Muller's species. It demonstrates that the diachronism of their FADs equals that of their LADs. We also note that more than 80% of the datums (FADs and LADs together) are diachronous.

12.3 Comparison with Californian Nannoplankton

To make it easier to compare the biochronologic dispersions of Sullivan's and Muller's nannofossils we have summarized in Fig. 12.5 the data from Figs. 11.9 and 12.4. The first two diagrams (Figs.12.5.a, 12.5.b) stress the quasi-identity of the relationships $D_a(x) = f[L_a(x)]$ and $D_d(x) = f[L_a(x)]$ in Muller's nannoplankton. This contrasts sharply with the marked difference in the dispersions of FADs and LADs observed in Sullivan's forms (Fig. 12.5.c).

Fig. 12.5.d emphasizes the fact that almost a quarter of Sullivan's species are diachronous on a regional scale, compared to about a seventh of the Mediterranean nannoplankton [sector D of the comparative diagram: $D_t(x)/L_a(x) \geqq 1$].

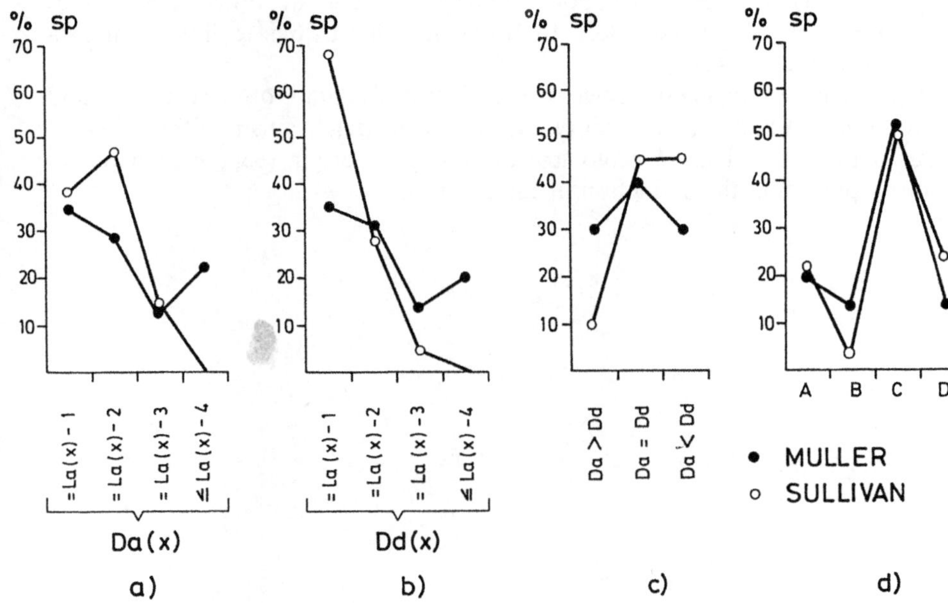

Fig. 12.5a–d. Comparison with Sullivan's data. **a** $D_a(x) = f[L_a(x)]$: percentage of species. **b** $D_d(x) = f[L_a(x)]$: percentage of species. **c** Relationship between D_a and D_d. **d** Percentage of species belonging to sectors A–D of the graph $D_t(x)/L_a(x)$

12.4 Conclusion

If the seven Pliocene sequences examined here are truly free of reworking (Muller 1978, p. 476), this suggests that the strong relative diachronism of the LADs of Californian nannofossils is due to continuous and subcontemporaneous lateral displacements that generated a local and artificial lengthening of the species' ranges (nondisjunctive reworking).

This interpretation could explain the relatively poor biochronologic resolution power of Sullivan's nannofossils compared to Muller's Mediterranean forms.

Berggren et al. (1985 a, p. 162) estimate that the interval NP-8 to NP-15 lasted about 10 million years. That interval corresponds to Sullivan's (1965) zones 1–4, which would then have a duration of about 2.5 million years each.

They also estimate (1985 b, p. 127) that the interval NN-12 to NN-21 lasted about 6 million years. That interval corresponds to biochronozones 3–7 (Pliocene-Quaternary) of Muller's sections (Sect. 12.2.1 and Fig. 12.3). Each of these five zones would then have a duration of about 1.2 million years.

If these estimates are correct, Muller's data generate zonal correlations that are twice as precise as Sullivan's. (Note that the subzones *A–H* described in Sect. 11.3 are ignored in the present discussion, for their lateral reproducibility is too poor). This greater precision would be due to the exceptionally good record of the Mediteranenan nannofossils examined here.

However, these explanations should be considered with caution, for two reasons:

1. Martini's (1971) "standard" zones (in particular the interval NP-8 to NP-15) are not very reliable (see Sect. 16.4) and it is hazardous to link them to a geochronometric scale.
2. The absence of disjunctive reworking in Sullivan's data is quite surprising and we cannot exclude the possibility that the strong diachronism of the LADs of his nannofossils has an ecological or biological origin (e.g., greater actual rarity preceding the extinction of the species).

Chapter 13
Age of Tethyan Radiolarites

13.1 A Difficult Geological Problem

To date, the unitary association method has been applied to a limited number of important and previously unsolved geological problems. One of these applications led to particularly spectacular conclusions concerning some aspects of the paleotectonic evolution of the Tethys (Baumgartner 1984a, b; 1987). It will be useful to review that study here.

13.2 Radiolarian Zones in the Tethyan Realm

The stratigraphic interval studied by Baumgartner covers the Middle Jurassic to the Lower Cretaceous. He established a zonation using unitary associations based on the analysis of the stratigraphic relationships of 110 species in 43 sections located in the Central Atlantic and the Tethys. These abundant data are published in Baumgartner 1984b (pp. 809–816): they are not reproduced here.

Processing these data with the DV-86 program produces an initial sequence of 15 unitary associations (Fig. 13.1).

The interpretation of that sequence led Baumgartner (1984b) to define nine biochronozones (denoted A_0-E_2) that can be correlated over a large area and tied to the standard stages by means of other coexisting fossil groups.

13.3 Biochronologic Correlations and Geological Interpretation

The geological and paleotectonic conclusions Baumgartner drew from these biochronologic correlations are of particular interest.

To summarize these results, we will examine the most significant stratigraphic sequences he studied in various locations of the Tethys (Fig. 13.2). These sequences illustrate the relationships among the different sedimentary environments (seamounts, deep basins, ophiolitic substratums, etc.) and the age of the radiolarites in the different paleogeographic domains. The slightly simplified diagram shows several highly interesting facts:

1. The oldest radiolarites (U.A.0 and U.A.1: Bathonian) were deposited in basins formed prior to the opening of the Central Tethys and which were sites of deep water sedimentation as Early as the Middle Triassic and up to the Jurassic (Pindos Triassic Bassins, Asklipion-Argolide nappe and some of the Hawasina nappes).

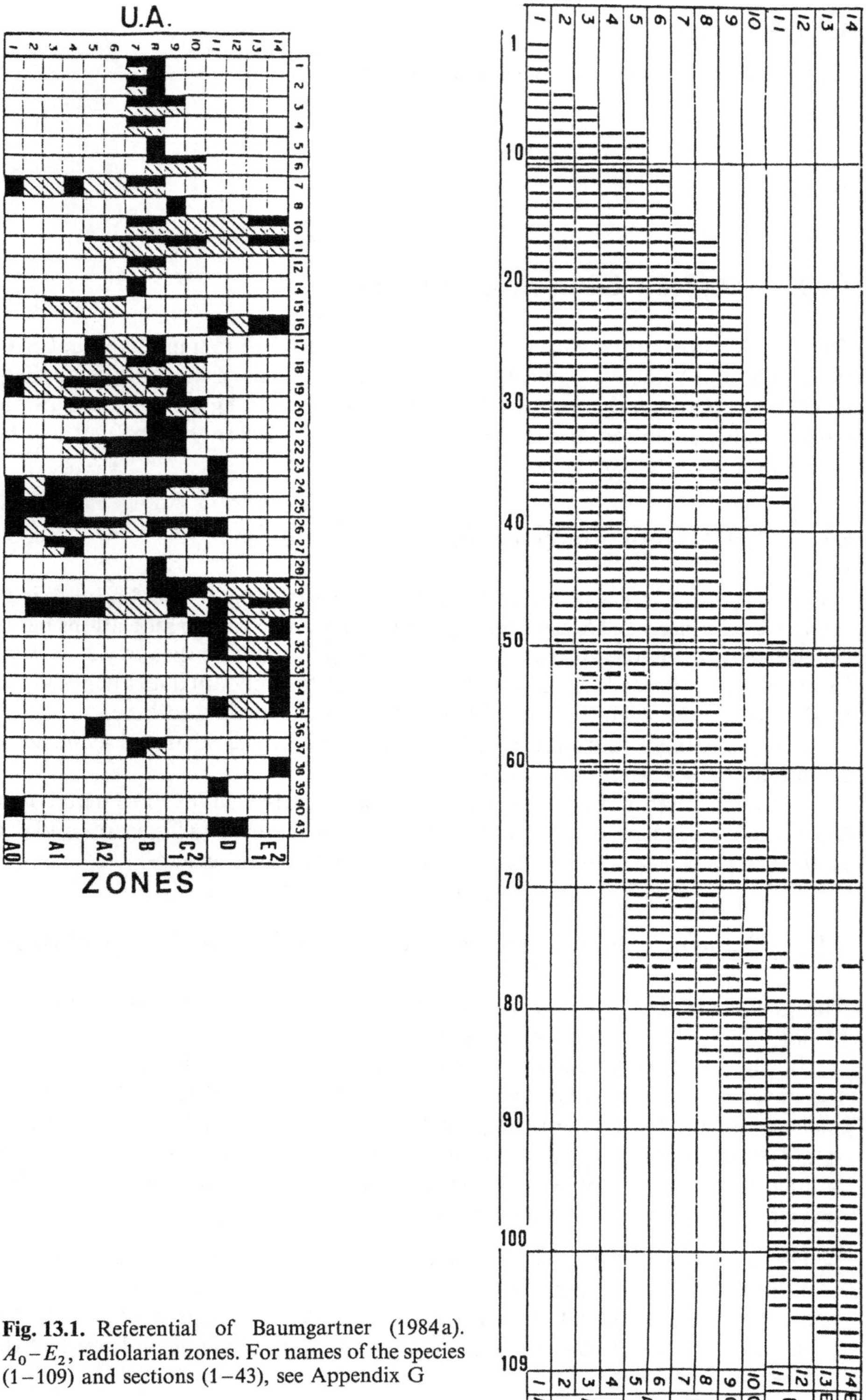

Fig. 13.1. Referential of Baumgartner (1984a). $A_0 - E_2$, radiolarian zones. For names of the species (1–109) and sections (1–43), see Appendix G

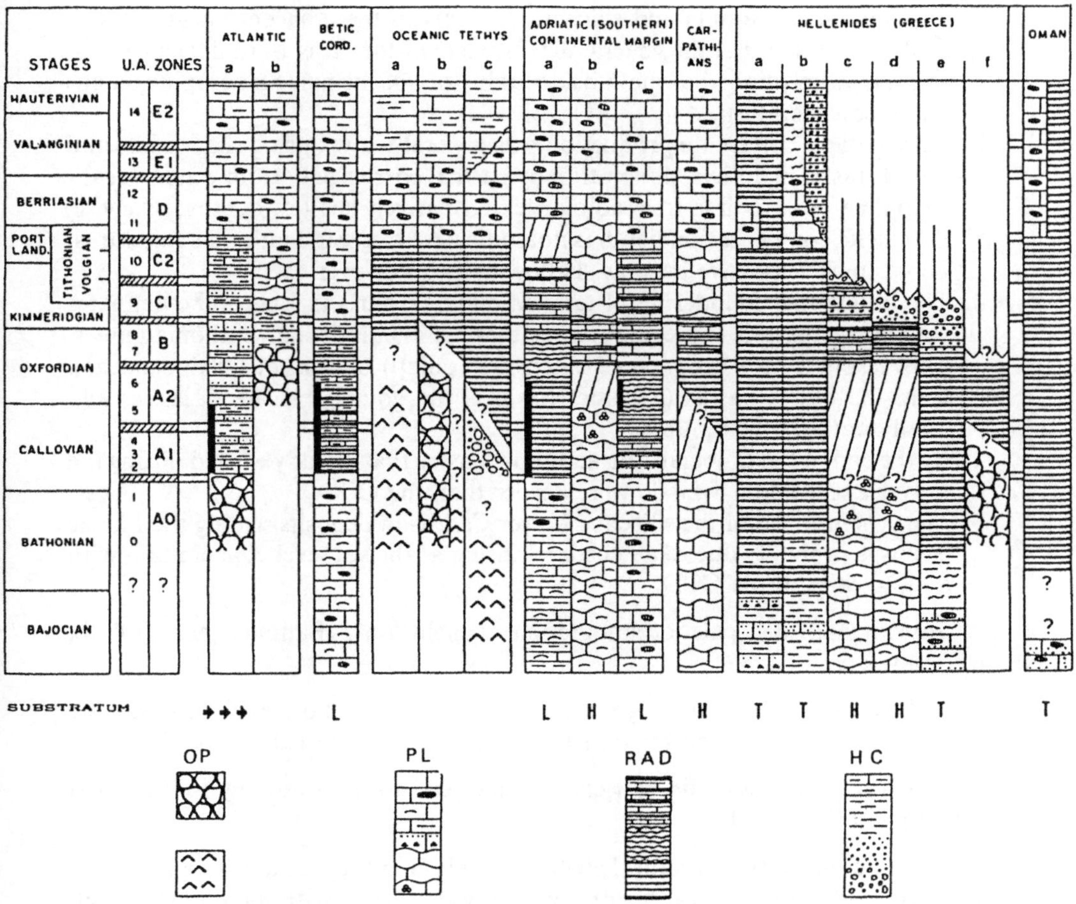

Fig. 13.2. Correlation between the pelagic sequences of the Atlantic and the Tethys for the Bajocian-Hauterivian interval. Ante-Bajocian substratum: Triassic basins (*T*); Liassic basins (*L*); pelagic seamounts (*H*). Lithology: ophiolites (*OP*); pelagic limestones (*PL*); radiolarites (*RAD*); hemipelagic clastics (*HC*). (After Baumgartner 1984 b)

2. Beginning in the Lower Callovian, radiolarites were deposited in pelagic basins (Liassic Basins) formed during the rifting phase affecting the shallow margins during the Lower and Middle Lias (e.g., Subbetic, Lombardy, Umbria).
3. Seamounts and plateaus of Lower and Middle Jurassic age (Argolis, Trento Plateau, Carpathians) received radiolarite sedimentation since the Oxfordian.
4. Radiolarites deposited on oceanic crust (Elba, Liguria, Vardar Ocean) are not older than Middle Jurassic.

The above fundamental observations strongly support Baumgartner's (1984 b) conclusions:

1. Radiolarite accumulation is basically controlled by the relative bathymetry of the basins: this is clearly demonstrated by the contemporaneity of carbonate

deposits on seamounts and siliceous deposits in the adjacent basins. It also explains the fact that the general diachronism of the onset of radiolarite deposition is strictly linked to the subsidence history of the paleogeographic realms under consideration.

2. Since pelagic carbonate production was generally low during the Middle and Late Jurassic in Tethys, dissolution by basinal corrosive bottom waters resulted in lime-poor to lime-free radiolarites at relatively shallow depths. Likewise, the calcite compensation depth is not indicative of great depth; in some Jurassic basins it may have been as shallow as 1 km or even less.
3. The differential foundering of tethyan margins starts in Lias and culminates with the formation of oceanic crust in that realm in Bajocian-Bathonian times. This resulted in a progressive extension (onlap) of siliceous sedimentation from the oldest basins (Triassic) towards the younger basins (Liassic) and finally onto the seamounts.
4. By the end of the Bajocian, intracratonic basins that already existed during the Lower and Middle Triassic were deeper than the CCD.
5. The synchronism of the siliceous Lower Callovian deposits sealing the Liassic basins is probably due to an eustatic raising of the sea level concomitant with a rapid raise of the CCD.

Introducing his paleooceanographic conclusions, Baumgartner (1984b, p. 749) writes:

«The present radiolarian ages demonstrate fundamental chronostratigraphic relationships which are no longer left to the imagination of the geologist.»

This opinion recalls the indirect homage paid to paleontology by the great Trümpy (1971, p. 295):

«Some enthusiastic structural geologists will claim they are able to unravel the structure and the structural history of mountain chains without a single fossil. This is quite true. provided that there are no fossils.»

13.4 Rates of Faunal Turnover

The great richness of biostratigraphic data on the Bathonian-Tithonic interval compiled by Baumgartner (1984b) can be used to characterize the rate of faunal turnover among radiolarians during this period. To give a simple picture of this variation, we have constructed a bivariate diagram (Fig. 13.3.a) showing the number of species that appear and disappear in the different unitary associations (U. A.1 – U. A.13).

This diagram clearly shows a high diversification rate during the Middle Jurassic (U. A.1 – U. A.6) followed by a high extinction rate during the Late Jurassic (U. A.7 – U. A.10) and near the Jurassic-Cretaceous boundary (interval U. A.10 – U. A.12).

For testing the significance of these apparent turnover rates, we have studied the variation of the "distance" (or dissimilarity) between each adjacent unitary association (Baumgartner's U. A.0 – U. A.14). For each pair k_i, k_j of U. A., we

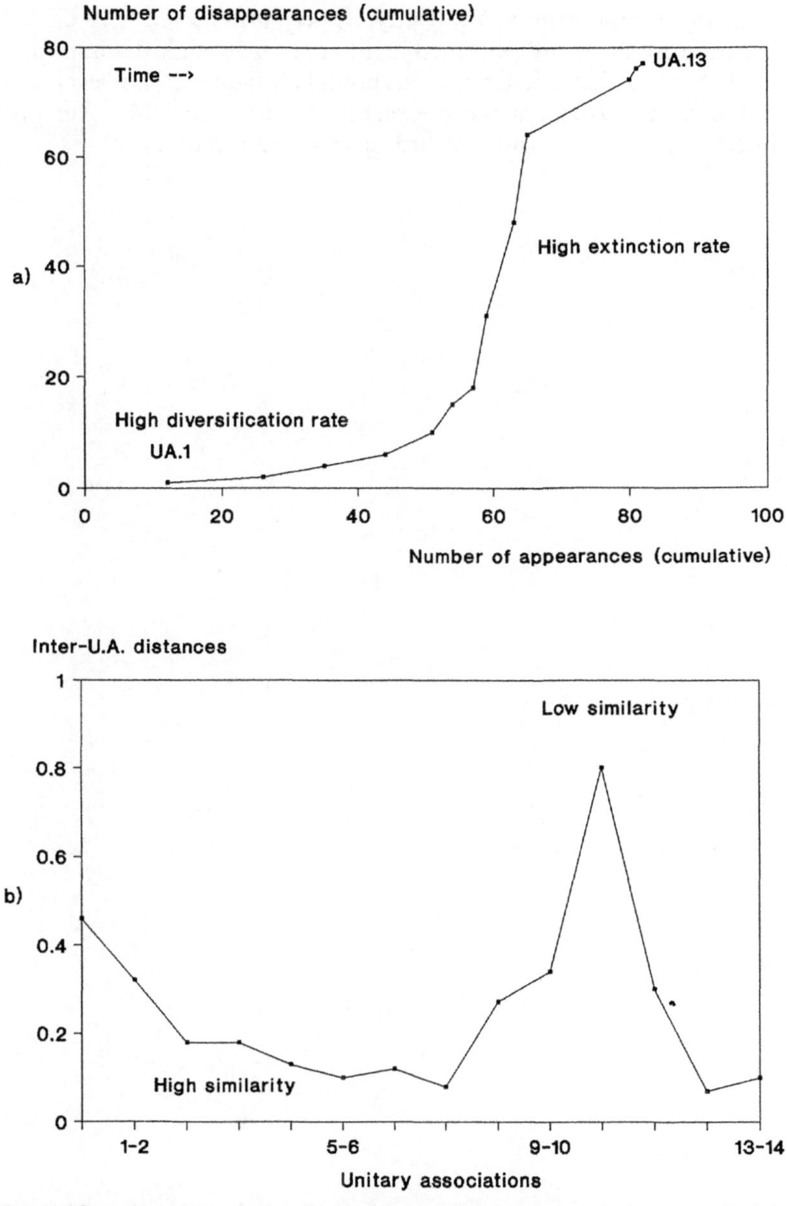

Fig. 13.3 a, b. Rate of faunal turnover during the Bathonian-Tithonic interval. A_0-C_2, radiolarian zones (U.A. O–U.A. 14 in Baumgartner 1984 b)

calculate a distance D defined as follows:

$$D = (|k_i - k_j|/|k_i|) + (|k_j - k_i|/|k_j|) .$$

The diagram in Fig. 13.3.b shows the variation of the inter-U. A. distance in the interval U. A.0–U. A.14. The small distances between the unitary associations during the Middle and Early Upper Jurassic (U. A.2–U. A.9) indicate an accurate fossil record within that interval. The high dissimilarity between the U. A. near the Jurassic-Cretaceous boundary indicates, however, that the data are not sufficient to give a reliable picture of the faunal turnover rate during that interval (gaps in the fossil record or insufficent data).

Chapter 14
Unitary Associations and Ammonite Zones

14.1 Introduction

Ammonites have a well-deserved reputation as an almost ideal fossil group for correlating rocks over vast geographic areas. Specialists of this group have always constructed their zonations empirically, and – except for taxonomic mistakes – their correlations are rarely contradicted by later discoveries.

We have often been asked whether biochronologic scales constructed by the unitary association method are better than those constructed empirically by ammonite workers. The question is legitimate, for we know from experience that the intuition of a good stratigrapher (i.e., of classical Oppelian inspiration) needs no help from computers as long as the data are sufficiently simple and coherent.

The best way to answer the question is of course to compare empirical solutions by ammonite workers with those we obtain with our theoretical model.

As an example, we will reexamine here the data of an excellent study published by Dommergues and Meister (1987) on the Carixian zonation (Middle Lias). Ammonite sequences of this stage are well known and have been described in detail for many parts of the northwest European realm.

14.2 Original Data

The biostratigraphic data analyzed by Dommergues and Meister were compiled from recent monographs on the Carixian in seven northwest European regions: (1) Portugal (Dommergues 1987); (2) Great Britain (Phelps 1983); (3) Spain (Comas Rengifo 1985); (4) Quercy (Cubaynes et al. 1984); (5) Causses (Meister 1986); (6) Burgundy (Dommergues 1979) and (7) southern Germany (Schlatter 1980). These data are given in Fig. 14.1. They are presented as composite sections in which only the maximal horizons are considered.

Of the genera and species described in these regional monographs, Dommergues and Meister considered 74 taxa with large geographic distribution (taxa 1–74: Appendix H.1). They grouped some taxa in order to extend the geographic range of some closely related genera and species (e.g., taxon 38 represents *Vicinodiceras* + *Liparoceras* and taxon 17 represents *Polymorphites* gr. *polymorphus* + *pseudodubari*). These unions of taxa have an important effect on the biochronologic solution that we will discuss in detail later.

Finally we note that most of the species belonging to the genus *Acanthopleuroceras* seem to coexist in southern Germany (possible condensation). These strati-

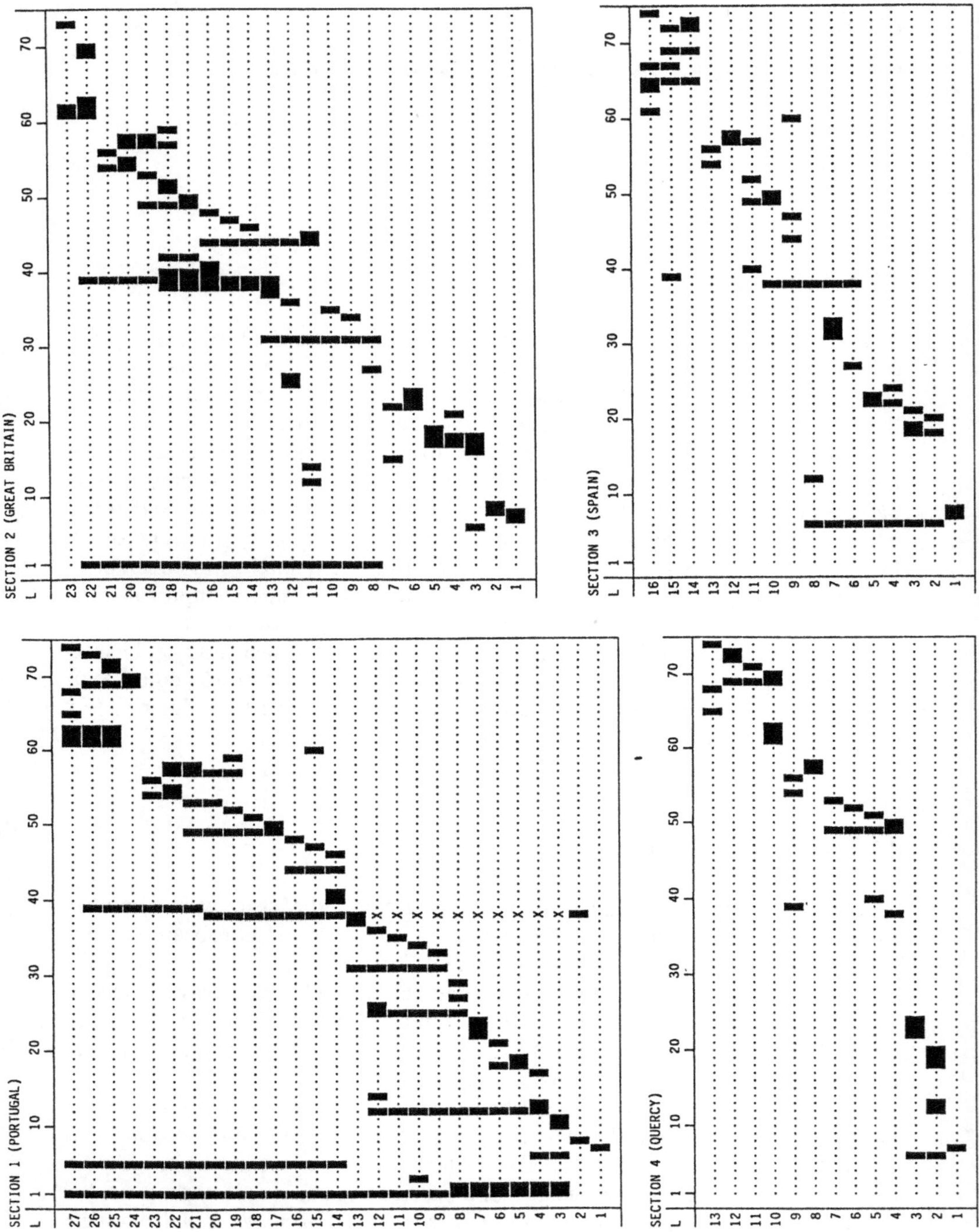
SECTION 2 (GREAT BRITAIN)
SECTION 3 (SPAIN)
SECTION 1 (PORTUGAL)
SECTION 4 (QUERCY)

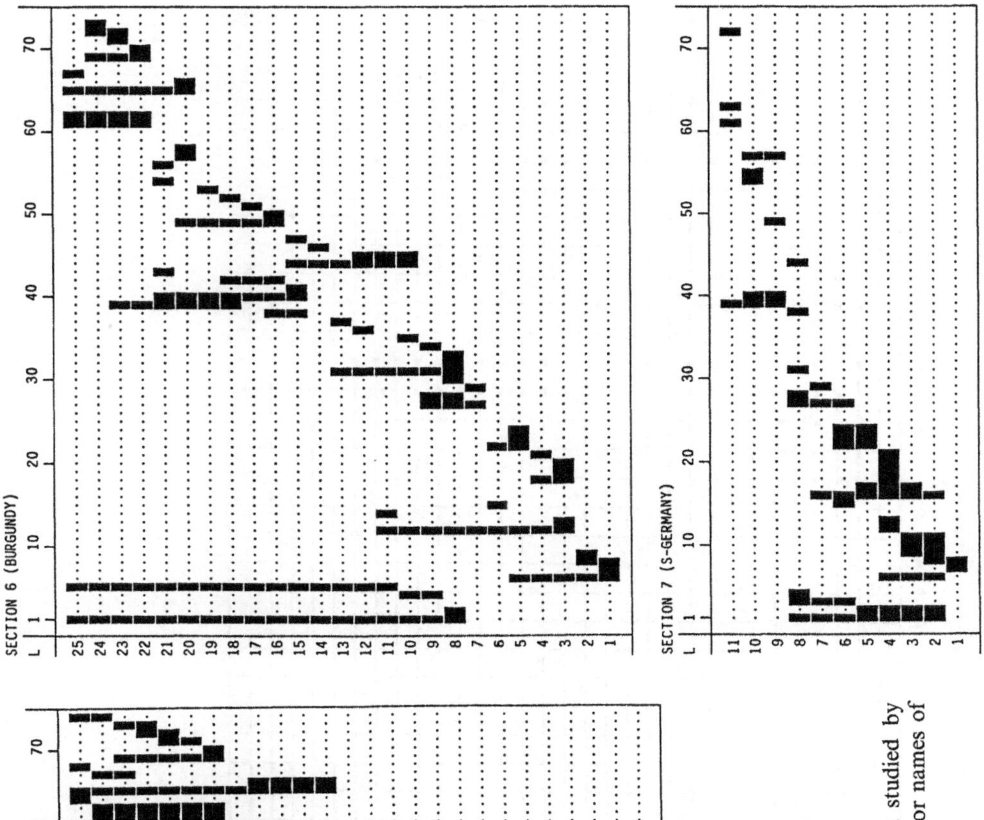

Fig. 14.1. Local stratigraphic ranges of the 74 taxa studied by Dommergues and Meister (1987) in seven regions. For names of species and genera, see Appendix H.1

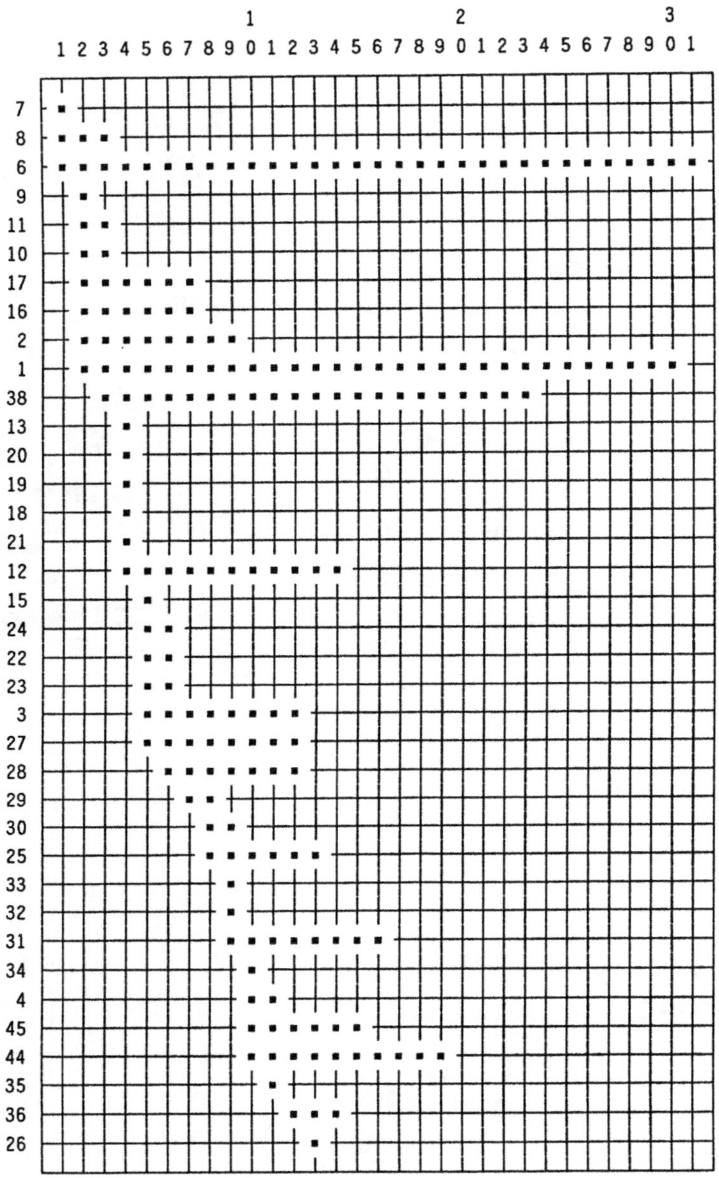

Fig. 14.2

graphic data, considered as dubious by Dommergues and Meister, are omitted in their original compilation.

14.3 Zonation

Analyzing the above raw data produces the range chart of Fig. 14.2. Using this as a starting point, Dommergues and Meister have set the reproducibility

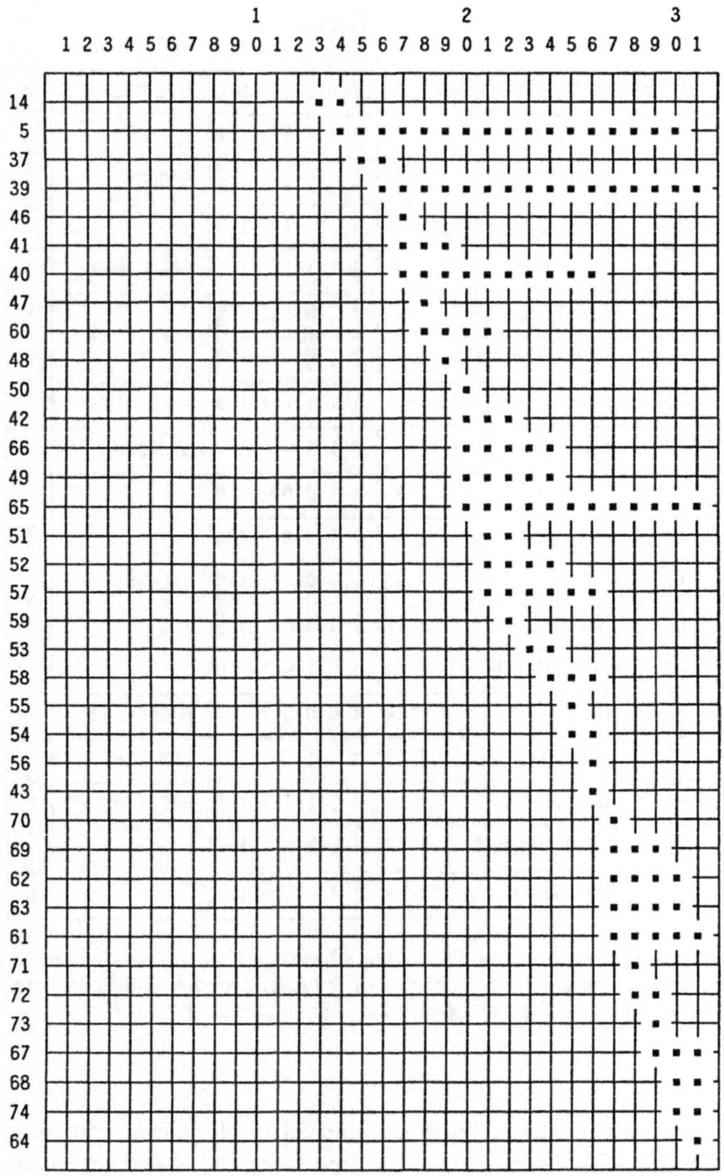

Fig. 14.2. Synthesis of data listed in Fig. 14.1 (output of the DV-86 program; after Dommergues and Meister 1987)

threshold at four regions out of seven (4/7) for defining biochronozones that are significant in the northwest European realm. These biochronozones are denoted Z1–Z18 in Fig. 14.3 (column D & M).

Note that Dommergues and Meister did not consider the possibility of dividing their *Capricornus* empirical subzone (Z6) into two widely identifiable biochronozones: (1) U. A.21–22, characterized by *P.rectiradiatum* and *A.latae-*

U.A.=>	BIOCHRONOZONE	D & M	SECTIONS 1 2 3 4 5 6 7	EMPIRICAL SZ	Z
30-31	DEPRESSUM	Z 1	1 0 1 1 1 0 0	SUBNODOSUS ?-?-?-?-?-?-?	
29	NITESCENS	Z 2	1 1 1 1 1 1 █		
28	MONESTIERI	Z 3	1 0 0 1 1 1 █	STOCKESI	MARGARI
27	OCCIDENTALE	Z 4	1 1 0 1 1 1 0		
25-26	FIGULINUM	Z 5	1 1 1 1 1 1 1	FIGULINUM	
23-24	CRESCENS	Z 6	1 1 █ 1 1 1 █	CAPRICORNUS	DAVOEI
21-22	CAPRIC-HYBRIDA		1 1 █ 1 1 1 █		
20	MACULATUM	Z 7	1 1 1 1 1 1 0	MACULATUM	
17-19	CRASSUM	Z 8	1 1 1 0 1 1 0	LURIDUM	
15-16	LEPIDUM	Z 9	1 1 0 0 1 1 0		
12-14	ACTAEON	Z 10	1 1 0 0 1 1 0		IBEX
11	VALDANI	Z 11	1 1 0 0 1 1 █	VALDANI	
10	MAUGENESTI	Z 12	1 1 0 0 1 1 █		
9	ARIETIFORME	Z 13	1 0 1 0 1 1 0		
7-8	MASSEANUM	Z 14	1 0 0 0 1 1 1	MASSEANUM	
5-6	JAMESONI	Z 15	1 1 1 1 1 1 1	JAMESONI	
4	BREVISPINA	Z 16	1 1 1 1 1 1 1	BREVI+POLYM	JAMESON
2-3	NODOFI-BIRUGA	Z 17	1 1 0 0 0 1 1	TAYLORI	
1	NODOGIGAS	Z 18	1 1 1 1 0 1 1		

Fig. 14.3. Comparison of the zonation based on U.A. analysis (*column D & M*: zones Z_1-Z_{18}) and the classical carixian subzones of the seven sections studied by Dommergues and Meister. *Black rectangles,* strictly identified unions of zones

costa and by the coexistence interval of *A.hybrida* and *A.capricornus*; and (2) U. A.23–24, characterized by *A.crescens* and *A.samontaensis.*

14.4 Precision of the Method

Correlations generated by the empirical zonation are shown in Fig. 14.4. That figure allows us to compare the U. A. solution with these correlations. It shows that our method enables us to approximately double the precision of the empirical correlations.

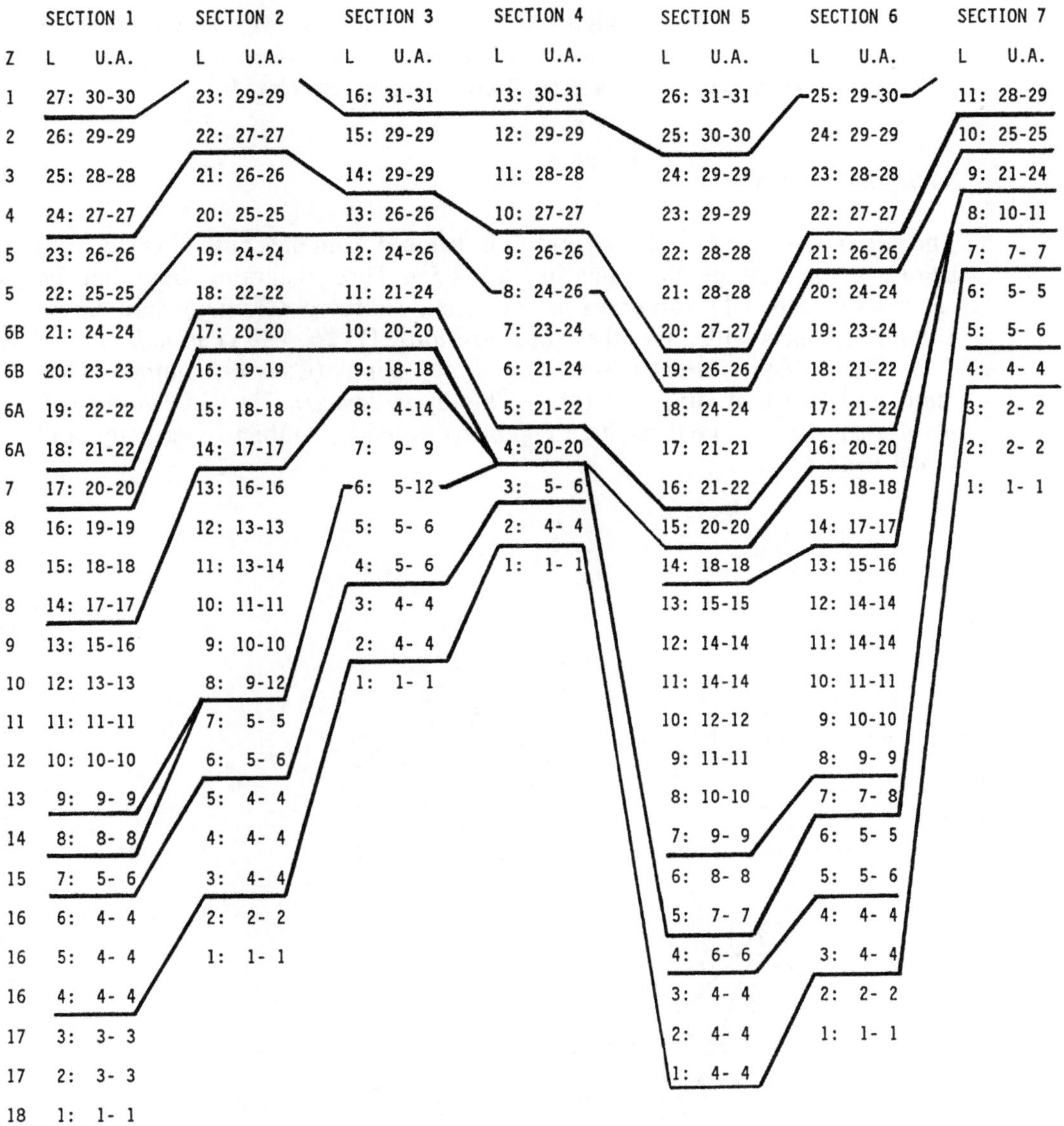

Fig. 14.4. Comparison of the empirical correlations (*solid lines*) and the U.A. correlations

14.5 Problematic Virtual Coexistences

14.5.1 Nature of the Problem

The DV-86 program detects 519 S_3 and Z_4'' circuits and three Z_4 circuits. Adding 24 virtual edges to the biostratigraphic graph studied by Dommergues and Meister is sufficient to destroy all these conflicting stratigraphic relationships. Dommergues and Meister (loc.cit., p. 929, Fig. 6) consider that almost all the virtual coexistences deduced by our algorithm are justified. However, they contest the validity of our conclusion that taxa 8 (*Phricodoceras* gr.*taylori*) and 17 (*Polymorphites* gr.*polymorphus* + *pseudodubari*) are virtually associated.

14.5.2 Discussion

The virtual coexistence of taxa 8 and 17 is deduced from the conflicting superpositions of the Z_4'' type illustrated in Fig. 14.5.a. These diagrams show that the superposition 8 → 17 (observed in Portugal, southern Germany and Great Britain) conflicts with two other superpositions: (1) 16 → 38 (*Parinodiceras* → *Vicinodiceras*/*Liparoceras*) observed in three regions (Causses, southern Germany and Great Britain); and (2) 9 → 38 (*Hyperderoceras* → *Vicinodiceras*/*Liparoceras*) also observed in three regions (Burgundy, southern Germany, and Great Britain).

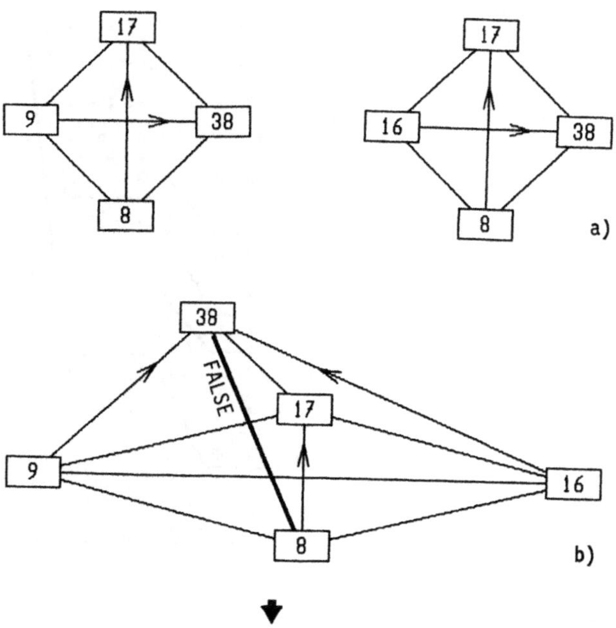

SEQUENCE OBTAINED AFTER REMOVING EDGE (8,38) (FALSE COEXISTENCE)

			17	38
	9	16	17	
8	9	16		

Fig. 14.5. Conflicting stratigraphic relationships of taxa 8, 9, 16, 17, and 38 resulting from the synonymy *Liparoceras*-*Vicinodiceras* (38)

It is easy to see two possible interpretations for these contradictions:

1. The six observed superpositions 16 → 38 and 9 → 38 are generated by unsufficient documentation and should be replaced by virtual coexistences of (16, 38) and (9, 38); or
2. The three observed superpositions 8 → 17 are meaningless (resulting from documentary gaps) and should be replaced by a virtual coexistence of 8 and 17.

Putting aside all subjective considerations, it is clear that the solution produced by our algorithm (i.e., replacement of the arc 8 → 17 by a virtual edge) is optimal, for it preserves the greatest possible number of observed superpositions.

However, Dommergues and Meister have good empirical reasons for criticizing this result. If we admit that it is incorrect, then we must assume that at least one of the coexistences represented in the subgraphs of Fig. 14.5.a is false.

Inspection of the local stratigraphic ranges of the taxa given in Fig. 14.1 shows that the vertical distribution of the composite taxon 38 (*Vicinodiceras* + *Liparoceras*) is highly discontinuous in Portugal. *Vicinodiceras* is recorded only in that region and its local range is limited to horizons 2–3, while *Liparoceras* does not appear before horizon 12.

From this we deduce that the artificial coexistence of *Liparoceras sensu lato* (38) and *Phricodoceras taylori* (8) accounts for the conflicting superpositions represented in Fig. 14.5. Removing the edge (8, 38) from these subgraphs eliminates the contradictions. In other words the disputed virtual coexistence would not have been constructed by our algorithm if *Liparoceras* and *Vicinodiceras* had not been grouped in the original compilation (Fig. 14.5.b).

Note that in this particular case, that slight taxonomic mistake has no effect on the correlations.

14.6 Paleontological Notes

It is interesting to note that some evolutionary variations among the Carixian ammonites studied by Dommergues and Meister seem to be entirely independent of any biogeographic control (very wide geographic distribution of the species without chronologic overlap among the morphotypes). The species' sequence of *Acanthopleuroceras* [*A.arietiforme (32)* → *A.maugenesti (34)* → *A.valdani (35)* → *A.actaeon (36)* → *A.lepidum (37)*] and that of *Aegoceras* [*A.maculatum (50)* → *A.lataecosta (51)* → *A.crescens (53)*] are of particular interest from that point of view.

By contrast, we remark that nine morphospecies of the genus *Platypleuroceras* (i.e., *P.brevispina; P.brevispinoides; P.aureum; P.muellensis; P.rotundum; P.oblungum; P.amplinatrix; P.tenuilobum; P.acanthobronni*) are strictly restricted to a single U. A. (U. A.4): they probably represent a single, highly variable, species.

Finally, we note that some contemporaneous groups like *Dayiceras* and *A.arietiforme*, *T.ibex*, *T.flandrini*, *T.stahli*, are biogeographically exclusive (see Dommergues and Meister, p. 930). Establishing the contemporaneity of such evident geographic exclusions is a by-product of our method.

Chapter 15
Quantitative Biostratigraphy

15.1 General Remarks

A number of quantitative techniques of biostratigraphic correlations have been published in the past 20 years. Most have been brilliantly summarized and analyzed by Edwards (1977, 1982a, b).

We agree entirely with her conclusions on several important points, notably the criticism of probabilistic approaches (see for example Guex, 1979). On others, we think that she is too kind, especially when she writes: «Most methods work well with good data and poorly with bad data ...» (1982a, p. 151).

A careful reading of her work suggests that she actually meant "Some methods ...", for some so-called quantitative methods produce false results from excellent data.

Jeletzky (1965) gets the credit for being the first to criticize correlation methods based on statistical concepts (similarities, etc.):

"Any attempt at the quantification of biochronological correlation is precluded by the fundamentally qualitative and nonstatistical nature of its most valuable data (index fossils)."

This opinion has often been attacked by statisticians (Cockbain 1966; McCammon 1970; Hay and Southam 1978; Brower 1985a).

The grounds for Jeletzky's criticism are, however, perfectly clear: it is the sampling that is to some extent random and not the fact that a species existed at a given moment. In other words, the essentially nonstatistical nature of the problem of interpreting biostratigraphic observations makes it impossible to establish "quantitative" and chronologically significant correlations using statistical models.

Most specialists interested in the problem of formalizing stratigraphic correlations are well acquainted with the UNESCO report, *Geological Correlation*, which summarizes various research projects of the International Geological Correlation Program (I. G. C. P.), especially projects number 1 (*Accuracy in Time*) and number 148 (*Quantitative Stratigraphic Correlation Techniques*).

The half-joking comment by Brower, published by Drooger (1980, p. 22) in the eighth volume of these reports is doubtless familiar to them:

"Since the inception of IGCP Project 148, most nonquantitative biostratigraphers have stubbornly resisted numerical methods. I believe that the lack of palatability is due to two causes. First, many of the quantitative methods are complex whereas most biostratigraphers are simple-minded. Secondly, most of the quantitative techniques utilise methodologies that are basically foreign to biostratigraphers."

This opinion, embraced by Drooger, is, however, ill-advised, and the resistance deplored by these specialists is actually due to the fact that most quantitative methods simply do not meet biostratigraphers' needs.

Before briefly discussing these methods, we draw the reader's attention to the fact that two elementary conditions must be satisfied for a biochronologic referential to be valid:

1. The characteristic elements (species or pairs of species) of its subdivisions cannot occupy contradictory stratigraphic positions in the sections where these subdivisions are identified.
2. A synthesis of complex biostratigraphic data must at the very least give an exact picture of all the observed coexistences of species when these coexistences do not result from reworking or condensation.

We will see below that several methods of quantitative biostratigraphy fail to meet these minimal conditions. If a geologist finds three species in the same bed of a given section, he has every right to doubt the conclusions of a statistician who points to his formulas to show that the three species occur in a chronologic sequence. To prove that a given method gives incorrect results, then, all we need to do is to show that it omits observed coexistences (Guex 1977, p. 315).

15.2 Multivariate Approaches

15.2.1 Introduction

It has long been recognized that multivariate techniques (derived from numerical taxonomy or from similarity analysis) do not produce reliable biochronologic correlations.

Millendorf and Heffner (1978, p. 313), referring to the multivariate approach in general (cluster analysis, factorial analysis, principal coordinates, principal components, multidimensional scaling) remarked judiciously:

"This approach ignores the effects of faunal gradation within an isochronous unit with respect to geographic position. Thus, if the faunal composition of such an isochronous unit changes across the study area, samples taken from distant points in the unit might be dissimilar enough not to cluster. Simply, the greater the lateral variation and the larger the distance between them, the less similar are two isochronous samples."

About the same time we independently expressed a similar opinion (Guex 1979, p. 198):

"The global composition of subcontemporaneous fossil assemblages is mainly controlled by ecological factors. Lateral variations in the composition of these assemblages often follow the lateral variations of the sedimentary facies. This trivial observation often enables us to prove that, in a group of consecutive fossil-bearing beds studied in one locality, the variation in the composition of successive fossil assemblages is also governed much more by ecological variations than by time. This means that the similarity (in the statistical sense of the word) between two locally consecutive fossil assemblages is governed primarily by the ecology and only secondarily by time. This simple remark shows that it is pointless to try to resolve the problem of constructing a biochronologic scale by using statistical methods like cluster analysis or multidimensional scaling."

In one way or another, all methods based on a global resemblance between fossil samples end in fixing the boundaries of statistical biofacies, and they do not make it possible to find, within a fossil assemblage, the species that are characteristic of the relative age of the deposits under study.

Four separate methods proposed in the recent literature are briefly discussed below.

15.2.2 Archaeological Seriations

Scott (1974) was the first to propose that the seriation techniques used by archaeologists (e.g., Kendall 1971, Wilkinson 1974, etc.) be applied to biostratigraphy.

These techniques have been applied by Millendorf et al. (1978), Brower and Burroughs (1982) and Brower (1985b, 1990).

The basic idea is the following. Start with a presence/absence (1's and 0's) table listing m species from n samples in p sections. The rows of this $n \times m$ table are permuted to minimize the number of 0's located between the 1's in the columns. The main algorithms used are due to Kendall (1971, using multidimensional scaling) and Wilkinson (1974); (for details, see Brower 1981 and 1985b).

These archaeological techniques are interesting in theory but when applied to stratigraphic problems they produce acceptable results only in extremely simple cases.

In a recent paper, Brower (1990) uses Drobne's alveolinid data (Fig. 3.21) as a case study for comparing the unitary associations, archaeological seriation, and probabilistic methods. Before processing these data, however, he disregards the sole theoretical interest of that problem by removing the reworked fauna from the database (see Sect. 3.8 and Chapter 7). Doing so, he transforms that relatively complex problem into a trivial one and concludes his discussion by stating that the three methods produce nearly identical correlation charts. We note, however, that Brower's statistical analysis (loc.cit., p. 414, Fig. 7) results in reversing the relative positions of alveolinid biochronozones 3 and 4 in several locations (see Sect. 7.4, Figs. 7.7 and 7.8).

A single example of applying archaeological methods to a relatively complex situation was published by Brower (1985b, pp. 103–106). It concerned the distribution of Deboo's (1965) ostracods and foraminifera. However, the correlations obtained in this particular case (Brower 1985b, Fig. 3) do not show any gap or transgressive-regressive cycle in Deboo's sections (Fig. 15.1).

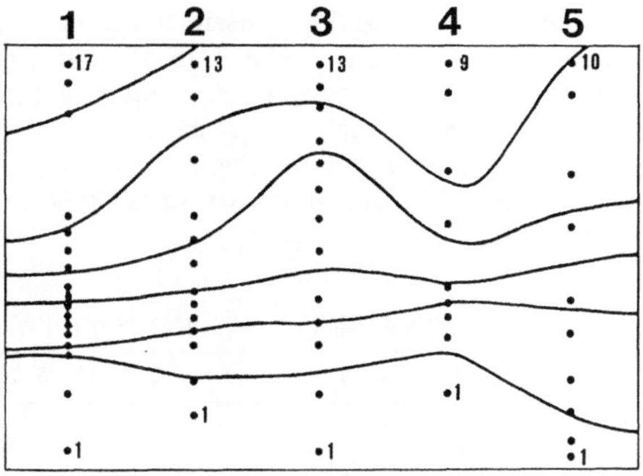

Fig. 15.1. Correlation of Deboo's sections (1 – 5) based on archaeological seriation (After Brower 1985b, p. 105, Fig. 3b)

15.2.3 Lateral Tracing

Millendorf and Heffner (1978) devised an interesting multivariate approach with the goal of detecting lateral series of subcontemporaneous biofacies that are vertically separated from each other by very distinct faunal discontinuities.

This method is based both on an algorithm of Sneath (1975a, b) known as lateral tracing and on an idea of Deboo (1965) for detecting diastems by looking for minima in the vertical variation of the faunal similarity of consecutive fossil-bearing samples in a given section.

The details of this elegant technique are beautifully explained by Millendorf and Millendorf (1982), therefore we will not repeat them here.

We simply note that this method works only in very special stratigraphic situations, clearly defined by Brower et al. (1978, p. 226):

"The only requirements for successful tracing are that the degree of lateral variation between the sections are less than the degree of vertical variation within the sections and that the sections be arranged along a linear transect."

Deboo's (1965) data, which happen to satisfy these very restrictive conditions, have also been analyzed by Millendorf et al. (1978, p. 229). The results produced by their algorithm are reasonably good (see loc.cit., Fig. 7) but not precise enough to justify any conclusions on the gaps and transgressive/regressive cycles demonstrated in our Chapter 10.

15.2.4 Principal Components

Hohn (1982) proposed using a modified version of the principal components technique to construct biochronologic scales. As an example, he (loc.cit., p. 109, Table 1) applied his method to some of the biostratigraphic data established by Palmer, and which we examined in this book in Section 2.4.

Hohn's result speaks for itself. His composite table (reproduced in Fig. 15.2) suggests that species *K.simplex* (No. 1 in our list), *C. beltensis* (No. 2) and *M.metra* (No. 12) occur in stratigraphic sequence. However, a glance at the data base mentioned above (our Fig. 2.6) shows that the three species coexisted in several sections over wide stratigraphic intervals (e.g., section 1). The same is true for species No. 14 (*C.australis*), No. 15 (*L.undulata*) and No. 16 (*M.*cf.*ariston*) observed to coexist in section 3.

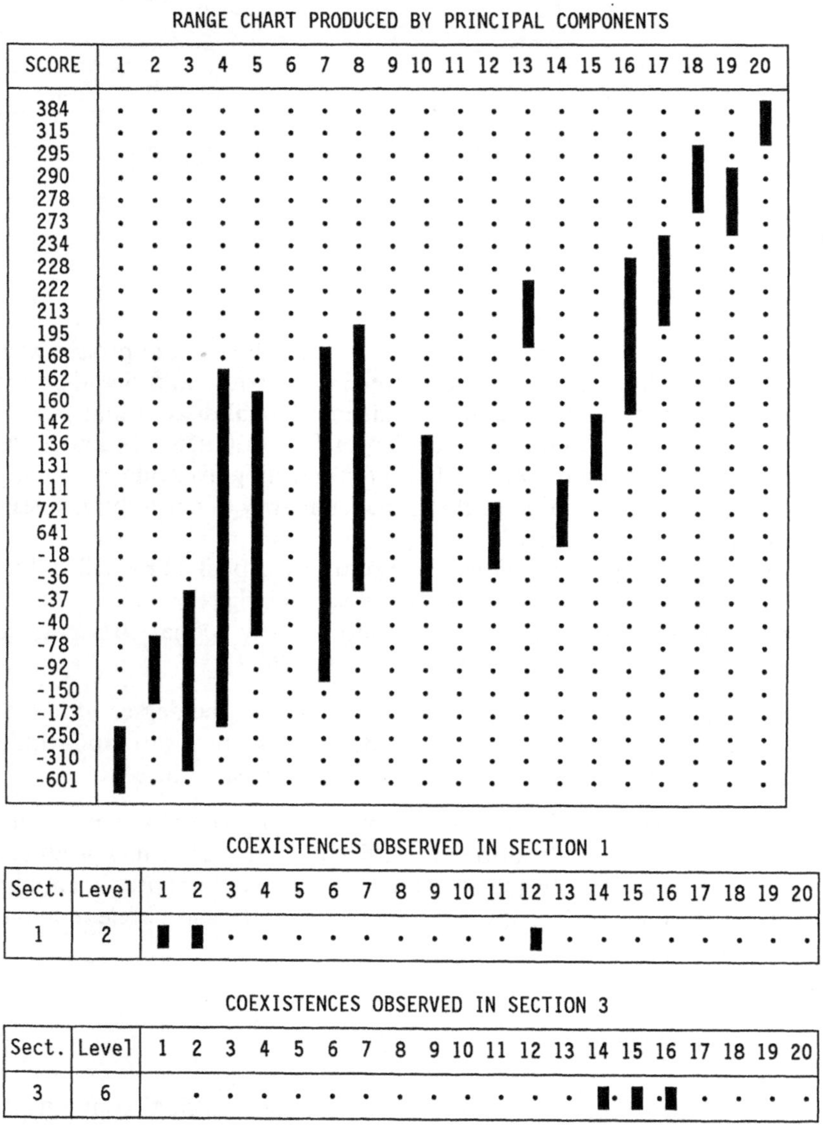

Fig. 15.2. Synthesis of Palmer's data (Fig. 2.6) produced by means of the principal components analysis. *Black bars*, stratigraphic distribution of species 1–20 (After Hohn 1982, p. 109. Table 1)

15.2.5 RBV Coefficients (Relative Biostratigraphic Value)

Some methods of multivariate analysis devised by Brower et al. (1978), Millendorf et al. (1978) and Brower (1984, 1985a) use a quantification of the concept of "index fossil": to each species is attributed a coefficient RBV (Relative Biostratigraphic Value of McCammon 1970) that simultaneously expresses its degree of facies independence, its vertical range and its geographic extension. Only species with a high RBV coefficient are used to establish correlations; the others are ignored.

The most important factor in establishing the Relative Biostratigraphic Value is the vertical range, which is estimated based on the local vertical extension of the taxon under consideration. However, it is clear that a species can have a short vertical range everywhere (for example because of strong facies dependence) while its actual existence duration is very long.

As a result, correlation techniques that use the RBV are mainly useful in delimiting biofacies that have a great lateral extension and are defined by a limited number of species (in some cases, these may well turn out to be more or less synchronous).

15.3 Probabilistic Approaches

The probabilistic method introduced by Hay (1972) is well known and has given rise to many developments (Southam et al. 1975; Worsley and Jorgens 1977; Hay and Southam 1978; Agterberg and Nel 1982a, b; Blank 1979; Blank and Ellis 1982; Doeven et al. 1982; Harper 1981, 1984; etc.).

The goal of Hay and his school is to establish sequences of datums (FADs and/or LADs) that are reliable from a statistical point of view. The method begins as follows:

For each pair (I, J) of biostratigraphic events (FAD and/or LAD) one calculates the probability P that the locally observed order in which these events occur is random.

If the value of $P' = 1 - P$ is greater than a certain threshold (0.8 or 0.9, etc.) one deduces that the sequence observed most frequently ($I \rightarrow J$ or $J \rightarrow I$) is not random and that, consequently, it is reliable. In other words, if I is observed below J ten times and above it only once, one deduces that the true relationship is probably I below J. Fig. 15.1.a illustrates, as an example, the situation in which I and J are two FADs and Fig. 15.1.b shows the situation in which I is a LAD and J is a FAD.

Although the goal of the probabilistic method is to produce a statistically reliable sequence of events, those who have seriously tried to apply Hay's method (original or modified) soon found that it produces range charts giving extremely truncated representations of the relative ranges of the species (Edwards 1977, 1982a; Baumgartner 1984a).

These artificial shortenings occur because the probabilistic method is based on false assumptions, as we can see from the examples in Fig. 15.3. In Fig. 15.3.a it

	OBSERVATION		PROB.CONCLUSION
a)	FAD I < FAD J = 10 X FAD J < FAD I = 1 X	== >>	SPECIES I APPEARS BEFORE SPECIES J

	OBSERVATION		PROB.CONCLUSION = FALSE
b)	LAD I < FAD J = 10 X LAD J < FAD I = 1 X	== >>	SPECIES I DISAPPEARS BEFORE SPECIES J

Fig. 15.3. **a** FAD *I* observed below FAD *J* ten times and above it once: from a probabilistic point of view, FAD *I* predates FAD *J*. **b** LAD *I* observed below FAD *J* ten times and above it once: from a probabilistic point of view, LAD *I* predates FAD *J*

is entirely reasonable to suppose that the relationship *I above J* is true, even if it is infrequent. It is obvious that, even in the most favorable cases, the true first appearance of a species is recorded only very rarely, if at all. In Fig. 15.3.b, it is equally obvious that (assuming no reworking) the relationship *I above J* is the true one.

In other words, the probabilistic philosophy leads one, in all critical situations, to choose either an arbitrary solution (case No.1) or the false solution (case No. 2).

An example of a spectacular shortening in the apparent ranges of species is given by Baumgartner (1984a, pp. 172–173) in a relevant analysis of probabilistic range charts. In this example (shown in Fig. 15.4) several species of radiolarians that coexist in several sections appear in chronologic sequence in the probabilistic referential.

Another interesting example is given by Agterberg (1985, pp. 253–255) in a probabilistic analysis of the stratigraphic relationships of Drobne's alveolinids (Sect. 3.8; Fig. 3.21). In this study Agterberg used the RASC method (Ranking and Scaling) of Agterberg and Nel (1982a).

The initial range chart he obtained gives an extremely truncated picture of the relative ranges of the species and it places in an apparent sequence several species whose coexistence has actually been observed in original samples (Fig. 15.5.a).

To reconcile these contradictions between observed facts and his probabilistic solution, Agterberg (1985, p. 255) proceeds as follows:

"According to the original scaling model, events in sections are normally distributed about their average position with standard deviation equal to $\sigma = 0.7071$. Consequently, the observed highest occurrence of a fossil in a section would occur with a probability of 95% below its RASC value decreased by 1.645 $\mathrm{x}\sigma = 1.16$. This value provides a more reasonable estimate of the true highest occurrence or last appearance datum (LAD) than the original RASC value. Likewise 1.16 can be added to the RASC distance estimated for a lowest occurrence or first appearance datum (FAD) along the RASC scale."

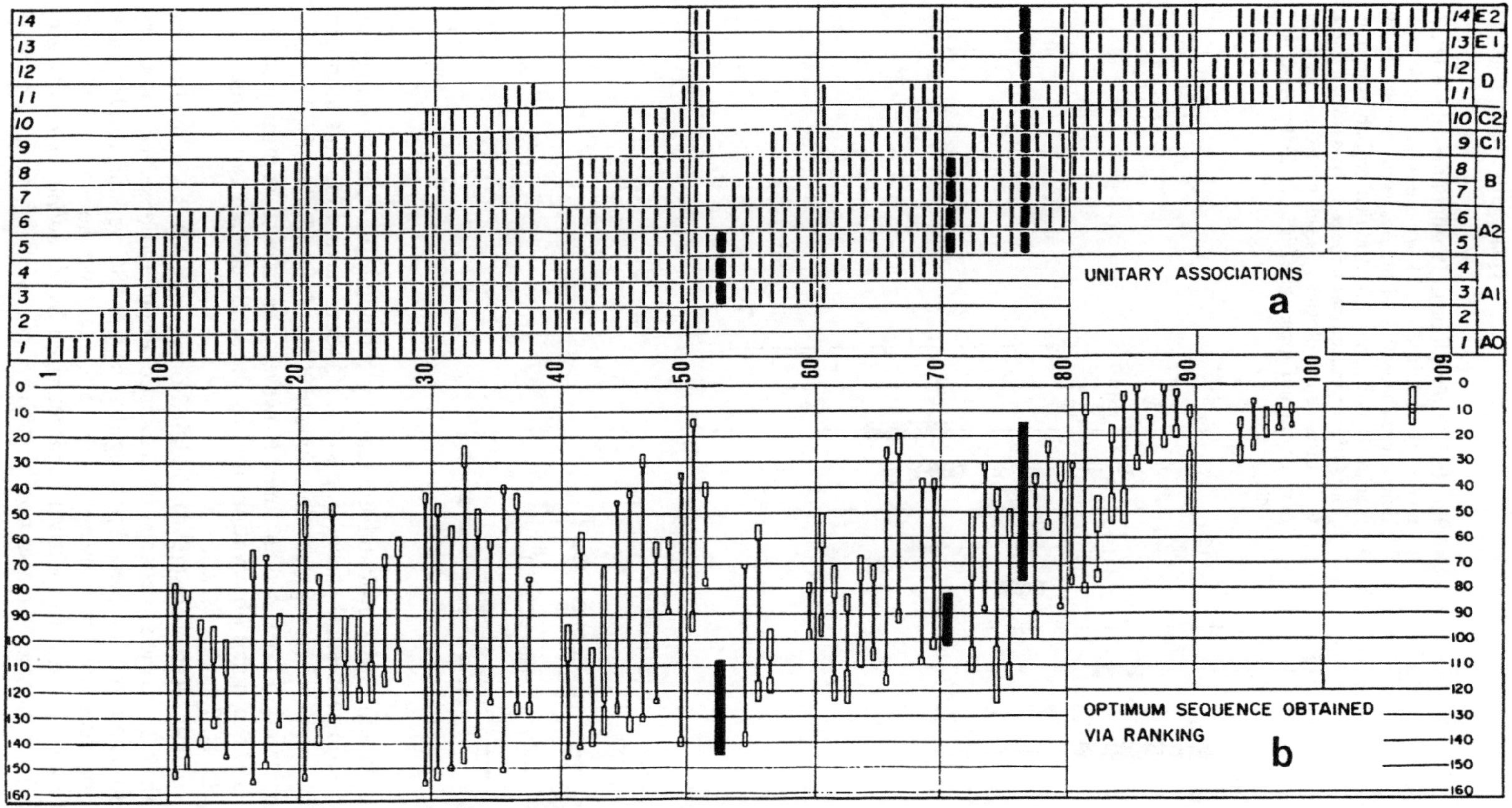

Fig. 15.4. **a** Synthesis of biostratigraphic data established by Baumgartner (1984a) on the distribution of the radiolarians from the western Tethys. The relative ranges of the species are expressed in terms of unitary associations. **b** The same data processed using the RASC probabilistic model: the *black bars* represent species whose coexistence is observed in several sections but which appear in a sequence in the probabilistic synthesis [After Baumgartner 1984a (modified)]

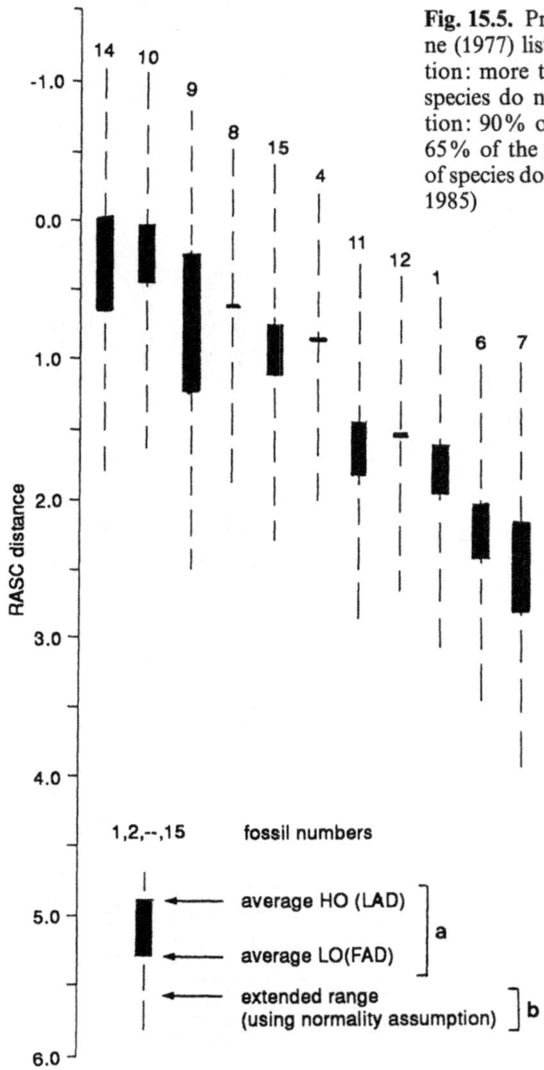

Fig. 15.5. Probabilistic treatment of the data of Drobne (1977) listed in Fig. 3.21. *a* Pure probabilistic solution: more than 70% of the observed coexistences of species do not appear. *b* Modified probabilistic solution: 90% of the species appear to be associated and 65% of the chronologically significant superpositions of species do not appear (see Fig. 7.7); (After Agterberg 1985)

Thus he modifies the probabilistic scale by extending the relative ranges of species by a constant length, according to the standard deviation. This modified range chart (Fig. 15.5.b) suggests that almost 90% of the 15 alveolinid species under consideration coexisted in time. However, this is also contradicted by Drobne's (1977) original observations and by the synthesis that can be made of these data (Sect. 7.4, Fig. 7.7.d).

More recently, Agterberg (1990) modified once again his RASC method to minimize the discrepancies between probabilistic output and observed co-occurrences of fossils in the samples. Applying that new RASC version to Drobne's example produces the scale given in Fig. 15.6. Here we note that Drobne's biochronozones 2 and 3 (respectively characterized by *Alveolina solida* (3) and *A.laxa* (13)) are merged. We also note that the coexistences of species 6 and 10

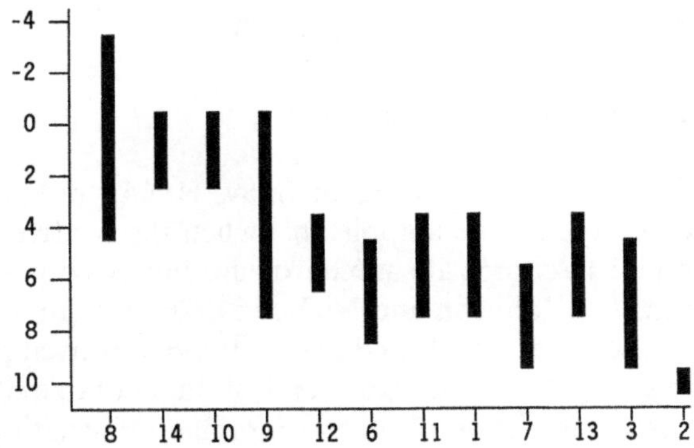

Fig. 15.6. Modified RASC treatment of Drobne's data (After Agterberg 1990, p. 299, Fig. 8.10)

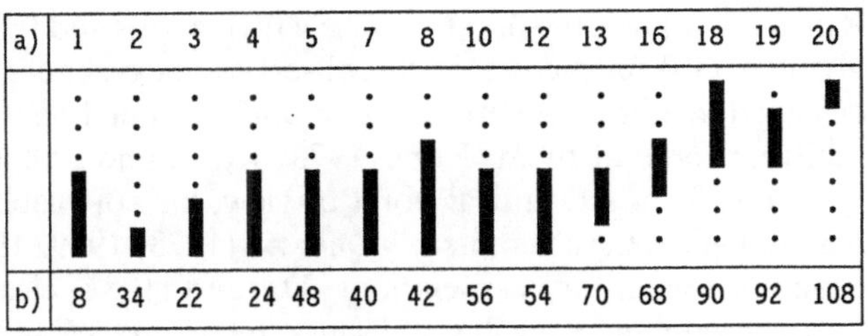

Fig. 15.7. Modified RASC treatment of Palmer's data (After Agtergerg 1990, p. 309, Fig. 8.13). *a* Species code numbers listed in Appendix A.1. *b* Species code numbers used in the original probabilistic treatment

(observed in Drobne's section 9, level 2) and of species 10 and 12 (observed in the same locality) are omitted.

Agterberg (1990) also applied the new RASC technique to Palmer's data examined in Chapter 2. This trivial example is useful in that the studied data set contains no initial contradictions (i.e., the biostratigraphic graph is an interval graph). From a graph theoretical point of view, such a problem has only one solution. Therefore, any method generating an output different from this unique solution is false. The new modified RASC solution to Palmer's problem is reproduced in Fig. 15.7. Comparing this probabilistic scale with the original data (Fig. 2.6) shows that the stratigraphic superpositions between the species *K.simplex* and *C.*cf.*albertensis* (1 → 13), *K.simplex* and *M.*cf.*ariston* (1 → 16), *K.variegata* and *C.*cf.*albertensis* (3 → 13), *M.metra* and *C.*cf.*albertensis* (12 → 13), *M.metra* and *M.*cf.*ariston* (12 → 16) are omitted without justification (no conflicting relationships).

Therefore, we are forced to contradict our friend Agterberg (1985, p. 255) when he concludes that the unitary association method produces results that are similar to those obtained with the modified probabilistic RASC model.

15.4 Deterministic Approaches

15.4.1 Geometric Methods

The geometric method of Shaw (1964), modified by Miller (1977), is rightly considered to be inapplicable when the relative positions of the datums used to establish composite standards are highly contradictory from one section to the next (see Johnson and Niebuhr 1976 and Guex 1979).

Edwards (1977, 1978, 1984, 1989) improved this method very considerably by relieving it of its statistical ballast. In favorable situations, this results in a productive combination of biostratigraphic information and lithologic data. An excellent example of this kind of application is given by Edwards 1989.

15.4.2 Data Set Reduction and Sequential Analysis

Rubel (1976) was the first to devise a formal method for constructing range charts giving a faithful picture of the observed coexistences of species. His method is designed to eliminate conflicting or undetermined relationships among species by omitting some of them (Rubel 1978; Rubel and Pak 1984; Pak 1984).

Thus, from a technical point of view, his approach is almost identical to the first independent attempts by Davaud (1978, 1979, 1982) to devise a computer program for unitary associations. Davaud (1982) clearly explains the details of this program and we will not discuss them here. We note only that this algorithm does not allow one to construct maximal compatible sets of species including the virtual coexistences that can be deduced from conflicting stratigraphic relationships (i.e., unitary associations in a strict sense). There are two reasons for this:

1. When the stratigraphic relationship of two species is undetermined, one of the species is omitted.
2. When two species occupy contradictory stratigraphic positions in different sections, one of them is also omitted.

As a result, this algorithm does not resolve the principal difficulties of the problem of biochronology (see Baumgartner 1984a. p. 167 and 1984b, p. 738). Except for trivial cases (Sect. 2.2.1), no virtual coexistence is taken into consideration.

Recent applications of Davaud's (1982) program demonstrate that, in complicated problems, this procedure of extreme simplification reduced the data to be processed by 60–70%. The best example is given by Schaaf (1985, p. 249) who applied the program to 111 species of Cretaceous radiolarians. Only 38 species of the 111 were kept in the final output.

To remedy this enormous loss of information, Schaaf (loc.cit.) combined the above algorithm and Hay's (1972) probabilistic approach (Sect.15.3): this mixed method is called "sequential analysis".

When applied to Schaaf's Cretaceous radiolarians, sequential analysis produced an extremely truncated picture of the species' total ranges (see Sect. 15.5.2).

This is not surprising, for that method combines the disadvantages of the two others (see Sect. 15.5.2).

15.5 Comparison of Range Charts

15.5.1 Statistical Methods Based on Bivariate Graphs

Statistically oriented biostratigraphers commonly use bivariate graphs to compare range charts or correlations produced by two different methods.

The *X*-axis of such graphs records the sequence of "events" (FADs and/or LADs) produced by method 1 and the *Y*-axis records the sequence of events produced by method 2.

The resulting *XY* plot is a line if the two methods produce identical results. If the two methods produce different results, the dots will be more or less scattered. In this latter case, the standard deviation of the events relative to the regression line is usually supposed to provide an estimate of the mismatch of the two sequences.

Note, however, that such statistical methods of comparison are worthless for a simple reason: false and correct syntheses of raw data often appear to be very similar when compared by means of bivariate graphs.

An example of this is given in Fig. 15.8. Suppose that we compare two range charts, *RC*1 and *RC*2, the first being an exact synthesis of biostratigraphic data and the second being false (i.e., the observed coexistences of species AB, BC and CD are not represented in *RC*2). In this hypothetical case, the bivariate scatter-

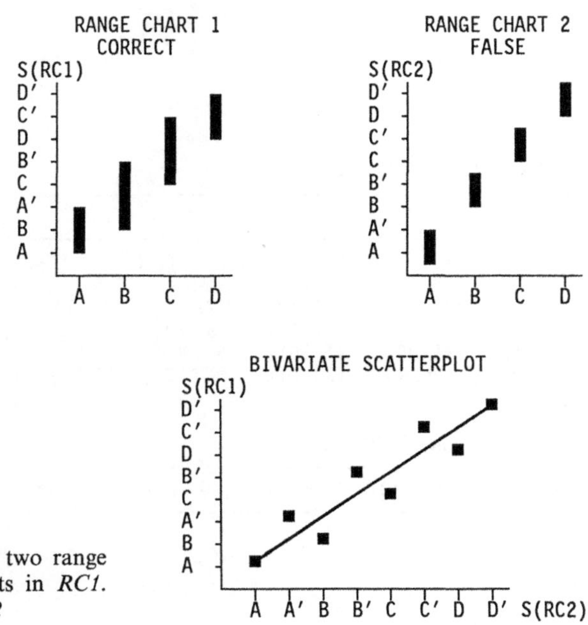

Fig. 15.8. Bivariate comparison of two range charts. *S*(*RC1*), sequence of events in *RC1*. *S*(*RC2*), sequence of events in *RC2*

plot is close to a line (Fig. 15.8); however, it would be absurd to conclude that *RC*1 and *RC*2 are similar.

15.5.2 Comparison of Adjacency Matrices

Any comparison of distinct range charts produced by different methods should, first of all, be based on common sense and should meet the following test: a correct biochronologic synthesis of a given data set must represent the taxonomic content of all the studied samples. Testing this can easily be done as follows:

1. We establish the adjacency matrix A of the coexistence graph (G) of the data on which a biochronologic synthesis is based;
2. We construct the (taxon-taxon) adjacency matrix A' of the synthesis itself.

Any observed coexistence recorded in A (i.e., $a_{ij} = a_{ji} = 1$) which is not recorded in A' represents an error.

As an example, we will examine Schaaf's (1985) cretaceous radiolarian zonation mentioned above (Sect. 15.4.2). Schaaf's original stratigraphic observations are given in Appendix I.2 (numerical database) and the zonation he obtained by means of his sequential analysis is reproduced in Fig. 15.9. (Note that we do not consider here the details of FAD/LAD relationships used in Schaaf's zonation, or those of Schaaf's taxa that are not listed in his original database).

Comparison of that semiprobabilistic range chart with the original data (Appendix I.2) shows that 56 coexistences are omitted and that 6 observed superpositions are reversed in Schaaf's synthesis: these contradictions are listed in Table 15.1.

Table 15.1. List of contradictions

Omitted coexistences
(7-9), (7-15), (7-21), (7-81), (7-89), (9-11), (9-23), (9-25), (9-70), (11-45), (12-70), (15-35), (15-45), (15-48), (15-57), (15-59), (15-72), (15-87), (15-103), (15-107), (19-45), (21-104), (22-71), (22-109), (28-83), (31-70), (34-45), (34-70), (34-71), (34-109), (37-45), (38-74), (38-81), (38-93), (38-110), (45-57), (45-68), (45-83), (45-88), (45-89), (45-100), (45-103), (45-110), (48-81), (65-70), (66-83), (68-109), (70-93), (79-104), (81-83), (81-87), (83-95), (83-110), (87-89), (93-104), (93-110)
Reversed superpositions
(38->70), (43->70), (70<-74), (70<-81), (70<-87), (70<-97)

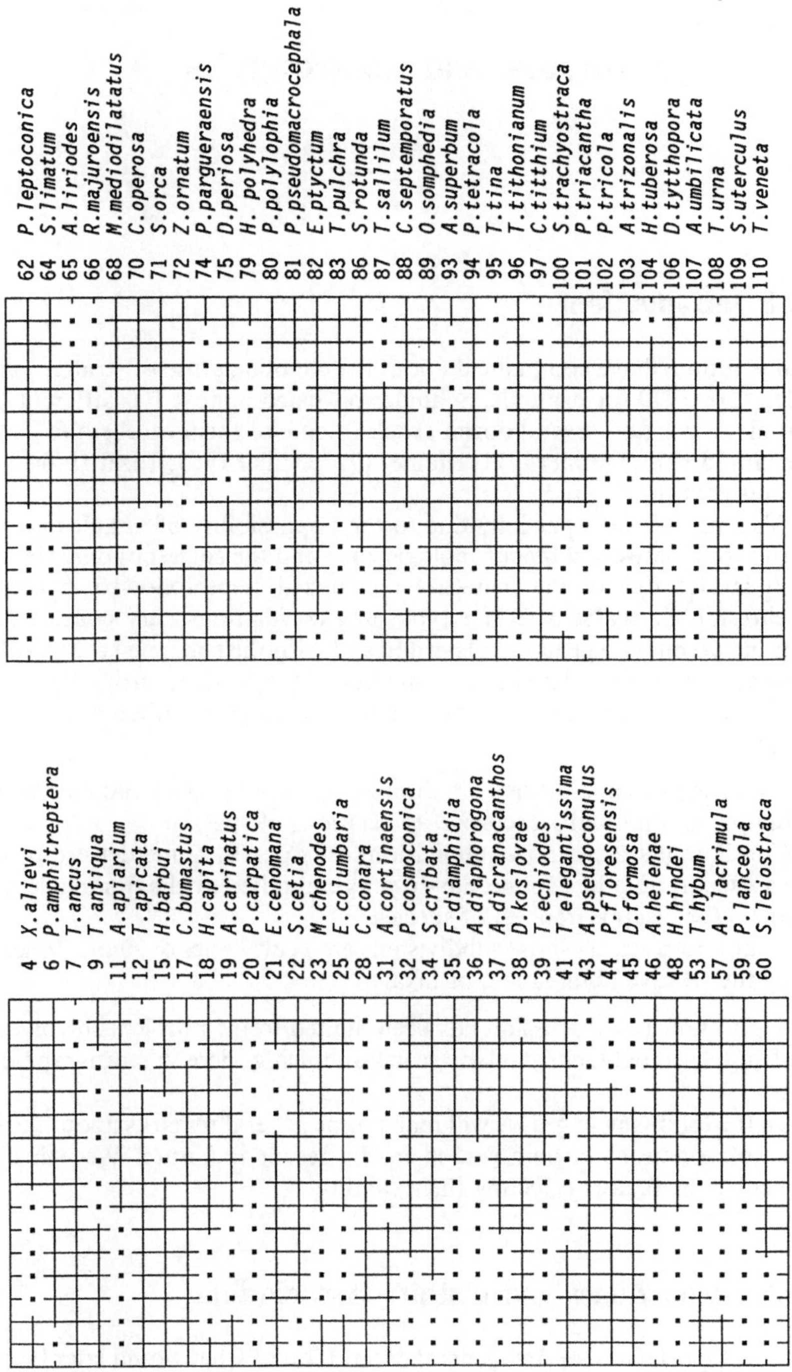

Fig. 15.9. Cretaceous radiolarian zonation generated by sequential analysis (After Schaaf 1985, p. 222, Fig. 50)

Chapter 16
Zones, Stratotypes, and Controversies

16.1 Introduction

Many authors have complained about the chaotic nature of stratigraphic terminology. In 1970 for example, Schindewolf listed almost 100 different biostratigraphic terms, and several dozen articles, some of them excellent (e.g., Eldredge and Gould 1977; Gabilly 1971; Mamet 1977; Miller 1965; Tozer 1967) discuss the famous concept of "zone".

This profusion is partially due to the impossibility of establishing a precise relationship between a biochronologic zone and the corresponding time interval.

It is also due to the fact that the logical steps followed in constructing biochronologic scales and in establishing correlations vary considerably both with the taxonomic group under study and with the concrete stratigraphic situation (nature of the sedimentation, quality of the fossil record, etc.).

Three major categories of biochronologic scales were differentiated in the first Chapter.

1. Scales whose subdivisions are discrete (noncontiguous) and characterized by unique and mutually exclusive associations of species.
2. Continuous scales whose subdivisions are Interval Zones defined by the intervals separating the evolutionary events embodied by the first phyletic appearances (or disappearances) of species.
3. Continuous scales whose subdivisions are Peak Zones or Acme Zones, defined by the relative abundances of taxa.

Up to now our discussion has been limited to the problems involved in constructing discrete biochronologic scales and using them in stratigraphic correlations.

In the following pages we will make some general remarks about the different kinds of zonations currently used in the recent literature. We will discuss in particular the factors affecting their validity.

16.2 Oppel Zones and Unitary Associations

In 1856, Oppel gave the first general description of what would later be called the Oppel Zone:

"Comparison has often been made between whole groups of beds, but it has not been shown that each horizon, identifiable in any place by a number of peculiar and constant species, is to be recognised with the same degree of certainty in distant regions."

"This task is admittedly a hard one, but it is only by carrying it out that an accurate correlation of a whole system can be assured. It necessarily involves exploring the vertical range of each separate species in the most diverse localities, while ignoring the lithological development of the beds; by this means will be brought into prominence those zones which, through the constant and exclusive occurrence of certain species, mark themselves off from their neighbors as distinct horizons. In this way is obtained an ideal profile, of which the component parts of the same age in the various districts are characterized always by the same species." (Oppel 1856, p. 3, translated in Arkell 1933, p. 16).

We note the following points about Oppel's discussion.

1. The petrographic nature of the strata is not taken into consideration when establishing the existence of a zone.
2. The zone is characterized by the constant and exclusive occurrence of certain species that are homotaxic over vast geographic ranges.
3. The end product of Oppelian analysis is an ideal sequence whose subdivisions are approximately the same age in all the regions where they are recognized.

These remarks show that a biochronozone constructed using unitary associations has the same attributes as an Oppel Zone.

1. It is defined and entirely characterized by the existence interval of one (or more) species and/or by the coexistence interval of *n* species.
2. This interval is unique and distinct from all others.
3. The zone is identifiable in a reproducible way by the species that characterize it.
4. Such a zone is at first an abstract construction and only secondarily a body of rock.

It is amusing to note that this last point still inspires polemics. Not so long ago Hancock (1977, p. 21) wrote:

"Some German stratigraphers (e.g., Schindewolf 1950) have even interpreted a zone as a time term to designate an interval during which was deposited the sediment that contains certain index fossils. Such a mystical inexactitude has dissatisfied twentieth-century American pragmatic stratigraphers."

In 1956 Arkell (p. 5) proposed a pragmatic solution to these mystical problems:

"British geologists have always envisaged a zone as a bed or stratum, a tangible object accessible to the hammer, though differing lithologically from place to place. Consequently some have thought it necessary to construct a parallel terminology to express the time units to which the various kinds of zones correspond."

"The need to keep time and rock distinct in our thoughts is obvious, and to the extent that this elaborate terminology has led to clarification of thought it has served a useful purpose. But beyond that it is unnecessary. No one uses it nor ever will. That Oppel himself fully appreciated the time element cannot be doubted (Schindewolf 1950). The fact that he nowhere defined a zone, nor made it clear whether his zones were stratigraphic or time intervals, may be taken to mean that he visualised zones from both aspects at once. The argument that a zone must be one or the other is sterile."

We hardly need mention that these words of Arkell have never been (to our knowledge) quoted in any Guide or Code of stratigraphic terminology.

16.3 Phylogenetic Seriations and Phylozones

Lacking precise information on the superpositional control between different fauna, some paleontologists who work on sparse and stratigraphically disconnected populations are forced to infer sequences of species (or assemblages of species) from an estimate of the relative degree of evolution of the individuals belonging to the species (see for example Thaler 1972 and Godinot 1981).

The criteria used to judge the more or less primitive character of a fauna depend largely on the modes of morphologic transformation known in its zoological group:

1. Increase of hypsodonty among micro-mammals;
2. More or less gradual increase in size among many organisms;
3. Progressive elongation of the shell among some large foraminifera, etc.

Chronologic sequences deduced from the (presumed) degree of evolution of different populations are sometimes verified later; but this is far from being always the case.

A good example of a false dating resulting from an incorrect phylogenetic interpretation is that of the *Nejdia* beds in Saudi Arabia. For a long time this ammonite genus was thought to have evolved from the Phymatoceratinae; as a consequence, researchers in the field deduced that *Nejdia* was from the Upper Toarcian. However, recent discoveries in Morrocco proved that *Nejdia* is actually from the lower Toarcian and that it derived from *Bouleiceras* (Guex 1973).

We note incidentally that the unitary association method can be quite useful in detecting and interpreting the contradictions that can result from applying such phylogenetic seriations. We simply replace the notion of "biostratigraphic" graph with that of "phylogenetic" graph in which the arcs represent the (hypothetical) relationships primitive → advanced (instead of representing superpositions of species).

The use of phylogenetic criteria to establish chronologic sequences has however been criticized often, and with justification. There are several reasons for this.

The most important criticism is the inherent risk of circular reasoning. It is in fact impossible to establish the validity of a model of morphologic transformations that affect a fossil group through time without first proving that these transformations do not simply reflect reversible responses to local ecological variations (Fig. 16.1). However, such a proof necessarily requires correct biochronologic correlations.

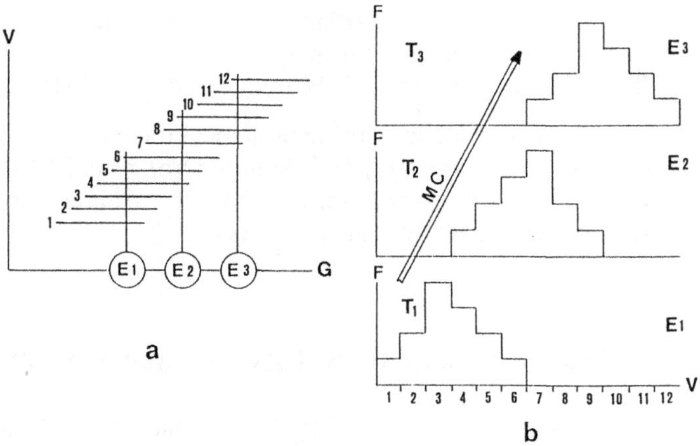

Fig. 16.1. a Distribution of 12 morphologic variants (*V*) of the same species in different environments ($E_1 - E_3$). *G*, geographic and ecological gradient. **b** Frequency (*F*) of the different variants in three fossiliferous beds ($T_1 - T_3$) in a single stratigraphic section where the environments $E_1 - E_3$ are superposed. *MC*, apparent morphocline. From Guex 1979, p. 213

As Eldredge and Gould (1977, p. 35) wrote, in the context of a similar discussion:

"We shall always be able to document differences among successive population samples in a vertical sequence, but these will, at most, represent adaptive shifts to minor environmental changes rather than a dominant evolutionary force inexorably leading to the creation of a new species."

We should also bear in mind that neontologists have a tendency to divide the anagenetic (i.e,. gradual) lineages that they observe into branches defined by the degree of evolution of the different species (biospecies) making up a lineage. These branches generally serve as a basis for defining phylozones. In practice, a sequence of phylozones is chronologically meaningful only if the succession of branches that define it is recognizable on a large scale.

These general observations lead us to note that the different morphotypes that make up a biospecies often have durations of existence that are not identical. If a fossil group is to have optimal biochronologic resolution, it is useful to consider each variant as a potential taxonomic entity. The purely neontologic doctrine of taxonomy, with the goal of giving a biological image of the paleontological species, is in theory justified.

However, when it is pushed to an extreme ("one bed, one name") it forces those who apply it dogmatically to treat as different species identical morphotypes of the same chronospecies, under the pretext that these morphotypes are found in populations whose variability is different depending on the stratigraphic levels.

This has the effect of obscuring precious information on the space-time distribution of different morphotypes of a supposed biospecies and on the ecological factors that affect the variability of the species. It also limits the efficient transmission of the raw biostratigraphic observations used in making correlations and constructing phylogenetic syntheses.

This is why we agree with Eldredge and Gould (1977, pp. 39–40):

"... we merely claim that the biological theory underlying most "modernist" approaches to biostratigraphy is simply wrong, and that the theory supporting "old-fashioned" biostratigraphy is correct. To date, evolutionary theory owes more to biostratigraphy than vice versa."

16.4 Oppel Zone, Interval Zones, and Datums

Over the past 30 years, the Oppel Zone has gradually been displanted by the concept of the Interval Zone. As we have seen, such a zone corresponds in principle to the interval separating two datums.

In its original sense the term "datum" means "first phyletic appearance" of a species. This implies that the appearance of a species, at least in one section, has been proven to be phyletic and is not simply fortuitous (e.g., sudden immigration of a species from another faunal realm). However, some biostratigraphers now use this term even when constructing zonations with groups in which the evolutionary relationships of the different species ("morphospecies") are not known. Thus the term datum has gradually lost its evolutionary connotation and has become, for many experts, synonymous with "first local appearance" and "first local disappearance" of a species. This is why we have deliberately used the word in this weaker (but objective) sense.

The practice of Interval Zones spread at the same time that datum lost its original meaning; this was encouraged by many authors of Guides and Codes who claim that the first phyletic appearance of a species is the best criterion for correlating stratigraphic sections.

Two arguments are advanced to justify this point of view.

The first says that in theory the true first appearance of a species is chronologically punctual. This is correct but has nothing to do with the stratigraphic reality. Let us assume that the descendence relationship between two species x and y is proved; in such a case, two stratigraphic relationships are possible for their respective FADs and LADs.

1. The proof of the relationship $x \rightarrow y$ is based on the existence of a transitional population (i.e., x and y coexisted more or less briefly within transitional populations). In this case, we find ourselves in the kind of difficulty described by Cita (1978, p. 672):

   "It is worth mentioning at this point how difficult it is to define the lower boundary of Zone N. 14, because of the inherent difficulties in pinpointing the first evolutionary occurrence of *G.nepenthes* in the slow and gradual evolution from *G.druryi.*"

   As we see it, the biochronologic interest of a phyletic relationship $x \rightarrow y$ lies in the use that can be made of the existence interval of the transitional population between x and y when making correlations: if this interval is short, it gives a precise criterion for the correlation.

2. A gap exists between the LAD of x and the FAD of y (lack of knowledge about transitional populations for documentary reasons). It is then clear that the FAD of y will behave just like any FAD (i.e., in the sense of the first local appearance of y). To test its reliability, it is necessary to prove that it is not diachronous. However, in the absence of external (and nonbiochronologic) arguments, this can be done only in the framework of an Oppelian zonation.

The second argument invoked by stratigraphers of the neontologists' school is the following: we know that the migration speeds of recent organisms can be very great (on the order of 10–100 km/year) and it is clear that the absolute time taken by a new group to colonize a newly opened environment is virtually nil compared to the resolution power of the biochronologic dating tool (see for example Berggren and Van Couvering 1978; Hottinger 1981).

However, this actualistic argument generally does not apply to fossils; to be convinced of this it is enough to remember that the diachronism of the FADs in some groups is sometimes greater than that of the LADs (Sect. 8.4.2.2).

Establishing a rule that FADs should be used to define the base of a zone will not inspire biostratigraphers to try to prove the chronologic homotaxy of their datums using Oppelian-type syntheses. On the contrary, it encourages them to be satisfied with stating (without any proof or using only pseudo-objective statistical criteria) that certain datums are chronologically significant, and with constructing scales whose subdivisions are precisely Interval Zones.

The accuracy of the correlations resulting from such zonations clearly depends on the diachronism of the datums used to define the limits between these zones. It is in fact obvious that an Interval Zone based on nonphylogenetic datums enables one to make correct correlations only if the amplitude of the biochronologic dispersion of the datums is zero (i.e., these characteristic datums must not be found in different Oppel Zones in different locations).

However, in the absence of physical criteria, these diachronisms (and the amplitude of the biochronologic dispersion of the species) can be detected only using the reproducible subdivisions of a correctly established range chart. In other words, the validity of an Interval Zone based on nonphylogenetic datums can be proven only if the Oppel Zones generated by the stratigraphic relationships of the species have been previously established.

To show that the use of nonphylogenetic Interval Zones easily leads to contradictions, we will follow the elementary logical procedure of comparing observed facts with the formal zonal definitions. This will naturally lead us to recognize the following kinds of situations:

1. Interval Zone *A* is defined by the interval separating the FAD of species x from the FAD of species y, while in fact the true first appearance of y precedes that of x.
2. Interval Zone *B* is defined by the interval that separates the LAD of species x from the FAD of species y while the observed facts show that x and y coexist in some locations (which raises the amusing question whether the existence interval of x intersects that of y or vice versa.).

As a concrete example, we will again look at the biochronologic distribution of the Paleogene nannoplankton from California (Chapter 11). Since 1965 no major revision of the stratigraphic data of Bramlette and Sullivan (1961) and of Sullivan (1964, 1965) has been made. Their magnificent work has however been used by all the nannoplankton workers who have published subsequent "syntheses" on the zonation of the Paleogene. The stratotype for the base of the (standard) zone NP-8 of Martini (1971) corresponds for example to the horizon number 1 from the Lodo section (section 10 in our Fig. 11.1), and the stratotype for the base of zone NP-15 corresponds to horizon number 11 of Media Aqua Creek (section 8 of our Fig. 11.1).

Sullivan's original observations have naturally been reinterpreted in terms of datums, and the zones proposed in the recent syntheses are all Interval Zones (Fig. 16.2). According to Martini (1971) and Martini and Muller (1986):

1. Zone NP-12 is defined by the interval separating the FAD of *D.lodoensis* (species No. 85 of our list; see Appendix E.1) from the LAD of *N.tribrachiatus* (No. 89).
2. Zone NP-13 is defined by the interval separating the LAD of *N.tribrachiatus* (No. 89) from the FAD of *D.sublodoensis* (No.96) and
3. Zone NP-14 is defined by the interval separating the FAD of *D.sublodoensis* (No. 96) from the FAD of *C.alatus* (No. 6).

However, Sullivan's original data (Figs. 11.2, 16.1) show that:

1. In sections 2 and 5 (Vaca Valley and Las Cruces), NP-12 completely or partially overlaps NP-14;
2. Species 89 and 96 (whose exclusion interval is supposed to define NP-13) coexist in three sections (Vaca Valley, Santa Anita, and Las Cruces; Figs. 11.2, 16.2).

The misuse of Interval Zones has given rise to exaggerated claims for the resolution power of biochronologic scales of this type. As stated by Fahreus (1986, p. 150):

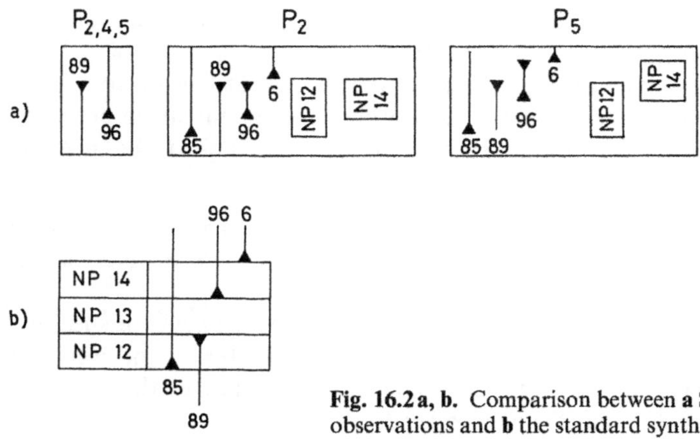

Fig. 16.2 a, b. Comparison between **a** Sullivan's (1965) original observations and **b** the standard synthesis of Martini (1971) (b)

"In recent years many biostratigraphic zonations proposed in the literature either imply or directly state such a fine biostratigraphic resolution that in many cases has to be considered beyond the point of practical reproducibility, i.e. correlations attempted back onto such zonations carry with them little or no precision."

Mamet (1977, p. 454) used his biting humor to attack this inflation:

"1800 Visean "species" are known in the literature, hence the possibility of erecting 1800 first occurrence zones, 1800 last occurrence zones, and 1800 acme zones. This is common practice and gives a maze of incomplete ranges inducing pseudocorrelations."

We will conclude our discussion of Interval Zones with the following remarks.

1. It is pointless to try to discover the time of phyletic appearance of a species before establishing as accurately as possible the chronologic relationships of the morphotypes whose evolutionary relationships one wishes to know. Proceeding in the opposite order leads to circular reasoning.
2. Constructing a relative biochronologic scale whose subdivisions are based on diachronous (and nonphylogenetic) datums can be compared to building a card castle with only one side: such structures do not stand up.

16.5 Standard Zones

The term "standard zone" is frequently used to describe the subdivisions of a biochronologic scale of supposed general value. Two drastically different views of the properties of standard zonations now exist: those of Tozer (1967) and Callomon (1965, 1984).

Discussing the Triassic standard zones in North America, Tozer (loc.cit. p. 10–11) writes:

"In most sections there are rocks between the zones that are devoid of characteristic fossils. Those parts of the section without diagnostic fossils cannot be assigned to a zone. The alternative procedure – to include beds lacking the characteristic fauna in one or other of the adjacent zones – practised by some contemporary workers (e.g., Donovan 1956, p. 198) is not followed. From this it should be clear that the zones, as arranged, are not regarded as forming a continuum."

"Future discoveries will probably reveal distinctive faunas that will form the basis for new zones between those now recognized. If and when such new zones are discovered, a decision may be necessary regarding the placing of the zone within a hierarchy of stages. Such decisions should be arbitrary. A new zone found between the Sverdrupi and Romunderi Zones might be regarded as either Dienerian or Smithian, depending upon the affinity of its fauna. At present it seems undesirable to adopt the alternative policy advocated by Callomon (1965), namely to define inflexibly the base of a stage such as the Smithian as the base of the Romunderi Zone."

We see that Tozer rightly considers that the subdivisions of a standard biochronologic scale are discrete when based on discontinuous biostratigraphic data.

According to Callomon (1984), a standard biochronologic scale must have the following properties:

1. It must be continuous.
2. Its subdivisions (zones) must be contiguous (no gap, no overlap).
3. The lower limit of each standard zone must be defined by a "Golden Spike" in a stratotype.
4. The Golden Spike automatically defines the top of the subjacent zone, even when faunal criteria are lacking.

Callomon's concept of standard scales may be criticized on three grounds:

1. It fails to take into account both the nature of the biostratigraphic data on which the scale is based, and the analytical methods used to establish the correlations (no distinction between continuous vs discrete scales).
2. It ignores the fact that strongly discontinuous paleontological data (ammonites, radiolarians, nannoplankton etc.) necessarily generate discrete biochronologic scales: imposing a continuum on such scales is equivalent to overlooking that fundamental property.
3. It neglects the essential requirement that a stratotype must locally record the superpositional control between two consecutive discrete zones (i.e., the Golden Spike methodology does not require any superpositional control between two zones).

To this we add that the famous rule according to which "the base of a zone fixed by a Golden Spike in a stratotype automatically defines the top of the preceding zone" can be applied only if the oldest part of the higher zone is recorded in that stratotype. The rule obviously makes no sense if that condition is not satisfied. In other words, that convention can generate confusion and has no place in a codification of biochronologic procedures.

In summary, we can describe a biochronologic scale as "standard" only if it is widely applicable and results from a correct synthesis. The quality of such a scale depends entirely on the method used to establish it. Each type of zonation (continuous vs discrete) requires its own procedure for zonal definition. Whatever the procedure, the definition must not contradict observed facts.

16.6 Acme Zones and Peak Zones

The local stratigraphic interval in which a species shows its maximum development can be considered independently of its total local stratigraphic range. If we can prove that such an interval is always located between other chronologically significant intervals, we can conclude that it is also chronologically significant. This interval can then be considered the acme of the species, and it can be used to define a zone (the "Acme Zone").

Remane's (1963) work on the Calpionellid is an excellent example of biochronologic correlations based on acme zones. The acme of *Calpionella alpina*, characteristic of the zone of the same name, is located between the acmes of the genera *Crassicolaria* and *Tintinnopsella*. The widely identifiable *C.alpina* zone is an acme zone with solidly established validity. Nevertheless we note that Remane's zonation is good because the preservation of the Calpionellid is not selective vis-à-vis the different species he used. In addition, the stratigraphic sections he studied were the site of relatively continuous sedimentation, so relative abundances of these taxa here are a reliable biochronologic indicator.

Very precise correlations can also be obtained using Peak Zones defined by pollen diagrams, biologs, or other morphometric variograms on the scale of a basin or a province.

The problems of interpretation posed by these abundance or morphometric variograms are very similar to those faced by the geophysicists who use electrical logging to establish lithostratigraphic correlations. In some geological contexts (the Quaternary, for example) these methods no doubt produce the most reliable correlations, but they do not come into the scope of this book. Readers interested in these approaches can refer to Reyment (1980) and Malmgren and Kennet (1981).

16.7 Stratotypes

The nature of the boundary between two consecutive biochronologic zones is different for continuous scales (Peak Zones, Acme Zones, Interval Zones) than for discrete scales (Oppel Zones). In the former, the (theoretical) limit is a surface, and in the latter it is an interval of separation.

For practical reasons, it is useful to delineate the boundary between two continuous zones, using a Golden Spike in a stratotype.

Also for practical reasons (notably the instability of scales) it is useful to designate a stratotype in which the superposition of two discrete (noncontiguous) zones is well documented. For Oppelian scales, however, it is inappropriate to place a Golden Spike in the reference section ["unless the spike is rather thick..." as Mamet (1977, p. 455) puts it so nicely].

Consequently, the stratotype for a discrete biochronologic boundary differs from that of a chronostratigraphic limit (e.g., the base of a stage) which necessarily is a surface.

An amusing anecdote illustrates this necessity.

In 1976 Guex and Taylor published a little note concerning the Hettangian-Sinemurian boundary in the northwest European, Tethyan and North American realms. It dealt more precisely with the correlation of the *Schlotheimia marmorea* zone (Tethyan) and the *Badouxia canadensis* zone (North American equivalent to the *Marmorea* zone) with the Lower Jurassic standard zonation proposed by Dean et al. (1961) for the northwest European realm. The zonal terminology for the relevant interval near the Hettangian/Sinemurian limit is provided in Fig. 16.3.

Stage	Zone	Subzone
SINEMURIAN		*Conybeari* sz
HETTANGIAN	*Angulata* Zone	*Complanata* sz
		Extranodosa sz
	Liasicus Zone	*Laqueus* sz
		Portlocki sz
	Planorbis Zone	*Johnstoni* sz
		Planorbis sz

Fig. 16.3. Zones and subzones (*sz*) of the boreal Hettangian

Until very recently the base of the Sinemurian was defined paleontologically by the first abundant appearance of the Arietitidae (*Metophioceras* in particular) above the beds with *Schlotheimia complanata* (index of the subzone of that name).

In 1983, Bloos made a convincing case that this definition was incorrect. His recent discoveries showed in fact that the Arietitidae first became common in the *Complanata* subzone (undisputed topmost Hettangian).

The *Marmorea* zone was initially considered as strictly equivalent to the northwest European *Angulata* zone, and it was not until 1942 that Spath planted doubts in some minds by affirming that the holotype of the alpine species (*Schlotheimia marmorea* Oppel) was of Sinemurian age. Since then, authors have used exclusively *Schlotheimia angulata* to characterize the last zone of the "boreal" Hettangian (Fig. 16.3), without, however, correlating the alpine *Marmorea* zone with the northwest European lower Sinemurian.

Field work in North America subsequently led us to consider that the *Marmorea* and *Canadensis* zones are actually younger than previously thought. We concluded that those zones should be assigned to the lowermost Sinemurian and not to the Hettangian.

That conclusion was sharply criticized by Bloos (1983), who cited recent discoveries he made in southern Germany.

Before examining Bloos's arguments, it is useful to note that most specialists agree that the Dorset coast (Lyme Regis area) provides the best stratotype for delineating the base of the Sinemurian stage. That splendid stratigraphic sequence has been studied by a number of authors; the most detailed description is given by Lang (1924) and Spath (1924) who planted the historic Golden Spike in their bed 16 (Fig. 16.4).

The last Hettangian ammonite-bearing beds of that section yield *Schlotheimia pseudomoreana*, a species that is very close to *S.complanata* (see Bloos 1979), the index of the *Complanata* subzone. These latest Hettangian beds are separated from the first beds containing typically Sinemurian fauna by an unfossiliferous stratigraphic interval; the first ammonites of uncontested Sinemurian age (e.g., *Metophioceras conybeari*) appear in bed 19 of this section (Palmer 1972).

Bloos's research in southern Germany led him to introduce, in 1979, a new horizon (characterized by *Schlotheimia depressa*), located between the *Complanata* subzone and the *Metophioceras longidomum* horizon (generally considered

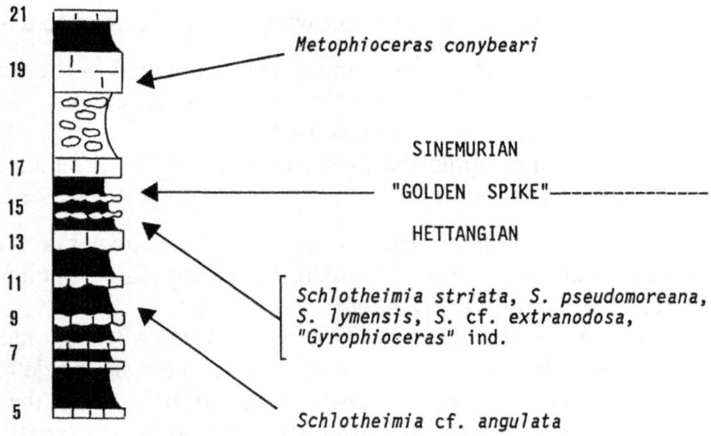

Fig. 16.4. Hettangian-Sinemurian boundary in Dorset (Lang 1924; Spath 1924) and stratotype of the *Conybeari* subzone (lower Sinemurian)

equivalent to the *Conybeari* horizon of the stratotype). More recent work (Bloos 1983) convinced him that *S.marmorea* had a limited range in the *Depressa* horizon. He concluded that all the authors who claimed to have found chronologically characteristic species of the *Marmorea* zone in beds of undisputed Sinemurian age (i.e., post-*Conybeari*) probably made mistakes in their identifications and that as a consequence the conclusions of Guex and Taylor (1976) were incorrect (see Taylor 1986 for a detailed discussion)

Bloos's 1983 article is interesting in that the recent discoveries on which he based his conclusions demonstrate unequivocally that *S.marmorea* is of post-*Complanata* age. On this point he involuntarily confirmed the deductions of his opponents.

In fact, Bloos merely sidestepped the main problem that we broached in 1976: that of the age of the *Canadensis* zone. This North American equivalent to the *Marmorea* zone was for a long time attributed by some experts to the Early or Middle Hettangian, a major error that makes any reasonable interpretation of the phylogeny of the Lower Jurassic ammonites impossible.

Whether the *Schlotheimia marmorea* beds should be attributed to the topmost Hettangian or to the lowermost Sinemurian is purely a problem of convention: this problem in fact disappears if we accept the boundary set by Lang (1924) and Spath (1924) in the stratotype.

16.8 Integrated Biochronology and Calibration

The term "integrated biochronology" refers to the procedure that consists of both studying the stratigraphic relationships among taxa belonging to different fossil groups (subclasses, classes, etc.) and using these relationships to construct a composite biochronologic scale.

Four conditions must be met when constructing such a scale:

1. The different fossil groups must be documented in the same way;
2. They must come from the same stratigraphic sections;
3. The nomenclature must be consistent;
4. The biostratigraphic databases established by different specialists must be published *in extenso*.

This last condition is often not met; the literature contains abundant examples of proposed "zonations" in which the authors fail to publish the raw data on which their zonations are based.

This has two troublesome consequences: (1) when a zonation turns out to be inapplicable by other researchers, there is no way to determine whether the contradictions arise from the new observations or from the original syntheses; and (2) a zonation must be constantly modified by new biostratigraphic data, but modifications are possible only if one can consider all the previously established data.

We note that the nonintegrated study of stratigraphic relationships among different fossil groups generally leads to establishing zones with different boundaries. An example can be found in the contradictory correlations that result from studying ammonites and conodonts in the lower Triassic in Pakistan (Chapter 9). Correlating biochronologic scales based on different zoological groups is a procedure of calibration and not integration.

In general it is more interesting to analyze the contradictions in correlations based on different groups than to try to integrate biostratigraphic databases that have totally different structures (conodonts vs ammonites, for example).

Finally, we should mention once more the problem of constructing biochronologic scales based on zoological groups affected by considerable provincialism, for biogeography is a by-product of correlations and not the reverse (see Baumgartner et al. 1980, p. 48). Once correct biogeographic provinces have been delimited, it is sometimes necessary (and productive) to construct parallel biochronologic scales for the different biogeographic realms in order to establish faunal migrations between these provinces. This is a common practice among ammonite workers.

16.9 Validity of a Zone

All Guides and Codes of stratigraphic terminology have one glaring omission in common: they never tell how to construct biochronologic zones. This omission leads to another: they never discuss the criteria governing the validity of such zones.

It is however obvious that the validity of a biochronologic zone depends neither on its formal designation in a particular stratotype, nor on the arbitrary choice of a phyletic relationship. Instead it depends entirely on its identifiability outside its type-section. This identifiability of a zone is contingent on the correctness of its definition, which in turn depends entirely on the way the zone is constructed.

In all cases with complex biostratigraphic data (strongly discontinuous sedimentation and fossil record), the use of discrete zones to correlate terrains is imperative, for the relative abundances of different taxa are not reliable. The main criterion for the validity of these Oppelian zones is their reproducibility and mutual superpositional control.

Note that discrete zones constructed by means of the U. A. method do not need to be formally defined: their correlation is a mere by-product of the construction of the referential. Within the framework of that methodology, new biostratigraphic data can easily be added to an old database for reprocessing and updating correlations.

Interval Zones differ in that their recognition in stratigraphic sequences depends entirely on their definition (i.e., they cannot be identified by means of characteristic taxa only: absences of some taxa are also required.).

At least two standard situations occur that should invalidate such zones:

1. An Interval Zone defined by the interval between the LAD of species x and the FAD of species y when x and y coexist (Sect. 16.4);
2. An Interval Zone defined by the interval between the FAD of a species x and the FAD of a species y when the FAD of y predates that of x (the same is true for LADs).

Appendix A
Palmer's (1965) Database

A.1 Code Numbers of Species

1 *Kormagnostus simplex*
2 *Coosella beltensis*
3 *Kinsabia variegata*
4 *Spicule B*
5 *Pseudagnostus* (?) *nordicus*
6 *Arcuolimbus convexus*
7 *Opisthotreta depressa*
8 *Tricrepicephalus coria*
9 *Tricrepicephalus texanus*
10 *Kingstonia pontotocensis*
11 *Llaonaspis modesta*
12 *Meteoraspis metra*
13 *Coosia* cf. *albertensis*
14 *Crepicephalus australis*
15 *Llaonaspis undulata*
16 *Maryvillia* cf.*ariston*
17 *Llaonaspis peculiaris*
18 *Dictyonina perforata*
19 *Raaschella ornata*
20 *Labiostria conveximarginata*
21 *Angulotreta digitalis*

A.2 Code Numbers of Sections and Levels

Section 1: Morgan Creek

Level 1 = height 419; level 2 = height 446; level 3 = height 453; level 4 = height 464; level 5 = height 467; level 6 = height 469; level 7 = height 472; level 8 = height 475; level 9 = height 483; level 10 = height 485; level 11 = height 494; level 12 = height 504; level 13 = height 516; level 14 = height 529; level 15 = height 530; level 16 = height 532; level 17 = height 538; level 18 = height 547; level 19 = height 561; level 20 = height 570

Section 2: White Creek

Level 1 = height 628; level 2 = height 653; level 3 = height 655; level 4 = height 677; level 5 = height 695; level 6 = height 706; level 7 = height 725; level 8 = height 744; level 9 = height 750; level 10 = height 775

Section 3: James River

Level 1 = height 42; level 2 = height 120; level 3 = height 132; level 4 = height 144; level 5 = height 163; level 6 = height 201;

level 7 = height 218; level 8 = height 221; level 9 = height 223;
level 10 = height 247; level 11 = height 248

Section 4: Little Llano River

Level 1 = height 450; level 2 = height 468; level 3 = height 486;
level 4 = height 507; level 5 = height 537; level 6 = height 548;
level 7 = height 557; level 8 = height 558; level 9 = height 559;
level 10 = height 560

Section 5: Lion Mountain

Level 1 = height 85; level 2 = height 125; level 3 = height 155;
level 4 = height 180; level 5 = height 190; level 6 = height 200;
level 7 = height 215

Section 6: Pontotoc

Level 1 = height 395; level 2 = height 425; level 3 = height 541;
level 4 = height 594; level 5 = height 602; level 6 = height 609;
level 7 = height 618; level 8 = height 622; level 9 = height 647;
level 10 = height 650

Section 7: Streeter

Level 1 = height 105; level 2 = height 183; level 3 = height 287;
level 4 = height 295; level 5 = height 325; level 6 = height 369;
level 7 = height 387; level 8 = height 393

Section 8: Threadgill Creek

Level 1 = height 655; level 2 = height 656; level 3 = height 660;
level 4 = height 673; level 5 = height 693; level 6 = height 702;
level 7 = height 732; level 8 = height 736; level 9 = height 776;
level 10 = height 787; level 11 = height 805; level 12 = height 807;
level 13 = height 811; level 14 = height 815; level 15 = height 835

A.3 Numerical Database

Section 1

1 1 10, 2 1 3, 3 1 5, 4 4 12, 6 3 3, 7 2 11, 8 1 14, 9 1 1, 10 3 8, 11 7 9, 12 2 10,
13 13 13, 14 6 9, 15 12 12, 16 12 14, 17 13 13, 18 14 17, 19 14 16, 20 17 20

Section 2

1 1 3, 2 1 1, 3 1 2, 4 1 7, 5 1 1, 7 1 7, 8 1 6, 9 1 3, 10 1 6, 11 5 5, 12 1 4, 15 6 6,
16 6 6, 18 8 8, 19 8 9, 20 10 10, 21 8 9

Section 3

1 1 3, 2 2 3, 3 1 3, 4 4 9, 5 9 9, 6 2 3, 7 2 8, 8 2 7, 9 2 2, 10 6 9, 11 4 5, 13 6 9, 14 6 6, 15 6 6, 16 6 9, 17 7 7, 18 10 11, 20 11 11, 21 11 11

Section 4

1 1 1, 2 1 2, 3 1 1, 4 1 5, 5 1 1, 7 2 6, 8 2 7, 10 2 3, 12 4 4, 13 7 7, 14 4 4, 15 5 5, 16 5 7, 17 7 7, 18 9 10, 19 8 9

Section 5

1 2 2, 2 1 2, 5 2 2, 6 2 2, 7 1 1, 8 1 4, 9 1 1, 10 2 2, 12 2 2, 13 4 4, 16 3 4, 19 5 5, 20 6 7

Section 6

1 1 2, 2 2 2, 3 2 3, 7 4 4, 8 7 7, 13 4 6, 14 4 5, 16 6 7, 19 9 9, 20 10 10, 21 9 9

Section 7

1 1 1, 2 4 4, 3 2 3, 4 3 5, 5 6 6, 7 3 6, 8 6 6, 12 4 4, 13 6 6, 16 6 6, 17 6 6, 18 7 7, 19 7 8

Section 8

1 1 3, 3 1 4, 4 2 7, 5 2 10, 6 1 1, 7 2 8, 8 2 6, 9 2 2, 10 8 9, 11 6 6, 12 6 6, 13 10 10, 15 7 7, 16 7 9, 18 12 12, 19 11 12, 20 15 15, 21 13 14

Appendix B
Drobne's (1977) Database

B.1 Code Numbers of Species

1 *Alveolina moussoulensis*
2 *Alveolina aramea*
3 *Alveolina solida*
4 *Alveolina globosa*
5 *Alveolina avellana*
6 *Alveolina pisiformis*
7 *Alveolina pasticillata*
8 *Alveolina leupoldi*
9 *Alveolina montanarii*
10 *Alveolina aragonensis*
11 *Alveolina dedolia*
12 *Alveolina subpyreneica*
13 *Alveolina laxa*
14 *Alveolina guidonis*
15 *Alveolina decipiens*

B.2 Code Numbers of Sections and Maximal Horizons (M. H.)

Section 1: Fatji hrib

M. H.1 = level 11
M. H.2 = level 8
M. H.3 = level 7
M. H.4 = level 5

Section 2: Dane-Divaca

M. H.1 = level 7
M. H.2 = level 16
M. H.3 = level 20
M. H.4 = level 23

Section 3: Veliko Gradisce

M. H.1 = level 8
M. H.2 = level 12
M. H.3 = level 13
M. H.4 = level 14
M. H.5 = level 15
M. H.6 = level 19
M. H.7 = level 21

Section 4: Ritomece-Gradice

M. H.1 = level 6/2
M. H.2 = level 6/1
M. H.3 = level 5/1
M. H.4 = level 5/2

Section 5: Podgorje

M. H.1 = level 26
M. H.2 = level 48

Section 6: Podgrad-Hrusica

M. H.1 = level 7/1
M. H.2 = level 7/3
M. H.3 = level 7/4
M. H.4 = level 8/1

Section 7: Kozina-Socerb

M. H.1 = level 3
M. H.2 = level 6
M. H.3 = level 8
M. H.4 = level 4

Section 8: Golez

M. H.1 = level 20
M. H.2 = level 33
M. H.3 = level 34
M. H.4 = level 36
M. H.5 = level 46
M. H.6 = level 53

Section 9: Zbernica

M. H.1 = level 4
M. H.2 = level 12

Section 10: Dane, Istria

M. H.1 = level 1 – b
M. H.2 = level 7
M. H.3 = level 8

Section 11: Jelsane

M. H.1 = isolated sample (Drobne 1977, p.92)

B.3 Numerical Database

Section 1

2 1 1, 3 4 4, 4 3 3, 6 2 2, 7 3 3, 8 4 4, 12 3 3

Section 2

1 3 3, 3 1 1, 5 1 1, 6 2 3, 7 1 3, 9 3 4, 11 3 3, 12 3 3, 13 2 2, 14 4 4

Section 3

2 1 1, 3 2 2, 6 2 3, 7 2 3, 9 6 7, 10 7 7, 11 5 5, 12 4 4, 13 3 3, 14 6 7, 15 4 6

Section 4

1 1 2, 4 3 3, 7 1 1, 8 2 2, 9 1 4, 10 3 4, 11 1 2, 12 2 2, 14 3 4, 15 4 4

Section 5

4 2 2, 5 1 1

Section 6

1 2 2, 4 3 3, 7 1 2, 9 3 3, 10 4 4, 13 1 1, 14 4 4

Section 7

2 1 1, 3 3 3, 5 2 2, 7 3 4, 13 4 4, 15 4 4

Section 8

2 1 1, 5 1 1, 6 3 3, 7 2 2, 8 5 5, 10 6 6, 13 4 4, 14 5 6

Section 9

6 2 2, 10 1 2, 12 2 2, 15 1 1

Section 10

1 1 1, 6 1 1, 8 3 3, 10 2 2, 11 1 1, 12 1 1

Section 11

9 1 1, 10 1 1, 14 1 1

Appendix C
Sweet's (1970) Database

C.1 Code Numbers of Species

3 *Ellisonia clarki*
4 *Ellisonia delicatula*
5 *Ellisonia gradata*
6 *Ellisonia robusta*
8 *Ellisonia torta*
9 *Ellisonia triasica*
10 *Ellisonia* sp.ind.
12 *Neogondolella elongata*
13 *Neogondolella jubata*
14 *Neospathodus homeri*
16 *Neospathodus dieneri*
18 *Neospathodus triangularis*
19 *Neospathodus pakistanensis*
21 *Neospathodus timorensis*
22 *Neospathodus spathi*
23 *Neospathodus waageni*
24 *"Prioniodella" prioniodellides*
26 *Xaniognathus deflectens*
27 *Xaniognathus elongatus*
28 *Xaniognathus* sp.ind.

C.2 Code Numbers of Samples

Landu: 1 = K12–21 ; 2 = K12–1 to K12–12
Narmia: 1 = K1–21 ; 2 = K1–36/37 ; 3 = K1–50 + T63–167
Zaluch: 1 = K13–9 ; 2 = K13–11 ; 3 = K13–13
Nammal: 1 = K6–8D to K6–8F; 2 = K6–8G ; 3 = K6–27–A;
4 = K6–30; 5 = K6–31B; 6 = K6–36/37
Chhidru: 1 = K3–15; 2 = K4–CS; 3 = K4–UCL

C.3 Numerical Database

Section 1

4 2 2, 5 2 2, 8 2 2, 9 1 2, 10 2 2, 12 1 2, 13 2 2, 14 2 2, 18 2 2, 22 2 2, 23 1 1, 27 2 2

Section 2

3 3 3, 4 2 2, 5 1 3, 9 1 2, 12 2 3, 13 2 2, 14 2 3, 18 2 3, 19 1 1, 21 3 3, 22 2 2, 26 1 1, 27 2 3, 28 2 3

Section 3

4 1 3, 5 1 3, 9 1 2, 13 3 3, 18 3 3, 22 3 3, 23 1 2

Section 4

3 6 6, 4 4 4, 5 1 6, 9 1 6, 10 6 6, 12 1 5, 13 4 6, 14 5 6, 16 1 2, 18 4 6, 19 1 2, 24 1 6, 26 1 1, 27 4 5, 28 3 3

Section 5

4 2 2, 5 1 3, 6 1 2, 9 1 3, 16 1 2, 19 1 1, 23 2 3, 24 1 1, 26 1 1

Appendix D
Deboo's (1965) Database

D.1 Code Numbers of Samples

Deboo's (1965) original numbering of samples is given from top to bottom. For technical reasons, they are renumbered from bottom to top in the present book.

D.2 Code Numbers of Sections

1 Wayne and Clark Counties (composite section)
2 St Stephen Quarry
3 Little Stave Creek
4 Jackson
5 Perdue Hill

D.3 Code Numbers of Species (Benthic Foraminifera and Ostracods)

1 *Alabamina wilcoxensis*
2 *Angulogerina byramensis*
3 *Angulogerina danvillensis*
4 *Angulogerina vicksburgensis*
5 *Anomalina cocoaensis*
6 *Anomalina danvillensis*
7 *Astacolus danvillensis*
8 *Bolivina lazanensis*
9 *Bolivina dalli*
10 *Bolivina gracilis*
11 *Bolivina mississippiensis*
12 *Bolivina retifera*
13 *Bolivina striatellata*
14 *Bulimina jacksonensis*
15 *Cancris cocoaensis*
16 *Cibicides cocoaensis*
17 *Cibicides pippeni*
18 *Cibicides pseudoungerianus*
19 *Cibicides* sp.1
20 *"Darbyella" danvillensis*
21 *Discorbitura dignata*
22 *Eponides byramensis*
23 *Eponides jacksonensis*
24 *Eponides mariannensis*
25 *Eponides* sp.1
26 *Fursenkoina dibollensis*
27 *Guttulina problema*
28 *Hanzawaia mississippiensis*
29 *Hanzawaia* sp.1
30 *Karreriella advena*
31 *Lankesterina frondea*
32 *Lanticulina convergens*
33 *Liebusella turgida*
34 *Marginulina cocoaensis*
35 *Massilina decorata*
36 *Palmula caelata*
38 *Planulina cocoaensis*
38 *Planulina cooperensis*

39 *Planulina lobatulus*
40 *Pseudoclavulina cocoaensis*
41 *Pseudogaudryina jacksonensis*
42 *Pseudogaudryina* sp.1
43 *Pyrulina* sp.
44 *Ramulina* sp.
45 *Robulus carolinianus*
46 *Robulus cultratus*
47 *Robulus limbosus*
48 *Robulus rectidorsatus*
49 *Robulus vicksburgianus*
50 *Saracenaria ornatula*
51 *Sigmomorphina costifera*
52 *Sigmomorphina jacksonensis*
53 *Siphonina advena*
54 *Siphonina eocenica*
55 *Siphonina danvillensis*
56 *Spiroloculina* sp.1
57 *Spiroplectammina alabamensis*
58 *Stilostomella cocoaensis*
59 *Stilostomella jacksonensis*
60 *Textularia adalta*
61 *Textularia dibollensis*
62 *Textularia subhauerii*
63 *Textularia tumidulum*
64 *Textularia* sp.2
65 *Uvigerina cocoaensis*
66 *Uvigerina glabrans*
67 *Uvigerina jacksonensis*
68 *Uvigerina topilensis*
69 *Uvigerina vicksburgiensis*
70 *Vaginulina lalickeri*
71 *Vulvulina advena*
72 *Eponides ouachitaensis*
73 *Textularia haerii*
74 *Subcarinata quinqueloba*
75 *Valvulinaria octomerata*
76 *Flabellina lanceolata*
77 *Flabellina* sp.1
78 *Frondicularia tenuissima*
79 *Anomalin bilateralis*
80 *Discorbis cocoaensis*
81 *Globulina alabamensis*
82 *Globulina gibba*
83 *Globulina inaequalis*
84 *Guttulina irregularis*
85 *Hanzawaia* sp.2
86 *Liebusella byramensis*
87 *Marginulina hantkeni*
88 *Marginulina multiplicata*
89 *Eponides obesa*
90 *Robulus inusitatus*
91 *Saracenaria hantkeni*
92 *Spiroplectammina mississippiensis*
93 *Textularia porrecta*
94 *Textularia* sp.1
95 *Uvigerina dumblei*
96 *Uvigerina gardnerae*
97 *Globobulimina ovata*
98 *Rectoglandulina ovata*
99 *Angulogerina* sp.1
100 *Asterigerina gallowayi*
101 *Acanthocythereis* n.sp.1
102 *Actinocythereis dacyi*
103 *Actinocythereis gibsonensis*
104 *Actinocythereis* n.sp.1
105 *Actinocythereis* n.sp.2
106 *Ambocythere* n.sp.1
107 *Argilloecia hiwanneensis*
108 *Bairdia* sp.1
109 *Bairdopillata ocalana*
110 *Brachycythere mississippiensis*
111 *Buntonia israelskyi*
112 *Buntonia* n.sp.1
113 *Bythocypris gibsonensis*
114 *Clithrocytheridea caldwellensis*
115 *Clithrocytheridea garretti*
116 *Clithrocytheridea grigsbyi*
117 *Cushmanidea* n.sp.1
118 *Cyamocytheridea watervalleyensis*
119 *"Cythereis" dohmi*
120 *"Cythereis" hysonensis*
121 *Cytherella* sp.1
122 *Cytherelloidea cocoaensis*
123 *Cytheretta jacksonensis*
124 *Cytheropteron danvillensis*
125 *Digmocythere russelli*
126 *Digmocythere watervalleyensis*
127 *Echinocythereis jacksonensis*
128 *Echinocythereis mcguirti*
129 *Eucythere woodwardensis*
130 *Haplocytheridea ehlersi*
131 *Haplocytheridea montgomeryensis*
132 *Haplocytheridea* n.sp.1

133 *Hemicythere kniffeni*
134 *Henryhowella florienensis*
135 *Isocythereis couleycreekensis*
136 *Jugosocythereis bicarinata*
137 *Jugosocythereis vicksburgensis*
138 *Konarocythere spurgeonae*
139 *Krithe hiwanneensis*
140 *Krithe* n.sp.1
141 *Loxoconcha concentrica*
142 *Loxoconcha creolensis*
143 *N.gen.*n.sp.1
144 *Occultocythereis broussardi*
145 *Paracypris rosefieldensis*
146 *Paracytheridea belhavenensis*
147 *Paracytheridea woodwardensis*
148 *Propontocypris mississippiensis*
149 *Pteryogocythereis ivani*
150 *Pteryogocythereis murrayi*
151 *Trachyleberidea blanpiedi*
152 *Trachyleberis montgomeryensis*
153 *Trachyleberis* n.sp.1
154 *Trachyleberis* n.sp.2
155 *Triginglymus* n.sp.1

D.4 Coefficients (Benthic Foraminifera Only)

X	$D_a(x)$	$D_d(x)$	$G_r(x)$	$L_a(x)$	$I_t(x)$
1	2	0	1	4	62
2	2	1	1	4	83
3	1	1	0.8	3	6
4	1	1	0.8	2	98
5	0	0	1	1	64
6	0	0	1	1	134
7	0	0	1	1	73
8	0	0	1	1	10
9	0	1	1	2	17
10	1	0	0.8	2	105
11	0	1	1	2	123
12	1	1	0.8	3	89
13	0	1	1	2	53
14	1	0	1	2	9
15	0	0	0.8	1	101
16	1	1	1	3	23
17	1	0	1	3	59
18	1	1	1	3	54
19	0	1	1	2	48
20	0	0	0.8	1	72
21	0	1	0.6	2	81
22	2	0	1	3	5
23	0	1	1	3	25
24	1	0	1	2	33
25	0	0	0.6	1	80
26	1	3	0.8	4	162
27	1	1	0.8	2	113

X	$D_a(x)$	$D_d(x)$	$G_r(x)$	$L_a(x)$	$I_t(x)$
28	0	0	0.8	4	0
29	2	2	0.8	3	281
30	1	2	1	4	6
31	2	3	1	4	6
32	1	3	1	4	141
33	1	1	0.8	3	52
34	0	0	1	1	73
35	0	1	0.8	2	53
36	2	0	1	3	51
37	1	0	0.8	2	128
38	1	0	1	2	9
39	2	3	1	4	6
40	0	1	0.8	2	72
41	0	0	1	1	85
42	0	1	0.8	2	57
43	2	2	0.6	3	315
44	1	2	0.8	3	35
45	0	0	1	3	39
46	1	2	1	3	103
47	1	1	1	3	83
48	0	0	0.8	1	39
49	0	0	0.6	1	31
50	0	0	1	1	63
51	0	0	1	1	218
52	0	0	1	1	73
53	1	0	1	2	33
54	0	0	0.8	1	149
55	1	1	1	3	6
56	1	1	1	2	53
57	2	2	1	3	31
58	0	0	1	1	71
59	0	0	1	3	71
60	0	1	0.8	2	540
61	0	0	1	1	152
62	1	1	1	2	41
63	0	0	0.8	1	98
64	0	1	1	2	109
65	1	1	1	3	6
66	1	1	1	3	157
67	1	0	1	2	9
68	1	1	1	3	25
69	0	1	1	2	66
70	0	1	1	2	7

X	$D_a(x)$	$D_d(x)$	$G_r(x)$	$L_a(x)$	$I_t(x)$
71	0	0	1	1	55
72	1	1	0.8	2	119
73	0	1	0.6	2	118
74	0	2	0.8	3	11
75	1	0	1	3	21
76	2	2	0.6	3	215
77	0	0	0.6	1	48
78	0	0	0.6	1	120
79	2	0	1	4	0
80	0	1	1	2	44
81	2	1	0.8	4	134
82	1	0	1	4	0
83	0	1	0.6	2	260
84	0	0	1	4	0
85	2	1	0.6	3	8
86	1	2	1	4	0
87	0	0	0.6	1	45
88	0	1	0.6	2	43
89	1	1	1	3	1
90	1	2	1	4	0
91	1	1	1	4	23
92	2	2	1	4	26
93	0	1	0.6	2	316
94	0	0	1	1	218
95	0	0	0.6	1	88
96	1	2	1	3	16
97	1	2	0.8	4	6
98	0	0	0.6	1	11
99	0	1	0.4	2	304
100	0	0	0.4	1	75

D.5 Numerical Database

Section 1

1 12 15 2 11 15 3 4 8 5 7 8 6 4 8 7 3 8 8 4
6 9 4 11 10 4 8 11 11 17 13 4 8 14 1 10 15 3
10 16 1 8 17 4 17 18 15 15 19 1 3 20 4 4 21 14
15 22 5 17 23 1 10 24 14 17 26 2 6 27 16 17 28
2 17 30 3 14 31 1 2 32 2 2 34 3 8 35 3 15 36
16 17 38 3 10 39 1 3 40 4 7 41 3 11 43 16 16 44
4 8 45 4 16 46 3 9 47 3 9 48 4 6 50 4 7 51 1

2 52 3 8 53 15 17 54 1 2 55 3 14 56 4 8 57 14
14 58 3 8 59 3 7 60 1 1 61 1 1 62 12 17 63 16
16 64 1 1 65 1 8 66 2 12 67 6 10 68 1 8 69 11
15 70 7 14 71 4 6 72 5 8 73 14 14 74 4 9 75 10
17 78 3 9 79 11 17 80 4 10 81 14 14 82 1 17 83
1 3 84 1 17 85 1 3 86 4 14 87 4 8 89 11 12 90
1 16 91 4 17 92 1 17 93 1 3 94 1 2 96 1 11 99
2 2 101 4 11 102 10 17 103 1 7 104 2 3 105 1 3
107 4 15 110 1 1 111 4 14 112 4 8 113 4 17 115
1 3 116 1 3 118 1 2 119 1 8 120 1 3 121 15 17
122 1 1 123 1 3 125 1 17 126 1 1 127 1 3 128 4
17 129 12 17 130 4 8 131 1 10 132 1 1 133 15 17
134 1 8 135 1 4 137 15 17 139 4 17 142 4 9 143
1 3 144 1 1 145 12 17 148 15 17 149 4 17 150 4
6 151 12 17 152 4 6 153 9 10 155 1 2

Section 2

1 3 10 2 9 10 3 2 10 4 9 10 5 4 6 6 6 8 7 3
4 8 4 8 9 4 10 10 2 8 11 9 10 12 4 7 13 3 8
14 3 8 15 4 8 16 3 8 17 8 13 18 7 10 19 2 6
20 4 4 21 9 10 22 9 12 23 1 7 24 9 13 25 12 13
27 9 12 28 2 13 29 1 2 30 1 8 31 2 2 32 3 13
33 3 9 34 3 6 35 3 10 36 9 13 37 3 7 38 3 7
39 1 2 40 4 7 41 3 7 42 3 7 43 4 4 44 6 13 45
3 13 46 3 12 47 4 11 48 4 7 49 13 13 50 3 7 51
2 2 52 3 4 53 9 13 54 1 2 55 3 11 56 5 8 57
2 8 58 3 8 59 3 7 60 1 2 61 1 1 62 9 13 63 13
13 64 2 3 65 3 11 66 3 10 67 1 7 68 3 8 69 9
13 70 4 4 71 3 7 72 8 8 74 5 10 75 7 13 76 3
7 77 6 7 79 3 13 80 4 7 81 8 13 82 3 13 84 1
13 85 11 11 86 1 13 88 4 7 89 5 13 90 1 13 91
3 12 92 1 7 93 1 1 94 2 2 95 4 7 96 1 10 97
2 8 98 8 8 99 2 3 101 3 7 102 7 13 103 1 4 104
3 4 109 1 1 111 3 7 114 1 2 115 1 2 116 1 2 118
1 2 119 1 6 120 2 2 121 12 13 122 1 2 123 1 1
125 2 13 126 1 2 127 1 2 128 9 13 131 1 6 134 1
8 135 1 6 136 1 1 137 12 13 138 1 1 139 7 12 140
3 12 141 1 7 142 1 2 143 1 2 144 2 3 146 1 1 147
13 13 148 12 13 149 3 13 151 8 13 152 5 10 154
8 8

Section 3

1 4 11 2 5 12 3 1 6 4 8 11 5 3 5 6 3 5 7 3
4 8 3 5 9 5 6 10 4 5 11 6 12 12 2 8 13 3 5
14 3 5 15 3 5 16 1 8 17 7 13 18 6 10 19 1 2
20 3 5 21 8 11 22 5 13 23 1 6 24 6 13 25 11 13

26 5 12 27 11 13 28 2 13 29 1 10 30 3 12 31 2
13 32 3 11 33 5 13 34 3 4 35 3 7 36 4 13 37 3
5 38 3 5 39 2 13 40 3 5 41 3 5 42 3 10 43 11
3 44 6 13 45 3 13 46 3 13 47 3 5 48 3 5 49 11
13 50 3 4 51 1 2 52 3 3 53 6 13 54 1 2 55 2
5 56 4 7 57 1 2 58 3 3 59 3 4 60 1 2 61 1 1
62 8 13 64 1 2 65 1 7 66 3 7 67 2 5 68 3 7 69
8 12 70 3 8 71 3 5 73 9 12 74 5 13 75 6 13 77
4 5 78 3 5 79 2 13 80 3 5 81 6 11 82 1 13 83
2 84 1 13 86 2 13 87 3 5 88 3 5 89 5 12 90
2 10 91 2 10 92 1 10 93 1 2 94 1 2 95 3 5 96
3 5 97 1 13 98 3 5 100 3 3 101 3 6 102 6 13 103
1 3 104 3 3 105 1 2 106 13 13 107 10 13 108 1 13
109 1 2 113 3 13 114 1 2 115 1 2 116 1 2 117 1
2 118 1 2 119 1 5 120 1 2 121 6 13 123 1 2 125
1 13 126 1 2 127 1 2 128 3 13 131 1 4 134 1 5
135 1 5 136 1 2 137 11 13 139 6 13 142 1 2 143
1 2 144 1 4 145 6 13 147 10 13 148 9 13 149 3 13
150 4 4 151 6 13 152 6 8

Section 4

1 1 9 2 6 8 3 1 6 4 6 7 5 2 2 6 1 1 7 2 4
8 3 4 9 3 4 10 4 4 11 8 9 12 2 6 13 2 4 14
3 4 15 2 4 16 2 3 17 5 9 18 5 7 19 1 1 20 2
4 22 5 9 23 1 2 24 5 9 26 4 8 27 5 6 28 1 9
29 5 9 30 2 4 31 5 9 32 3 7 33 5 8 34 2 4 35
2 8 36 2 9 37 1 4 38 1 4 39 5 8 40 2 6 41 2
4 42 2 4 44 2 9 45 2 9 46 6 6 47 1 4 48 2 4
50 2 3 51 1 1 52 2 4 53 5 9 55 1 7 56 5 5 57
1 7 58 2 4 59 2 4 60 1 2 61 1 1 62 5 8 64 1
1 65 3 5 66 3 4 67 3 4 68 3 4 69 6 9 70 4 8
71 4 4 72 5 7 73 5 5 74 3 7 75 4 9 76 2 3 77
4 4 78 2 4 79 2 9 80 2 4 81 1 9 82 1 9 84 1
9 85 2 8 86 2 9 87 4 4 88 3 6 89 4 9 90 2 9
91 2 8 92 2 9 94 1 1 95 2 4 96 1 1 97 2 4 98
3 4 101 3 6 102 3 9 103 1 1 105 3 4 107 5 9 108
1 9 109 1 1 110 1 1 111 3 4 113 3 9 114 1 1 115
1 1 116 1 1 117 1 5 119 1 5 120 1 1 123 1 1 125
1 9 126 1 1 128 6 9 129 7 8 131 1 5 134 3 4 135
1 4 136 1 1 137 8 9 138 1 1 139 5 9 143 1 1 145
5 5 148 5 8 149 3 9 151 5 9 152 3 6

Section 5

1 4 8 2 8 10 4 10 10 5 5 6 6 5 6 7 5 6 8 5
5 9 5 6 11 8 10 12 6 7 13 5 7 14 5 6 16 3 5
17 7 10 18 7 10 19 1 4 22 9 10 23 1 6 24 7 10

25 9 10 26 1 4 29 1 7 30 3 6 31 4 7 32 5 8 33
5 8 34 5 6 36 7 10 37 5 6 38 3 6 39 3 4 41 5
5 42 5 8 45 7 10 46 5 6 47 5 5 49 10 10 50 5
6 51 1 4 52 5 6 53 7 10 54 2 4 55 5 6 57 1 4
58 5 6 59 5 6 61 1 4 62 9 10 63 9 10 64 3 4
65 5 6 66 5 7 67 1 6 68 4 6 69 7 10 70 5 5 71
5 6 72 5 5 75 7 10 76 9 9 79 4 10 80 5 7 82
2 10 83 4 4 84 1 10 86 5 8 89 5 10 90 5 6 91
5 7 92 8 8 94 3 4 96 4 5 97 5 5 100 5 6 101
5 6 102 5 9 103 1 5 104 3 6 107 8 10 110 1 2 111
3 3 113 3 10 114 1 7 115 1 4 116 1 4 118 1 4 119
1 6 120 1 3 122 1 5 123 3 3 125 3 10 127 1 4 128
5 8 129 5 8 131 1 2 134 1 6 135 1 6 136 2 4 137
9 10 138 1 1 139 5 10 142 1 6 143 1 4 144 2 6
145 9 10 146 1 6 148 8 10 149 6 10 150 6 6 151
7 10 152 2 2 155 1 2

Appendix E
Sullivan's (1964, 1965) Database

E.1 Code Numbers of Species

1 *Chiphragmalithus cristatus*
2 *Chiphragmalithus acanthodes*
3 *Chiphragmalithus calatus*
4 *Chiphragmalithus dubius*
5 *Chiphragmalithus protenus*
6 *Chiphragmalithus quadratus*
7 *Coccolithus bidens*
8 *Coccolithus californicus*
9 *Coccolithus expansus*
10 *Coccolithus grandis*
11 *Coccolithus solitus*
12 *Coccolithus staurion*
13 *Coccolithus gigas*
14 *Coccolithus delus*
15 *Coccolithus consuetus*
16 *Coccolithus crassus*
17 *Coccolithus cribellum*
18 *Coccolithus eminens*
19 *Cyclococcolithus gammation*
20 *Cyclococcolithus luminis*
21 *Discolithus pectinatus*
22 *Discolithus planus*
23 *Discolithus pulcher*
24 *Discolithus pulcheroides*
25 *Discolithus rimosus*
26 *Discolithus distinctus*
27 *Discolithus fimbriatus*
28 *Discolithus ocellatus*
29 *Discolithus panarium*
30 *Discolithus punctosus*
31 *Discolithus solidus*
32 *Discolithus vescus*
33 *Discolithus versus*
34 *Discolithus pertusus*
35 *Discolithus exilis*
36 *Discolithus duocavus*
37 *Discolithus inconspicuus*
38 *Cyclolithus robustus*
39 *Ellipsolithus macellus*
40 *Ellipsolithus distichus*
41 *Helicosphaera semilunum*
42 *Helicosphaera lophota*
43 *Lophodolithus nascens*
44 *Lophodolithus reniformis*
45 *Lophodolithus mocholophorus*
46 *Rhabdosphaera crebra*
47 *Rhabdosphaera morionum*
48 *Rhabdosphaera perlonga*
49 *Rhabdosphaera rudis*
50 *Rhabdosphaera scabrosa*
51 *Rhabdosphaera semiformis*
52 *Rhabdosphaera tenuis*
53 *Rhabdosphaera truncata*
54 *Rhabdosphaera inflata*
55 *Zygodiscus sigmoides*
56 *Zygodiscus adamas*
57 *Zygodiscus herlyni*
58 *Zygodiscus plectopons*
59 *Zygolithus concinnus*
60 *Zygolithus crux*
61 *Zygolithus distentus*
62 *Zygolithus junctus*
63 *Zygrhablithus simplex*
64 *Zygrhablithus bijugatus*
65 *Braarudosphaera bigelowi*
66 *Braarudosphaera discula*
67 *Micrantholithus flos*
68 *Micrantholithus inaequalis*
69 *Micrantholithus vesper*
70 *Micrantholithus basquensis*
71 *Micrantholithus crenulatus*
72 *Micrantholithus aequalis*

73 *Clathrolithus ellipticus*
74 *Rhomboaster cuspis*
75 *Polycladolithus operosus*
76 *Sphenolithus radians*
77 *Fasciculithus involutus*
78 *Discoaster barbadiensis*
79 *Discoaster binodosus*
80 *Discoaster deflandrei*
81 *Discoaster delicatus*
82 *Discoaster diastypus*
83 *Discoaster distinctus*
84 *Discoaster falcatus*
85 *Discoaster lodoensis*
86 *Discoaster multiradiatus*
87 *Discoaster nonaradiatus*
88 *Discoaster stradneri*
89 *Discoaster tribrachiatus*
90 *Discoaster cruciformis*
91 *Discoaster germanicus*
92 *Discoaster lenticularis*
93 *Discoaster martinii*
94 *Discoaster minimus*
95 *Discoaster septemradiatus*
96 *Discoaster sublodoensis*
97 *Discoaster helianthus*
98 *Discoaster limbatus*
99 *Discoaster mediosus*
100 *Discoaster perpolitus*
101 *Discoasteroides kuepperi*
102 *Discoasteroides megastypus*
103 *Heliolithus kleinpelli*
104 *Heliolithus riedeli*
105 *F.involutus* (acme of *77*)

E.2 Code Numbers of Sections and Maximal Horizons

Section 1: Pacheco

M. H.1 = A-6658
M. H.2 = A-6666
M. H.3 = A-6665
M. H.4 = A-6667
M. H.5 = A-6668
M. H.6 = A-6671
M. H.7 = A-6685
M. H.8 = A-6688
M. H.9 = A-6689

Section 2: Vaca Valley

M. H.1 = A-4781-A
M. H.2 = A-4781-B
M. H.3 = A-4781-J
M. H.4 = A-4781-O
M. H.5 = A-4782-L
M. H.6 = A-4782-N
M. H.7 = A-4782-S
M. H.8 = A-4783-A
M. H.9 = A-4783-E

Section 3 : New Idria

M. H.1 = B-4861-A
M. H.2 = B-4868
M. H.3 = B-4869
M. H.4 = B-4872
M. H.5 = B-4877
M. H.6 = B-4884
M. H.7 = B-4894
M. H.8 = B-4897
M. H.9 = B-4898

Section 4: Upper Canada De Santa Anita

M. H.1 = A-8932
M. H.2 = A-8935
M. H.3 = A-8939
M. H.4 = A-8945
M. H.5 = A-8946
M. H.6 = A-8947
M. H.7 = A-8948
M. H.8 = A-8950
M. H.9 = A-8951
M. H.10 = A-8952
M. H.11 = A-8953

Section 5: Las Cruces

M. H.1 = A-8855
M. H.2 = D-1017
M. H.3 = A-8857
M. H.4 = A-8862

Section 6: Simi Valley

M. H.1 = A-7617
M. H.2 = A-7628
M. H.3 = A-7629
M. H.4 = A-7634
M. H.5 = A-7636
M. H.6 = A-7639
M. H.7 = A-7641
M. H.8 = A-7642
M. H.9 = A-7643
M. H.10 = A-7644
M. H.11 = A-7648
M. H.12 = A-7652

Section 7: Tres Pinos

M. H.1 = B-7239
M. H.2 = B-7202
M. H.3 = B-7204
M. H.4 = B-7215
M. H.5 = B-7217
M. H.6 = B-7226
M. H.7 = B-7231
M. H.8 = B-7242
M. H.9 = B-7246
M. H.10 = B-7235

Section 8: Media Aqua Creek

M. H.1 = A-7079
M. H.2 = A-7077
M. H.3 = A-7071
M. H.4 = A-7070
M. H.5 = A-7067
M. H.6 = A-7063
M. H.7 = A-7089
M. H.8 = A-7088
M. H.9 = A-7057
M. H.10 = A-7052
M. H.11 = A-7051
M. H.12 = A-7050
M. H.13 = A-7049
M. H.14 = A-7048
M. H.15 = A-7046
M. H.16 = A-7009
M. H.17 = A-7020
M. H.18 = A-7025

Section 9: Upper Reliz Creek

M. H.1 = B-2769
M. H.2 = B-2771
M. H.3 = B-2772
M. H.4 = B-2773
M. H.5 = B-2775
M. H.6 = B-2779
M. H.7 = B-2780
M. H.8 = B-2782
M. H.9 = B-2784
M. H.10 = B-2786
M. H.11 = B-2788

Section 10: Lodo (Synthetic Section, Based on Bramlette and Sullivan 1961)

M. H.1 = 6 + 1
M. H.2 = 7
M. H.3 = 32
M. H.4 = 36
M. H.5 = 70
M. H.6 = OC-2
M. H.7 = OC-3
M. H.8 = OC-4
M. H.9 = GC-1
M. H.10 = GC-2
M. H.11 = GC-3

E.3 Coefficients

X	$D_a(x)$	$D_d(x)$	$G_r(x)$	$L(x)$	$I_t(x)$
1	1	0	0.9	2	179
2	0	0	0.7	1	123
3	0	1	0.9	2	123
4	2	3	1	4	14
5	1	1	0.8	2	191
6	0	0	1	1	103
7	0	0	0.8	1	116
8	0	0	0.5	1	108
9	0	1	0.6	2	87
10	0	2	1	3	74
11	1	1	0.7	3	81
12	1	1	0.6	2	128
13	0	0	0.5	1	112
14	1	2	1	4	0
15	1	2	1	4	51
16	0	2	1	3	95
17	0	1	0.8	2	139
18	0	0	1	1	298
19	1	2	1	3	99
20	0	1	0.3	2	131
21	0	2	1	3	82
22	1	3	1	4	0
23	1	2	0.9	4	62
24	1	1	0.9	4	1008
25	1	2	1	3	7
26	0	1	0.9	2	151
27	0	1	0.5	2	104
28	1	1	0.8	2	103
29	1	1	0.8	3	140
30	0	0	0.3	1	136
31	2	1	0.4	3	82
32	0	0	0.1	1	459
33	1	1	0.7	2	92
34	0	0	0.1	2	490
35	0	0	0.3	1	29
36	0	0	0.3	1	142
37	1	1	0.4	3	909
38	0	0	0.7	1	148
39	0	2	0.9	3	27
40	0	1	0.6	2	104
41	0	0	0.9	2	87

X	$D_a(x)$	$D_d(x)$	$G_r(x)$	$L(x)$	$I_t(x)$
42	1	0	0.9	3	80
43	1	1	1	3	10
44	0	1	0.9	2	89
45	1	1	0.8	3	143
46	1	1	0.7	2	87
47	1	1	0.6	2	87
48	0	1	0.7	2	93
49	1	1	0.5	2	139
50	0	0	0.4	2	345
51	1	1	0.8	2	103
52	1	1	0.8	3	139
53	1	1	0.6	2	129
54	0	1	0.3	2	161
55	0	1	0.6	2	104
56	0	1	0.3	2	148
57	0	0	0.2	1	361
58	0	1	0.2	2	140
59	2	1	0.7	3	1054
60	0	0	0.2	2	632
61	0	1	0.4	2	375
62	0	0	0.3	1	124
63	0	1	0.4	2	102
64	1	3	1	4	6
65	2	1	1	4	6
66	2	2	0.7	3	308
67	2	2	0.7	3	24
68	1	0	0.3	2	205
69	1	1	0.7	3	78
70	1	0	0.5	2	103
71	2	0	0.6	3	106
72	1	1	0.3	2	743
73	2	1	0.7	3	20
74	0	0	0.2	2	298
75	0	1	0.3	2	749
76	0	1	0.9	3	84
77	0	1	0.7	2	104
78	1	2	1	4	16
79	0	1	0.9	2	92
80	1	2	0.9	3	307
81	0	1	0.5	2	152
82	0	0	0.9	1	144
83	1	2	0.9	3	84
84	0	1	0.5	2	152

X	$D_a(x)$	$D_d(x)$	$G_r(x)$	$L(x)$	$I_t(x)$
85	0	1	0.9	3	84
86	1	0	0.5	2	152
87	1	1	0.8	2	124
88	1	1	0.8	3	116
89	0	1	0.9	2	96
90	1	0	0.6	2	103
91	0	1	0.6	2	94
92	1	2	0.4	3	146
93	1	1	0.3	2	318
94	0	1	0.7	2	224
95	0	2	0.4	3	75
96	1	0	0.4	2	140
97	0	1	0.3	2	652
98	0	0	0.3	1	410
99	0	1	0.2	2	345
100	0	0	0.2	1	620
101	0	1	1	3	83
102	0	1	0.5	2	582
103	0	0	0.3	1	40
104	0	1	0.7	2	580
105	0	0	1	1	681

E.4 Numerical Database

Section 1

1 7 9, 2 4 7, 3 2 5, 4 2 9, 6 8 9, 9 5 9, 10 2 9, 11 2 9, 12 2 4, 13 9 9, 14 2 9, 15 1 9, 16 2 9, 17 2 2, 19 2 8, 20 2 2, 21 2 9, 22 2 9, 23 2 7, 24 2 7, 25 2 7, 26 2 9, 27 2 8, 28 2 6, 29 2 9, 33 2 7, 35 6 6, 39 1 2, 40 1 1, 41 2 7, 42 2 9, 43 2 6, 44 2 3, 45 2 7, 46 2 7, 47 2 7, 48 2 6, 50 2 4, 51 2 7, 52 2 9, 55 1 1, 59 2 2, 63 1 1, 64 2 7, 65 1 7, 67 2 5, 69 2 9, 71 2 6, 72 1 1, 73 2 4, 76 2 7, 77 1 1, 78 2 9, 79 2 5, 80 2 9, 83 2 9, 85 2 9, 87 2 9, 88 2 9, 89 2 3, 91 2 7, 94 2 5, 95 2 9, 101 2 8, 102 1 1, 104 1 1, 105 1 1

Section 2

1 8 9, 3 1 2, 4 1 9, 6 8 9, 9 6 6, 10 1 9, 11 1 9, 12 9 9, 14 1 8, 15 1 6, 16 1 9, 17 1 7, 19 1 7, 20 1 8, 21 1 9, 22 1 7, 23 1 9, 24 1 9, 25 1 6, 26 4 4, 27 4 9, 28 2 6, 29 1 8, 31 3 7, 32 3 3, 33 1 7, 39 1 1, 41 1 7, 42 1 9, 43 1 7, 44 1 7, 45 1 8, 46 1 7, 47 3 6, 48 1 7, 50 1 7, 51 8 9, 52 1 9, 53 3 7, 60 1 5, 64 1 9, 65 2 9, 66 5 6, 67 2 7, 68 3 7, 69 1 9, 70 3 6, 71 3 7, 73 1 6, 75 3 6, 76 1 9, 78 1 8, 79 3 6, 80 4 6, 83 1 9, 85 1 9, 87 8 8, 88 1 9, 89 1 7, 90 3 7, 91 1 7, 92 1 1, 93 7 7, 94 3 4, 95 1 9, 96 3 7, 101 1 8

Section 3

3 2 9, 4 2 9, 5 2 9, 7 1 1, 10 2 9, 11 7 9, 14 2 9, 15 1 9, 16 2 9, 17 2 8, 18 1 1, 19 2 9, 21 2 9, 22 2 9, 23 2 9, 24 2 9, 25 2 9, 28 4 9, 30 2 9, 33 2 9, 36 2 7, 37 2 5, 39 1 9, 40 1 1, 41 2 9, 42 2 9, 43 2 9, 44 2 9, 46 2 9, 47 2 6, 48 2 9, 49 2 9, 50 8 9, 53 2 9, 55 1 1, 59 1 9, 64 2 9, 65 1 9, 66 2 3, 67 3 9, 68 2 9, 69 2 9, 71 9 9, 73 3 9, 76 2 9, 77 1 1, 78 2 9, 79 2 9, 80 2 7, 82 2 6, 83 3 9, 85 2 9, 86 2 2, 88 2 9, 89 2 9, 90 6 9, 91 5 6, 92 2 3, 94 5 5, 95 4 4, 101 2 9, 104 1 1, 105 1 1

Section 4

1 7 11, 2 7 11, 3 4 6, 4 4 11, 5 4 4, 7 1 2, 8 1 3, 9 11 11, 10 5 11, 11 4 11, 12 7 11, 14 4 11, 15 1 11, 16 4 11, 17 5 6, 18 1 3, 19 5 11, 21 4 11, 22 4 11, 23 4 11, 24 4 11, 25 1 10, 26 6 11, 27 8 9, 28 9 10, 29 6 10, 33 7 8, 37 2 8, 38 1 3, 39 1 11, 40 1 2, 41 5 8, 42 5 11, 43 1 11, 44 5 7, 45 7 7, 46 9 9, 47 8 8, 48 4 8, 49 5 5, 51 7 10, 52 8 9, 53 7 8, 54 7 9, 55 1 2, 56 1 2, 57 1 2, 59 4 8, 61 1 2, 62 1 2, 63 1 2, 64 4 11, 65 2 9, 66 9 9, 69 6 9, 70 9 9, 71 8 9, 73 6 6, 76 4 11, 77 1 3, 78 4 11, 79 4 10, 80 4 11, 82 4 5, 83 5 10, 84 1 3, 85 4 11, 86 3 5, 87 10 10, 88 4 10, 89 4 11, 90 6 6, 91 4 9, 94 4 8, 96 8 11, 97 1 2, 101 4 9, 102 1 1, 104 1 3, 105 1 3

Section 5

1 4 4, 2 2 3, 3 1 3, 4 2 4, 6 4 4, 10 1 4, 11 3 3, 14 1 3, 15 1 2, 16 1 3, 19 2 4, 22 1 3, 23 1 3, 24 1 3, 25 1 3, 26 2 3, 41 1 3, 42 1 4, 43 1 1, 44 1 1, 45 2 3, 47 3 3, 51 3 3, 52 2 4, 59 2 3, 64 1 3, 65 2 3, 76 1 4, 78 1 3, 79 1 2, 80 1 4, 82 1 1, 83 2 3, 85 1 3, 87 3 3, 88 2 3, 89 1 2, 94 1 1, 96 2 3, 101 1 3

Section 6

4 6 6, 5 4 7, 7 1 6, 8 1 6, 10 11 11, 14 4 10, 15 1 12, 16 8 11, 18 2 7, 19 8 8, 22 4 7, 24 7 12, 25 1 5, 28 9 9, 31 9 9, 34 2 11, 37 2 10, 38 2 4, 39 1 5, 40 1 7, 43 4 11, 55 1 7, 56 2 7, 57 2 7, 58 2 7, 59 1 12, 61 1 5, 62 1 5, 63 1 7, 64 2 7, 65 2 12, 67 3 7, 72 10 10, 73 2 9, 77 1 7, 78 6 10, 81 2 7, 82 10 11, 84 2 7, 92 4 6, 98 2 7, 99 5 6, 100 4 7, 101 12 12, 102 2 2, 103 1 1, 104 2 2, 105 1 7

Section 7

1 10 10, 3 4 8, 4 4 8, 5 4 7, 6 9 10, 10 3 10, 14 4 7, 15 1 8, 16 4 10, 17 4 4, 18 2 2, 19 7 10, 21 5 7, 22 6 6, 23 4 5, 24 4 8, 25 4 6, 39 1 7, 41 4 8, 42 8 10, 43 4 8, 44 4 5, 55 1 1, 59 4 7, 64 4 6, 65 5 8, 67 5 5, 76 3 10, 77 1 2, 78 3 9, 79 4 8, 80 4 9, 81 2 3, 82 3 7, 83 3 8, 84 2 3, 85 4 8, 86 2 3, 88 4 8, 89 4 8, 101 4 8, 103 1 1, 104 1 2, 105 1 2

Section 8

1 16 18, 2 10 15, 3 2 7, 4 2 16, 5 2 2 , 6 16 18, 7 1 1, 8 1 1, 9 14 15, 10 2 17, 11 7 17, 13 17 18, 14 2 15, 15 2 15, 16 7 18, 17 2 14, 18 1 1, 19 2 15, 21 3 16, 22 3 15, 23 2 15, 24 2 15, 25 2 15, 26 10 15, 28 3 5, 29 9 14, 31 15 15, 33 3 15, 36 2 6, 37 7 9, 38 1 1, 39 1 7, 41 3 15, 42 7 18, 43 2 12, 44 3 7, 45 10 15, 46 4 12,

48 6 11, 49 3 6, 51 10 15, 52 11 15, 53 3 3, 59 7 10, 61 1 1, 64 5 15, 65 3 14, 66 3 8, 67 3 13, 68 7 13, 69 3 15, 70 12 12, 71 12 15, 72 3 5, 73 3 6, 75 5 6, 76 2 18, 77 1 1, 78 2 16, 79 2 14, 80 2 15, 81 1 1, 82 2 5, 83 5 15, 84 1 1, 85 3 15, 86 3 6, 87 14 14, 88 9 15, 89 2 11, 90 9 15, 91 7 9, 93 18 18, 94 2 9, 96 10 15, 97 1 1, 101 2 15, 102 1 1, 104 1 1, 105 1 1

Section 9

3 1 7, 4 2 11, 5 2 4, 10 1 11, 12 11 11, 14 1 11, 15 1 11, 16 1 11, 17 2 11, 19 5 11, 21 2 11, 22 4 11, 23 1 11, 24 1 9, 25 2 11, 26 9 11, 28 1 1, 29 10 11, 30 1 2, 33 6 6, 39 1 5, 41 2 11, 42 8 11, 43 1 10, 44 1 8, 46 6 6, 48 2 8, 49 6 8, 53 1 3, 64 1 11, 65 2 9, 66 7 7, 69 11 11, 76 1 11, 78 1 11, 79 1 11, 80 4 9, 82 1 6, 83 2 2, 85 1 11, 88 3 11, 89 1 8, 90 8 11, 94 3 3, 101 1 11

Section 10

1 6 11, 2 6 8, 3 4 5, 4 4 10, 5 3 3, 6 9 11, 7 1 2, 9 6 10, 10 4 11, 11 5 11, 12 10 10, 13 11 11, 14 3 9, 15 1 5, 16 5 6, 17 4 5, 18 1 1, 19 5 6, 21 4 10, 22 4 9, 23 3 8, 25 2 8, 26 6 8, 27 7 8, 28 5 8, 29 6 9, 30 4 5, 31 3 6, 33 4 6, 35 7 8, 36 4 5, 38 2 2, 39 1 5, 40 1 5, 41 5 8, 42 5 9, 43 3 6, 44 5 5, 45 6 8, 46 5 6, 47 5 6, 48 4 6, 49 5 5, 50 5 6, 51 6 9, 52 6 10, 53 4 5, 54 7 9, 55 1 5, 56 3 5, 58 3 5, 60 5 7, 61 3 5, 62 1 1, 63 1 5, 64 4 9, 65 3 6, 66 3 6, 67 5 6, 69 5 6, 71 3 6, 73 3 6, 74 3 5, 75 5 6, 76 4 9, 77 1 5, 78 4 11, 79 4 5, 80 4 5, 81 2 5, 82 4 5, 83 4 10, 84 2 5, 85 4 11, 86 3 5, 87 6 10, 89 4 5, 90 5 6, 91 4 6, 92 3 5, 93 11 11, 95 5 10, 97 1 5, 98 3 3, 99 3 4, 101 4 6, 102 2 5, 104 1 5, 105 1 3

Appendix F
Müller's (1978) Database

F.1 Code Numbers of Species

1 *Braarudosphaera bigelowi*
2 *Ceratolithus cristatus*
3 *Ceratolithus rugosus*
4 *Ceratolithus tricorniculatus*
5 *Coccolithus abisectus*
6 *Coccolithus pelagicus*
7 *Coronocyclus nitescens*
8 *Coronocyclus cf. nitescens*
9 *Cyclococcolithus floridanus*
10 *Cyclococcolithus leptoporus*
11 *Cyclococcolithus macintyrei*
12 *Cyclococcolithus rotula*
13 *Discoaster assymetricus*
14 *Discoaster aulakos*
15 *Discoaster cf. bollii*
16 *Discoaster brouweri*
17 *Discoaster calcaris*
18 *Discoaster cf. calcaris*
19 *Discoaster challengeri*
20 *Discoaster deflandrei*
21 *Discoaster druggii*
22 *Discoaster exilis*
23 *Discoaster icarus*
24 *Discoaster cf. hamatus*
25 *Discoaster kugleri*
26 *Discoaster pentaradiatus*
27 *Discoaster perplexus*
28 *Discoaster quinqueramus*
29 *Discoaster cf. quinqueramus*
30 *Discoaster subsurculus*
31 *Discoaster surculus*
32 *Discoaster tamalis*
33 *Discoaster triradiatus*
34 *Discoaster variabilis*
35 *Discolithina japonica*
36 *Discolithina multipora*
37 *Discosphaera tubifera*
38 *Emiliana huxleyi*
39 *Gephyrocapsa oceanica*
40 *Gephyrocapsa* sp. (small)
41 *Helicosphaera ampliaperta*
42 *Helicosphaera carteri*
43 *Helicosphaera perch-nielseniae*
44 *Helicosphaera sellii*
45 *Helicosphaera walbersdorfensis*
46 *Holodiscolithus macroporus*
47 *Lithostromation perdurum*
48 *Oolithotus fragilis*
49 *Pontosphaera syracusana*
50 *Pontosphaera* sp.
51 *Pseudoemiliania lacunosa*
52 *Reticulofenestra pseudoumbilica*
53 *Rhabdosphaera clavigera*
54 *Rhabdosphaera stylifer*
55 *Scapholithus fossilis*
56 *Scyphosphaera apsteini*
57 *Scyphosphaera conica*
58 *Scyphosphaera globulata*
59 *Scyphosphaera intermedia*
60 *Scyphosphaera procera*
61 *Scyphosphaera pulcherrima*
62 *Scyphosphaera recurvata*
63 *Sphenolithus abies*
64 *Sphenolithus belemnos*
65 *Sphenolithus heteromorphus*
66 *Syracosphaera pulchra*
67 *Thoracosphaera heimi*
68 *Thoracosphaera saxea*
69 *Triquetrorhabdulus rugosus*
70 *Umbellosphaera tenuis*
71 *Umbilicosphaera mirabilis*
72 *Umbilicosphaera jafarii*

F.2 Code Numbers of Sections

1 = DSDP site 371
2 = DSDP site 372
3 = DSDP site 374
4 = DSDP site 375
5 = DSDP site 376
6 = DSDP site 378
7 = DSDP site 378A

F.3 Code Numbers of Samples

Only the first and last levels of the schematic sections given in Fig. 12.3 are related to Müller's original numbers. The intermediate levels are numbered in a continuous manner in our compilation (see Müller, 1978, Tables 3–9).

Section 1 (see Müller 1978, Table 3):

Level 1 = Sample 8–2, Interval 91–92 cm
Level 31 = Sample 1–1,top

Section 2 (see Müller 1978, Table 4):

Level 1 = Sample 46–3, Interval 46–47 cm
Level 91 = Sample 1–2, Interval 109–110 cm

Section 3 (see Müller 1978, Table 5):

Level 1 = Sample 11–1, Interval 86–87 cm
Level 38 = Sample 1–1, Interval 40–41 cm

Section 4 (see Müller 1978, Table 6):

Level 1 = Sample 13–1, Interval 135–136 cm
Level 42 = Sample 1 CC (red)

Section 5 (see Müller 1978, Table 7):

Level 1 = Sample 17–1, Interval 121–122 cm
Level 35 = Sample 1–1, Interval 51–52 cm
NB: Samples 6–1 to 63 (Interval 95 cm) ignored

Section 6 (see Müller 1978, Table 8):

Level 1 = Sample 11 CC
Level 45 = Sample 1–1, Interval 130–131 cm

Section 7 (see Müller 1978, Table 9):

Level 1 = Sample 3–6, Interval 125–126 cm
Level 15 = Sample 1–1, Interval 110–111 cm

F.4 Coefficients

X	D_a	D_d	L	X	D_a	D_d	L
1	5	4	6	37	1	1	2
2	2	1	3	38	0	0	1
3	1	3	4	39	0	0	2
4	0	1	2	40	4	4	5
5	0	1	2	41	0	0	1
6	2	4	7	42	2	0	7
7	0	0	1	43	0	0	1
8	0	0	2	44	0	0	1
9	0	0	1	45	1	1	3
10	2	0	6	46	1	2	6
11	2	1	6	47	1	4	6
12	1	1	4	48	3	1	4
13	0	2	3	49	3	4	5
14	1	1	3	50	1	3	5
15	0	1	2	51	1	0	3
16	1	0	4	52	2	2	5
17	0	0	2	53	2	0	4
18	0	1	2	54	5	1	7
19	1	0	2	55	2	1	6
20	0	1	2	56	2	1	6
21	0	0	1	57	1	1	3
22	0	1	2	58	0	2	4
23	1	0	2	59	2	3	7
24	–	–	–	60	1	0	2
25	0	0	1	61	1	1	5
26	1	2	4	62	2	3	6
27	2	2	3	63	2	1	4
28	1	1	2	64	0	0	1
29	0	0	1	65	0	1	2
30	1	0	2	66	1	0	6
31	0	1	3	67	1	1	5
32	1	2	4	68	1	0	2
33	2	2	3	69	0	0	2
34	2	0	3	70	1	0	2
35	1	0	6	71	1	0	3
36	1	1	4	72	1	1	3

F.5 Numerical Database

Section 1

1 7 31, 2 30 30, 3 5 8, 4 2 2, 6 1 31, 10 15 31, 11 1 22, 13 5 16, 16 2 22, 19 2 2, 26 2 16, 31 6 16, 32 3 16, 34 3 3, 35 2 31, 37 31 31, 38 26 31, 39 23 31, 40 26 30, 42 1 31, 46 4 25, 47 6 20, 49 24 30, 50 2 10, 51 15 25, 52 1 14, 53 15 31, 54 25 30, 55 5 30, 56 8 31, 57 3 11, 59 3 29, 60 3 16, 61 5 24, 62 10 20, 63 1 7, 66 5 31, 67 2 31, 70 27 31, 71 25 31

Section 2

1 38 86, 3 84 84, 4 71 83, 5 1 41, 6 1 91, 7 1 32, 8 29 61, 9 5 27, 10 44 91, 11 22 89, 12 24 72, 13 85 85, 14 32 66, 15 68 68, 16 46 87, 18 47 55, 19 68 84, 20 1 45, 21 9 12, 22 22 69, 23 77 77, 25 41 70, 26 80 87, 28 80 80, 29 79 80, 30 32 65, 31 77 87, 32 80 81, 34 44 85, 35 85 91, 36 1 85, 39 90 91, 40 90 90, 41 9 21, 42 1 91, 43 5 21, 44 87 89, 45 27 69, 46 57 89, 47 33 77, 49 84 90, 50 45 80, 51 86 91, 52 6 85, 53 87 91, 54 28 91, 55 86 90, 56 86 90, 59 40 86, 61 39 89, 62 45 90, 63 8 85, 65 15 53, 66 37 91, 67 86 86, 68 23 59, 69 49 82, 71 90 91, 72 30 62

Section 3

1 31 38, 2 31 36, 3 6 29, 4 2 10, 6 1 38, 10 2 36, 11 2 28, 12 1 10, 13 11 14, 16 1 28, 19 1 9, 26 1 28, 31 1 28, 32 12 14, 33 27 28, 34 1 10, 35 13 38, 37 36 38, 38 37 38, 39 29 38, 40 33 36, 42 2 38, 44 14 25, 46 13 38, 47 2 23, 48 37 38, 49 19 38, 50 4 29, 51 14 30, 52 1 15, 53 24 38, 54 15 38, 55 13 38, 56 22 35, 57 2 24, 58 6 24, 59 5 17, 61 4 28, 62 4 18, 63 1 10, 66 11 38, 67 5 38, 70 35 38, 71 35 38

Section 4

1 17 21, 3 42 42, 4 41 42, 5 1 5, 6 1 41, 7 7 7, 10 9 42, 11 6 42, 12 9 42, 14 9 37, 15 15 26, 16 15 42, 17 16 36, 18 20 36, 19 17 41, 20 1 1, 22 6 17, 23 11 37, 25 8 13, 26 16 42, 27 6 6, 28 30 30, 29 24 37, 31 30 42, 32 15 35, 34 6 37, 35 10 42, 36 7 42, 41 1 1, 42 1 42, 45 16 34, 46 11 42, 47 11 37, 48 42 42, 49 21 21, 50 19 42, 52 2 42, 53 42 42, 54 8 42, 55 9 42, 56 16 42, 57 42 42, 58 24 24, 59 7 42, 60 42 42, 61 13 42, 62 6 41, 63 6 42, 64 1 1, 65 2 5, 66 10 42, 68 10 14, 69 8 37, 72 11 37

Section 5

1 21 35, 2 24 35, 3 11 20, 4 5 14, 6 4 31, 10 2 35, 11 2 20, 12 2 16, 13 12 18, 16 4 20, 19 2 13, 26 5 20, 28 4 4, 30 4 4, 31 5 19, 32 7 18, 33 17 20, 34 3 13, 35 11 35, 36 3 15, 37 35 35, 38 32 34, 39 21 34, 40 22 35, 42 3 35, 46 14 35, 47 13 19, 48 27 34, 49 17 35, 50 7 20, 51 17 25, 52 2 16, 53 21 35, 54 7 35, 55 7 35, 56 5 35, 57 5 18, 58 5 19, 59 5 18, 60 9 18, 61 5 20, 62 5 26, 63 1 14, 66 5 35, 67 5 35, 70 23 35, 71 19 35

Section 6

1 13 45, 2 31 35, 3 4 42, 4 1 11, 6 1 44, 10 2 45, 11 1 34, 12 1 12, 13 9 12, 16 1 31, 19 6 8, 26 1 23, 27 2 7, 31 1 25, 32 9 21, 33 8 30, 34 2 5, 35 1 45, 36 1 12, 37 33 33, 39 34 45, 40 30 33, 42 2 45, 44 16 16, 46 2 45, 47 7 38, 48 19 44, 49 16 45, 50 4 38, 51 10 45, 52 1 12, 53 14 45, 54 5 40, 55 1 45, 56 6 45, 57 1 7, 58 7 42, 59 1 17, 60 2 12, 61 1 42, 63 1 12, 66 1 45, 67 7 45, 70 45 45, 71 24 45

Section 7

1 6 15, 2 11 11, 3 1 8, 4 1 10, 6 2 13, 10 6 15, 11 1 10, 12 1 9, 13 8 10, 16 2 10, 19 5 9, 26 1 9, 27 4 4, 31 1 10, 32 8 10, 33 6 9, 34 1 6, 35 1 15, 36 1 10, 37 14 14, 39 11 15, 40 8 8, 42 2 15, 46 6 15, 47 4 8, 48 14 14, 49 8 14, 51 10 15, 52 2 10, 53 15 15, 54 2 15, 55 2 15, 57 2 10, 58 6 6, 59 1 10, 61 1 10, 62 1 10, 63 3 10, 66 8 15, 67 1 14, 71 11 15

Appendix G
Baumgartner's (1984a) Database

G.1 Code Numbers of Sections

1 Dhimaina, Argolis
2 Angelokastron, Argolis
3 Prosimi, Argolis
4 Taxiarchis, Argolis
5 Kandhia, Argolis
6 Serrada, Trento High
7 Koliaki, Argolis
8 Theokafta, Argolis
10 Pinde, Central Greece
11 Marathos, Central Greece
12 Pojorita, Rumania
14 Piatra Soimului, Rumania
15 Gomielor, Rumania
16 Sivinita, Banat, Rumania
17 Besozzo II, Southern Alps
18 Mte Generoso, Southern Alps
19 Torre de Busi, Southern Alps
20 Valmaggiorre, Southern Alps
21 Besozzo I, Southern Alps
22 Sangiano, Southern Alps
23 Cava Rusconi, Southern Alps
24 Breggia, Southern Alps
25 Saltrio, Southern Alps
26 Torr. Bosso, Umbria
27 Monte Cetona, Umbria
28 Sta Anna, Sicily
29 Cape Verde, DSDP site 367
30 Blake Bahama Bank, DSDP site 534
31 Magellan Rise, DSDP site 167
32 Shatsky Rise, DSDP site 306
33 Shatsky Rise, DSDP site 307
34 W-Pacifique, DSDP site 195
35 W-Pacifique, DSDP site 196
36 Glasenbach, Austria
37 Point Salena, California
38 Veveyse, Switzerland
39 Blake Bahama Bank, DSDP site 5
40 Inuyama, Japan
43 Trattberg, Austria

G.2 Code Numbers of Species

1 *Trillus cf. seidersi*
2 *Unuma echinatus*
3 *Andromeda praepodbielensis*
4 *Stichocapsa* aff. *japonica*
5 *Andromeda praecrassa*
6 *Tetraditryma praeplena*
7 *Eurycitid* ind.
8 *Tricolocapsa plicarum*
9 *Mirifusus fragilis*
10 *Diacanthocapsa normalis*
11 *Guexella nudata*
12 *Gorgansium pulchrum*
13 *Napora pyramidalis*
14 *Higumastra imbricata*
15 *Theocapsomma cordis*
16 *Praeconocaryomma* (?) *hexacubica*
17 *Eucyrtidiellum unumaensis*
18 *Podobursa helvetica*
19 *Emiluvia premyogii*
20 *Higumastra* aff. *inflata*
21 *Stichocapsa convexa*
22 *Tetraditryma corralitosensis*
23 *Emiluvia sedecimporata*
24 *Stichocapsa japonica*

25 *Tritrabs hayi*
26 *Paronaella bandyi*
27 *Paronaella mulleri*
28 *Parvicingula dhimenaensis*
29 *Pseudocrucella adriani*
30 *Acanthocircus suboblongus*
31 *Hsuum brevicostatum*
32 *Tetratrabs zealis*
33 *Protunuma costata*
34 *Tetraditryma pseudoplena*
35 *Homeoparonaella argolisensis*
36 *Triactoma jonesi*
37 *Tritrabs rhododactylus*
38 *Spongocapsula palmerae*
39 *Bernoullius cristatus*
40 *Archaeohagiastrum munitum*
41 *Hagiastrid* sp. *A*
42 *Hsuum maxwelli*
43 *Andromeda podbielensis*
44 *Eucyrtidiellum pustulatum*
45 *Tritrabs casmaliaensis*
46 *Triactoma blakei*
47 *Ristola altissima*
48 *Perispyridium ordinarium*
49 *Saitoum pagei*
50 *Emiluvia salensis*
51 *Sethocapsa leiostraca*
52 *Triactoma tithonianum*
53 *Stylocapsa oblongula*
54 *Monotrabs plenoides*
55 *Mirifusus guadalupensis*
56 *Bernoullius dicera*
57 *Angulobracchia purisimaensis*
58 *Pseudocrucella sanfilippoae*
59 *Emiluvia* (?) sp.
60 *Staurosphaera antiqua*
61 *Napora bukryi*
62 *Napora deweveri*
63 *Homoeoparonaella elegans*
64 *Paronaella kotura*
65 *Triactoma cornuta*
66 *Eucyrtidiellum ptyctum*
67 *Mirifusus major*
68 *Homoeoparonaella giganthea*
69 *Emiluvia hopsoni*
70 *Tritrabs ewingi*
71 *Tritrabs exotica*
72 *Ristola procera*
73 *Paronaella broennimanni*
74 *Tetratrabs bulbosa*
75 *Diboloachras chandrika*
76 *Napora lospensis*
77 *Acaenotyle diaphorogona*
78 *Podobursa spinosa*
79 *Emiluvia pessagnoai*
80 *Mirifusus chenodes*
81 *Emiluvia orea*
82 *Archaeodictyomitra apiara*
83 *Paronaella* (?) *hipposidericus*
84 *Podocapsa amphitreptera*
85 *Paronaella* (?) *diamphidia*
86 *Acanthocircus dicranacanthos*
87 *Sethocapsa cetia*
88 *Acaeniotyle umbilicata*
89 *Triactoma echiodes*
90 *Mirifusus minor*
91 *Syringocapsa lucifer*
92 *Pantanellium* (?) *berriasianum*
93 *Ristola cretacea*
94 *Parvicingula cosmoconica*
95 *Obesacapsula rotunda*
96 *Ditrabs sansalvatorensis*
97 *Angulobracchia* (?) *portmanni*
98 *Xitus cf.spicularius*
99 *Acaeniotyle diaphorogona*
100 *Obesacapsula rusconensis*
101 *Pseudodictyomitra depressa*
102 *Archaeodictyomitra excellens*
103 *Alievum helenae*
104 *Syringocapsa agolarium*
105 *Pseudodictyomitra carpatica*
106 *Holocryptocanium barbui*
107 *Thanarla elegantissima*
108 *Cecrops septemporatus*
109 *Sethocapsa uterculus*

Appendix H
Dommergues and Meister's (1987) Database

H.1 Code Numbers of Species and Genera

1 *Tragophylloceras*
2 *T.numismale*
3 *T.undulatum*
4 *T.ibex*
5 *T.loscombi*
6 *Radstockiceras*
7 *Apoderoceras nodogigas + leckenbyi*
8 *Phricodoceras* gr.*taylori*
9 *Hyperderoceras* gr.*ruga*
10 *Epideroceras nodofissum + trigonale*
11 *Coeloderoceras biruga*
12 *Metaderoceras*
13 *M.muticum + venustulum*
14 *M.*gr.*venarense*
15 *Coeloceras pettos*
16 *Parinodiceras*
17 *Polymorphites* gr.*polymorphus + pseudodubari*
18 *Platypleuroceras*
19 *P.brevispina + brevispinoides + aureum + muellensis*
20 *P.rotundum + oblungum*
21 *P.amplinatrix + tenuilobum +* cf.*oblongum + acanthobronni*
22 *Uptonia*
23 *U.jamesoni + lata*
24 *U.bronni*
25 *Dayiceras*
26 *D.polymorphoides*
27 *Tropidoceras*
28 *T.*gr.*flandrini*
29 *T.*gr.*masseanum*
30 *T.stahli*
31 *Acanthopleuroceras*
32 *A.arietiforme + subarietiforme*
33 *A.carinatum*
34 *A.maugenesti*
35 *A.valdani*
36 *A.acteon*
37 *A.lepidum*
38 *Liparoceras + Vicinodiceras*
39 *Becheiceras*
40 *Androgynoceras*
41 *A.geyeri*
42 *A.*gr.*hybrida*
43 *A.alloeotypus*
44 *A.(Beaniceras)*
45 *A.(B.) centaurus*
46 *A.(B.) rotundum*
47 *A.(B.) crassum*
48 *A.(B.) luridum*
49 *A.(Aegoceras)*
50 *A.(A.) maculatum*
51 *A.(A.) lateacosta*
52 *A.(A.) capricornus*
53 *A.(A.) crescens + samontaensis*
54 *A.(Oistoceras)*
55 *A.(O.) angulatum*
56 *A.(O.)* gr.*figulinum*
57 *Prodactylioceras*
58 *P.davoei*
59 *P.rectiradiatum*
60 *Reynesocoeloceras*
61 *Amaltheus*
62 *A.stockesi*
63 *A.bifurcus*
64 *A.*gr.*subnodosus*
65 *Protogrammoceras*
66 *P.psueodilectum*
67 *P.celebratum*

68 P.depressum + lusitanicum + normanianum
69 P.(Matteiceras)
70 P.(M.) occidentale
71 P.(M.) monestieri + diornatum
72 P.(M.) isselioides
73 P.(M.) nitescens
74 Fieldlingiceras

H.2 Numerical Database

Section 1

1 3 27, 2 3 8, 3 10 10, 5 14 27, 6 3 4, 7 1 1, 8 2 2, 10 3 3, 11 3 3, 12 4 12, 13 4 4, 14 12 12, 17 4 4, 18 5 6, 19 5 5, 21 6 6, 22 7 7, 23 7 7, 24 7 7, 25 8 12, 26 12 12, 27 8 8, 29 8 8, 31 9 13, 33 9 9, 34 10 10, 35 11 11, 36 12 12 , 37 13 13, 38 2 20, 39 21 26, 40 14 14, 41 14 14, 44 14 16, 46 14 14, 47 15 15, 48 16 16, 49 17 21, 50 17 17, 51 18 18, 52 19 19, 53 20 21, 54 22 23, 55 22 22, 56 23 23, 57 19 22, 58 21 22, 59 19 19, 60 15 15, 61 25 27, 62 25 27, 63 25 27, 65 27 27, 68 27 27, 69 24 26, 70 24 24, 71 25 25, 72 25 25, 73 26 26, 74 27 27

Section 2

1 8 22, 6 3 3, 7 1 1, 8 1 2, 9 2 2, 12 11 11, 14 11 11, 15 7 7, 16 3 3, 17 3 5, 18 3 5, 19 5 5, 21 4 4, 22 6 7, 23 6 6, 24 6 6, 25 12 12, 26 12 12, 27 8 8, 31 8 13, 34 9 9, 35 10 10, 36 12 12, 37 13 13, 38 13 18, 39 13 22, 40 16 18, 41 16 16, 42 17 18, 44 11 16, 45 11 11, 46 14 14, 47 15 15, 48 16 16, 49 17 19, 50 17 17, 51 18 18, 52 18 18, 53 19 19, 54 20 21, 55 20 20, 56 21 21, 57 18 20, 58 19 20, 59 18 18, 61 22 23, 62 22 23, 63 22 22, 69 22 22, 70 22 22, 73 23 23

Section 3

6 2 8, 7 1 1, 8 1 1, 12 8 8, 18 2 3, 19 3 3, 20 2 2, 21 3 3, 22 4 5, 23 5 5, 24 4 4, 27 6 6, 31 7 7, 32 7 7, 33 7 7, 38 6 10, 39 15 15, 40 11 11, 44 9 9, 47 9 9, 49 10 11, 50 10 10, 52 11 11, 54 13 13, 56 13 13, 57 11 12, 58 12 12, 60 9 9, 61 16 16, 64 16 16, 65 14 16, 67 15 16, 69 14 15, 72 14 15, 73 14 14, 74 16 16

Section 4

6 2 3, 7 1 1, 12 2 2, 13 2 2, 18 2 2, 19 2 2, 20 2 2, 22 3 3, 23 3 3, 24 3 3, 38 4 4, 39 9 9, 40 5 5, 49 4 7, 50 4 4, 51 5 5, 52 6 6, 53 7 7, 54 9 9, 56 9 9, 57 8 8, 58 8 8, 61 10 10, 62 10 10, 63 10 10, 65 13 13, 68 13 13, 69 10 12, 70 10 10, 71 11 11, 72 12 12, 73 12 12, 74 13 13

Section 5

1 4 21, 2 4 5, 3 7 10, 5 11 21, 6 1 26, 12 1 12, 13 1 1, 14 12 12, 16 3 3, 17 2 5, 18 2 3, 19 2 2, 20 2 2, 21 3 3, 22 4 4, 23 4 4, 24 4 4, 27 4 10, 28 4 10, 29 5 6, 30 6 6, 31 7 13, 32 7 7, 33 7 7, 34 8 8, 35 9 9, 36 10 11, 37 13 13, 38 5 16, 39 17 26, 40 14 19, 41 14 14, 42 15 17, 43 19 19, 44 8 14, 45 8 13, 47 14 14, 49 15 18, 50 15 15, 51 16 17, 52 17 18, 53 18 18, 54 19 19, 56 19 19, 57 17 19, 58 18 19, 60 17 17, 61 20 26, 62 20 25, 63 20 25, 64 26 26, 65 15 26, 66 15 18, 67 24 25, 68 26 26, 69 20 24, 70 20 20, 71 21 22, 72 22 23, 73 23 24, 74 25 26

Section 6

1 8 25, 2 8 8, 4 9 10, 5 11 25, 6 1 5, 7 1 1, 8 1 2, 9 2 2, 12 3 11, 13 3 3, 14 11 11, 15 6 6, 18 3 4, 19 3 3, 20 3 3, 21 4 4, 22 5 6, 23 5 5, 24 5 5, 27 7 9, 28 8 9, 29 7 7, 30 8 8, 31 8 13, 32 8 8, 33 8 8, 34 9 9, 35 10 10, 36 12 12, 37 13 13, 38 15 16, 39 18 23, 40 15 21, 41 15 15, 42 16 18, 43 21 21, 44 10 15, 45 10 12, 46 14 14, 47 15 15, 49 16 20, 50 16 16, 51 17 17, 52 18 18, 53 19 19, 54 21 21, 56 21 21, 57 20 20, 58 20 20, 61 22 25, 62 22 25, 65 20 25, 66 20 20, 67 25 25, 69 22 24, 70 22 22, 71 23 23, 72 23 24, 73 24 24

Section 7

1 2 8, 2 2 5, 3 6 8, 4 8 8, 6 2 4, 7 1 1, 8 1 2, 9 2 3, 10 2 3, 11 2 3, 12 4 4, 13 4 4, 15 6 6, 16 2 7, 17 3 5, 18 4 4, 19 4 4, 20 4 4, 21 4 4, 22 5 6, 23 5 6, 24 5 6, 27 6 8, 28 8 8, 29 7 7, 31 8 8, 38 8 8, 39 9 11, 40 9 10, 44 8 8, 49 9 9, 54 10 10, 55 10 10, 57 9 10, 61 11 11, 72 11 11, 63 11 11

Appendix I
Schaaf's (1985) Database

I.1 Code Numbers of Species

Only the taxa common to both, Schaaf's synthetic range chart (loc. cit., Fig. 50) and to its original database (loc. cit., p. 260) are listed below

4 *X.alievi*
6 *P.amphitreptera*
7 *T.ancus*
9 *T.antiqua*
11 *A.apiarium*
12 *T.apicata*
15 *H.barbui*
17 *C.bumastus*
18 *H.capita*
19 *A.carinatus*
20 *P.carpatica*
21 *E.cenomana*
22 *S.cetia*
23 *M.chenodes*
25 *E.columbaria*
28 *C.conara*
31 *A.cortinaensis*
32 *P.cosmoconica*
34 *S.cribata*
35 *F.diamphidia*
36 *A.diaphorogona*
37 *A.dicranacanthos*
38 *D.koslovae*
39 *T.echiodes*
41 *T elegantissima*
43 *A.pseudoconulus*
44 *P.floresensis*
45 *D.formosa*
46 *A.helenae*
48 *H.hindei*
53 *T.hybum*
57 *A. lacrimula*
59 *P.lanceola*
60 *S.leiostraca*
62 *P.leptoconica*
64 *S.limatum*
65 *A.liriodes*
66 *R.majuroensis*
68 *M.mediodilatatus*
70 *C.operosa*
71 *S.orca*
72 *Z.ornatum*
74 *P.pargueraensis*
75 *D.periosa*
79 *H. polyhedra*
80 *P.polylophia*
81 *P.pseudomacrocephala*
82 *E.ptyctum*
83 *T.pulchra*
86 *S.rotunda*
87 *T.sallilum*
88 *C.septemporatus*
89 *O.somphedia*
93 *A.superbum*
94 *P.tetracola*
95 *T.tina*
96 *T.tithonianum*
97 *C.titthium*
100 *S.trachyostraca*
101 *P.triacantha*
102 *P.tricola*
103 *A.trizonalis*
104 *H.tuberosa*
106 *D.tytthopora*
107 *A.umbilicata*
108 *T.urna*
109 *S.uterculus*
110 *T.veneta*

I.2 Numerical Database

Section 1

15 1 1

Section 2

28 4 5, 38 3 5, 93 5 5, 110 2 7

Section 3

15 1 1, 28 3 3, 38 3 3, 48 1 1, 74 3 3, 81 1 3, 93 3 3

Section 4

7 1 1, 93 2 2

Section 5

7 1 4, 28 1 1, 43 1 4, 93 2 2

Section 6

12 4 7, 28 4 12, 43 2 14, 44 2 14, 45 1 11, 65 13 14, 66 2 12, 74 6 11, 87 2 14, 93 2 11, 95 2 10, 97 4 9

Section 7

21 5 5, 28 5 5, 45 1 4, 66 5 5, 79 5 5, 81 2 4, 83 1 3, 104 5 5, 110 1 4

Section 8

11 3 10, 19 6 10, 34 2 10, 37 2 8, 45 1 17, 57 4 11, 68 1 7, 81 13 17, 83 3 14, 88 5 10, 100 4 10, 103 2 11

Section 9

45 1 4, 65 2 4, 66 4 5, 83 1 1, 93 3 3, 110 1 1

Section 10

12 8 8, 28 1 8, 43 7 12, 44 8 8, 45 1 12, 65 8 9, 66 1 10, 81 1 2, 83 1 2, 87 5 9, 93 5 9, 95 1 12, 97 7 9, 104 8 9

Section 11

43 1 1, 44 1 1, 45 1 4, 65 2 2, 66 1 5, 87 1 1, 93 1 4, 95 1 1, 104 1 1

Section 12

36 1 2, 57 1 1, 59 1 1, 62 1 2, 103 1 2, 107 1 2

Section 13

4 3 3, 20 3 4, 32 1 4, 35 4 4, 36 1 3, 37 1 2, 57 1 4, 59 1 4, 60 1 3, 62 3 4, 64 1 3, 80 1 4, 86 1 2, 88 1 3, 94 3 3, 100 1 3, 101 1 4, 102 3 4, 103 1 4, 106 3 3, 107 1 3

Section 14

4 2 5, 6 1 1, 19 2 5, 20 1 5, 22 1 1, 32 1 1, 35 1 5, 36 1 5, 37 1 5, 39 2 5, 46 2 5, 57 5 5, 59 1 5, 60 1 3, 62 2 5, 64 1 3, 80 1 3, 86 1 5, 88 2 5, 94 1 3, 96 1 1, 100 2 5, 101 1 5, 102 2 5, 103 1 5, 106 2 5, 107 1 5

Section 15

31 3 4, 70 4 6, 81 3 3, 93 6 6

Section 16

9 4 6, 11 2 6, 25 1 1, 70 3 3

Section 17

9 5 5, 11 5 7, 23 3 3, 25 1 5, 59 3 5, 70 3 3

Section 18

9 1 14, 11 1 13, 23 9 10, 25 2 11, 59 2 13, 70 6 8

Section 19

15 6 7, 18 2 2, 20 1 7, 36 1 5, 39 2 3, 46 1 4, 53 3 3, 57 1 4, 59 1 4, 71 1 4, 72 1 4, 75 5 5, 81 7 7, 88 1 1, 89 7 7, 101 2 4, 102 1 4, 103 1 6, 106 1 3

Section 20

4 1 2, 20 1 6, 36 1 5, 37 1 1, 39 1 2, 46 1 2, 53 1 2, 57 1 3, 59 1 3, 71 1 2, 72 1 2, 75 4 5, 88 2 2, 100 1 1, 101 1 2, 102 1 2, 103 1 4, 106 1 2, 107 1 5

Section 21

20 1 7, 36 1 8, 39 1 4, 46 1 4, 53 1 4, 57 1 3, 59 1 2, 71 1 2, 72 1 2, 101 1 1, 102 1 3, 103 1 4, 106 1 2, 107 1 8

Section 22

4 5 9, 6 1 2, 11 1 8, 15 12 20, 18 3 3, 19 8 8, 20 1 18, 22 1 3, 32 1 2, 35 1 11, 36 1 19, 37 1 2, 39 8 11, 46 1 10, 53 8 11, 57 4 9, 59 1 8, 60 1 2, 64 3 7, 68 1 2, 71 3 8, 72 4 9, 80 3 3, 86 1 2, 94 1 2, 96 1 2, 100 3 8, 101 1 8, 102 8 11, 103 1 12, 106 7 8

Section 23

4 5 7, 11 1 6, 15 8 10, 18 4 6, 19 6 7, 20 1 8, 22 1 3, 32 1 5, 35 6 6, 36 3 7, 37 1 7, 39 4 7, 46 4 7, 53 6 7, 57 5 8, 59 1 8, 60 2 5, 64 4 5, 68 1 5, 71 6 7, 72 6 8, 80 4 5, 86 4 5, 88 4 7, 94 5 5, 96 1 3, 100 4 6, 101 1 7, 102 6 7, 103 1 7, 106 6 7, 107 3 7

Section 24

38 2 5, 74 5 5, 87 4 5, 95 2 5, 108 3 5

Section 25

6 2 6, 22 4 10, 35 1 10, 37 6 8, 59 1 10, 68 6 10, 82 1 5, 86 7 10, 100 10 10, 101 1 10, 103 1 10, 109 10 10

Section 26

12 1 6, 38 1 3, 43 1 2, 65 4 7, 70 5 8, 74 1 3, 87 2 3, 97 1 2

Section 27

4 1 28, 9 29 32, 11 1 13, 15 31 31, 19 6 21, 20 3 32, 23 3 8, 25 1 22, 31 32 33, 34 1 5, 35 4 31, 36 2 29, 37 1 9, 46 1 15, 48 1 19, 53 17 26, 57 5 27, 59 2 26, 62 1 19, 70 1 11, 71 3 24, 83 1 14, 88 1 10, 100 2 6, 101 1 25, 102 2 14, 103 1 29, 106 6 23, 107 1 31, 109 3 16

Section 28

7 1 3, 20 1 3, 21 1 1, 31 1 2, 45 1 1, 66 2 3, 87 1 3, 89 1 3, 95 3 3

Section 29

7 1 1, 9 1 1, 15 1 1, 20 1 1, 21 1 1, 31 1 1, 45 1 1, 66 1 1, 81 1 1, 87 1 1, 89 1 1, 95 1 1

Section 30

7 8 9, 9 1 9, 15 5 9, 20 1 9, 21 5 9, 31 1 9, 45 1 9, 66 6 9, 81 4 9, 87 1 9, 89 1 9, 95 1 7

Bibliography

Agterberg FP (1985) Normality testing and comparison of RASC to unitary associations. In: Gradstein F, Agterberg F, Brower JC, Schwarzacher WS (eds): Quantitative Stratigraphy. Reidel, Dordrecht: 124–161

Agterberg FP (1990) Automated stratigraphic correlations. Elsevier, Amsterdam, 424 pp

Agterberg FP, Nel FP (1982a) Algorithms for the ranking of stratigraphic events. Comput Geosci 8/1:69–90

Agterberg FP, Nel FP (1982b) Algorithms for the scaling of stratigraphic events. Comput Geosci 8/2:163–189

Arkell WJ (1933) The Jurassic system in Great Britain. Oxford University, Oxford, 682 pp

Arkell WJ (1952) Jurassic ammonites from Jebel Tuwaiq, Central Arabia. Philos Trans Roy Soc Ser B 236:241–313

Arkell WJ (1956) Jurassic geology of the world. Oliver and Boyd, London 804 pp

Augustson JG, Minker J (1970) An analysis of some graph theoretical cluster techniques. J Assoc Comput Mach 17/4:571–588

Baumgartner PO (1984a) Comparison of unitary associations and probabilistic ranking and scaling as applied to mesozoic radiolarians. Comput Geosci 10/1:167–183

Baumgartner PO (1984b) A middle Jurassic-Early Cretaceous low-latitude radiolarian zonation based on unitary associations and age of Tethyan radiolarites. Eclogae Geol Helv 77/3:729–837

Baumgartner PO (1987) Age and genesis of Tethyan Jurassic radiolarites. Eclogae Geol Helv 80/3:831–879

Baumgartner PO, de Wever P, Kocher R (1980) Correlation of Tethyan Late Jurassic-Early Cretaceous radiolarian events. Cah Micropaléontol C.N.R.S, 2:24–74

Berge C (1973) Graphes et hypergraphes (2nd edn). Dunod, Paris, 576 pp

Berggren WA, van Couvering JA (1978) Biochronology. Am Assoc Pet Geol Stud Geol 6: 39–56

Berggren WA, Kent DV, Flynn JJ (1985a) Paleogene geochronology and chronostratigraphy. Geol Soc Mem 10:141–195

Berggren WA, Kent DV, van Couvering JA (1985b) Neogene geochronology and chronostratigraphy. Geol Soc Mem 10:211–250

Berger WH and Heath GR (1968) Vertical mixing in pelagic sediments. J Mar Res 26/2:134–143

Blank RG (1979) Applications of probabilistic biostratigraphy to chronostratigraphy. J Geol 87:647–670

Blank RG (1984) Comparison of two binomial models in probabilistic stratigraphy. Comput Geosci 10/1:59–68

Blank RG, Ellis CH (1982) The probable range concept applied to the biostratigraphy of marine microfossils. J Geol 90/4:415–433

Blieck A (1984) Les Hétérostracés Pteraspidiformes: Systématique, phylogénie, biostratigraphie, biogéographie. Cah Paléontol C.N.R.S, 2:199 pp

Bloos G (1979) Uber den Jura am Grossen Hassberg. Stutt Beitr Naturkd B/44:1–53

Bloos G (1983) The zone of Schlotheimia marmorea: Hettangian or Sinemurian? Newsl Stratigr 12/3:123–131

Bramlette MN, Sullivan FR (1961) Coccolithophorids and related nannoplankton of the early Tertiary in California. Micropaleontology 7/2:129–174

Brower JC (1981) Quantitative biostratigraphy, 1930–1980. In: Merriam (ed) (1981: Computer application in the earth sciences, Plenum, New York:63–103

Brower JC (1984) The relative biostratigraphic value of fossils. Comput Geosci 10/1:111–132

Brower JC (1985a) The index fossil concept and its application to quantitative biostratigraphy. In: Gradstein F, Agterberg FP, Brower JC, Schwartzacher WS (eds): Quantitative stratigraphy. Reidel, Dordrecht: 43–64
Brower JC (1985b) Archaeological seriation of an original data matrix. In: Gradstein F, Agterberg FP, Brower JC, Schwartzacher WS (eds): Quantitative Stratigraphy. Reidel Publ Company, Dordrecht: 95–108
Brower JC (1990) A case study for comparison of some biostratigraphic techniques using Paleogene alveolinids from Slovenia and Istria. Geol Survey Canada, Paper 89-9: 407–416
Brower JC, Burroughs WA (1982) A simple method for quantitative biostratigraphy. In: Cubbit JM, Reyment RA (eds): Quantitative stratigraphic correlation. John Wiley and Sons: 61–84
Brower JC, Millendorf SA, Dyman TS (1978) Quantification of assemblage zones based on multivariate analysis. Comput Geosci 4/3: 221–227
Callomon JH (1965) Notes on Jurassic stratigraphical nomenclature. Carpatho-Balkan Geol Assoc I/2: 81–85
Callomon JH (1984) Biostratigraphy, chronostratigraphy and all that – Again! International Symposium on Jurassic Stratigraphy, 1–8 Sept 1984 Erlangen. Geol Surv Denmark, Spec Pub, Copenhagen: 612–624
Callomon JH, Donovan DT (1971) A code of Mesozoic stratigraphical nomenclature. Mem. B.R.G.M. 75: 75–81
Carre B. (1979) Graphs and Networks. Clarendon Press, Oxford. 277 pp
Cheetham AH, Deboo PB (1963) A numerical index for biostratigraphic zonation in the mid-Tertiary of the eastern Gulf. Gulf Coast Assoc Geol Soc Trans 13: 139–147
Cita MB (1978) Biostratigraphy of miocene deep-sea sediments (sites 372 and 375). Init Repts 42/1: 671–685
Cockbain AE (1966) An attempt to measure the relative biostratigraphic usefulness of fossils. J Paleont 40/1: 206–207
Comas Rengifo MJ (1985) El pliensbachiense de la Cordillera Iberica. Thesis Univ Madrid, 591 pp
Cubaynes R, Boutet C, Delfaud J, Faure P (1984) La mégaséquence d'ouverture du Lias quercynois. Bull Centres Rech Explor Prod Elf Aquitaine 8/2: 333–370
Cubbit JM, Reyment RA (eds) (1982) Quantitative stratigraphic correlation. John Wiley and Sons, Chichester. 301 pp
Davaud E (1978) Un algorithme facilitant l'élaboration d'échelles biochronologiques. In: Davaud E, Guex J (1978: Traitement analytique "manuel" et algorithmique de problèmes complexes de corrélations biochronologiques. Eclogae Geol Helv 71/3: 598–610
Davaud E (1979) Automatisation des corrélations biochronologiques: un exemple d'application de l'informatique à la résolution d'un problème naturaliste complexe. Publ Spéciale Genève: 1–48
Davaud E (1982) The automation of biochronological correlation. In: Cubitt JM, Reyment RA (eds): Quantitative Stratigraphic Correlation. John Wiley and Sons. Chichester. pp 85–100
Davaud E, Guex J (1978) Traitement analytique "manuel" et algorithmique de problèmes complexes de corrélations biochronologiques. Eclogae Geol Helv 71/3: 581–610
Dean WT, Donovan DT and Howarth MK (1961) The liassic zones and subzones of the NW European province. Bull Brit Mus (NH) 4/10: 437–505
Deboo PB (1965) Biostratigraphic correlation of the type Shubuta Member of the Yazoo Clay and Red Bluff Clay with their equivalents in southwestern Alabama. Alabama Geol Surv Bull 80: 1–84
Doeven PH, Gradstein FM, Jackson A, Agterberg FP, Nel D (1982) A quantitative nonnofossil range chart. Micropaleontology 28/1: 85–92
Dommergues J-L (1979) Le Carixien bourguignon. Biostratigraphie, paléogéographie, approches paléontologiques et sédimentologiques. Thesis, Dijon University, 109 pp
Dommergues J-L (1987) L'évolution des Ammonitina au Lias moyen Carixien, Domérien basal) en Europe occidentale. Doc Lab Géol Lyon: 98–297 pp
Dommergues J-L, Meister C (1987) La biostratigraphie des ammonites du Carixien (Jurassique inférieur) d'Europe occidentale: un test de la méthode des associations unitaires. Eclogae Geol Helv 80/3: 919–938

Donovan DT (1956) The zonal stratigraphy of the Blue Lias around Keynsham, Somerset. Proc Geol Assoc London 66:182–212
Drobne K (1977) Alvéolines paléogènes de l'Istrie et de la Slovénie. Mem Suisses Paleontol 99:1–175
Drooger CW (1980) Accuracy in Time. In: Geological Correlation. Int Geol Correl Progr 8:22
Edwards LE (1977) Range chart as chronostratigraphic hypothesis. Unpublished thesis. Univ California, Riverside 188 pp
Edwards LE (1978) Range chart and no-space graphs. Comput Geosci 4/3:247–255
Edwards LE (1982a) Numerical and semi-objective biostratigraphy: review and prediction. North Amer Paleontol Conv Proc 1:147–152
Edwards LE (1982b) Quantitative biostratigraphy: the methods should suit the data. In: Cubitt JM, Reyment RA (eds) (1982: Quantitative stratigraphic correlation. John Wiley and Sons. Chichester p 45–60
Edwards LE (1984) Insight on why graphic correlation (Shaw's method) works. J Geol 92: 583–587
Edwards LE (1989) Supplemented graphic correlation: a powerful tool for paleontologists and nonpaleontologists. Palaios 4:127–143
Eldredge N, Gould SJ (1977) Evolutionary models and biostratigraphic strategies. In: Kauffman EG, Hazel JE (eds) (1977: Concepts and methods of biostratigraphy. Dowden, Hutchinson and Ross Inc, Stroudsburg, Pennsylvania: 25–40
Fahreus LE (1986) Spectres of biostratigraphic resolution and precision: rock accumulation rates, processes of speciation and paleoecological constraints. Newsl Stratigr 15/3:150–162
Fontaine P, Rey J (1984) Fluctuations phénotypiques, évolution et valeur chronostratigraphique d'alvéolines ilerdiennes du Minervois (Aude, France). Benthos (1983:207–212
Fulkerson DR, Gross OA (1965) Incidence matrices and interval graphs. Pac J Math 15/3:835–855
Gabilly J (1971) Méthodes et modèles en stratigraphie du Jurassique. Mém B. R. G. M. 75:5–16
Gallai T (1967) Transitiv orientierbare Graphe. Acta Math Acad Sci Hung 18:25–66
Ghouila Houri A (1962) Caractérisation des graphes non orientés dont on peut orienter les arêtes de manière à obtenir le graphe d'une relation d'ordre. C R Acad Sci Paris 254:1370–1371
Gilmore PC, Hoffman AJ (1964) A characterization of comparatibility graphs and of interval graphs. Canad J Math 16:539–548
Godinot M (1981) Usefulness and meaning of the mammalian specific lineages. Acta Geol Hispanica 16/1:249–258
Golumbic M (1980) Algorithmic graph theory and perfect graphs. Academic Press, New York 284 pp
Gradstein F, Agterberg FP, Brower JC, Schwarzacher WS (eds) (1985) Quantitative Stratigraphy. Reidel Publ Company, Dordrecht, Holland, 598 p
Guex J (1973) Sur l'âge et la position systématique du genre Nejdia Arkell. Bull Lab Géol Univ Lausanne 201:1–6
Guex J (1977) Une nouvelle méthode d'analyse biochronologique. Bull Lab Géol Univ Lausanne 224:309–322
Guex J (1978a) Le Trias inférieur des Salt Ranges (Pakistan): problèmes biochronologiques. Eclogae Geol Helv 71/1:105–141
Guex J (1978b) Influence du confinement géographique des espèces fossiles sur l'élaboration d'échelles biochronologiques et sur les corrélations. Bull Lab Géol Univ Lausanne 230: 115–124
Guex J (1978c) La zonation du Paléocène supérieur et de l'Eocène inférieur et moyen de Californie à l'aide du nannoplancton. Eclogae Geol Helv 71/3:581–598
Guex J (1979) Terminologie et méthodes de la biostratigraphie moderne: commentaires critiques et propositions. Bull Lab Géol Univ Lausanne 234:169–216
Guex J (1980) Calcul, caractérisation et identification des associations unitaires en biochronologie. Bull Lab Géol Univ Lausanne 249:111–126
Guex J (1981) Associations virtuelles et discontinuités dans la distribution des espèces fossiles. Bull Lab Géol Univ Lausanne 255:179–197

Guex J (1982) Remarques sur l'origine de la dispersion biochronologique du nannoplancton calcaire paléogène de Californie et sur la détection des remaniements. Bull Lab Géol Univ Lausanne 267:197–205

Guex J (1984a) Estimations numériques de la qualité de l'enregistrement fossile des espèces. Bull Lab Géol Univ Lausanne 274:79–89

Guex J (1984b) Recherche des associations unitaires: une nouvelle technique d'optimisation. Bull Lab Géol Univ Lausanne 279:63–66

Guex J (1987) Corrélations biochronologiques et Associations Unitaires. Presses Polytechniques Romandes, Lausanne 244 pp

Guex J (1988) Utilisation des horizons maximaux résiduels en biochronologie. Bull Lab Géol Univ Lausanne 300:135–142

Guex J, Davaud E (1982) Recherche automatique des associations unitaires en biochronologie. Bull Lab Géol Univ Lausanne 261:53–69

Guex J, Davaud E (1984) unitary associations Method: Use of Graph Theory and Computer Algorithm. Comput Geosci 10/1:69–96

Guex J, Davaud E (1986) Recherche des associations unitaires: option nouvelle et exemple d'application. Bull Lab Géol Univ Lausanne 287:195–205

Guex J, Taylor D (1976) La limite Hettangien-Sinémurien des Préalpes romandes au Nevada. Eclogae Geol Helv 69/2:521–526

Hancock JM (1977) The historic development of biostratigraphic correlation. In: Kauffman EG, Hazel JE (eds) (1977: Concepts and Methods of Biostratigraphy. Dowden, Hutchinson and Ross Inc, Stroudsburg, Pennsylvania:3–22

Hansen P (1980) Bornes et algorithmes pour les stables d'un graphe. In: Regards sur la Théorie des Graphes, Colloque de Cerisy, Presses Polytechniques Romandes, Lausanne:39–54

Harary F (1969) Graph Theory. Addison-Wesley Publication, Reading, 274 pp

Harper CW (1981) Inferring succession of fossils in time: the need for a quantitative and statistical approach. J Paleontol 55:442–452

Harper CW (1984) Ranking algorithms in quantitative biostratigraphy. Comput Geosci 10/1:3–15

Hay WW (1972) Probabilistic stratigraphy. Eclogae Geol Helv 65/2:255–266

Hay WW, Southam JR (1978) Quantifying biostratigraphic correlation. Ann Rev Earth Planet Sci 6:353–375

Hazel JE (1970) Binary coefficients and clustering in biostratigraphy. Geol Soc Amer Bull 81/11:3237–3252

Hazel JE (1977) Use of certain multivariate techniques in assemblage zonal biostratigraphy. In: Kauffman EG, Hazel JE (eds) (1977: Concepts and methods of biostratigraphy. Dowden, Hutchinson and Ross Inc, Stroudsburg, Pennsylvania:187–212

Hazel JE, Mumma MD, Huff WJ (1980) Ostracode biostratigraphy of the Lower Oligocene (Vicksburgian) of Mississipi and Alabama. Trans Gulf Coast Assoc Geol Soc 30:361–401

Hodson FR, Kendall DG, Tautu P (eds) (1971) Mathematics in the archeological and historical sciences. Edinburgh University Press, Edinburgh, 565 pp

Hohn ME (1982) Properties of composite sections constructed by least squares. In: Cubbit JM, Reyment RA (eds) (1982: Quantitative stratigraphic correlation). John Wiley and Sons: 107–117

Hottinger L (1981) The resolution power of the biostratigraphic clock based on evolution and its limits. Acta Geol Hispanica 16/1:233–242

Jeletzky JA (1965) Is it possible to quantify biochronological correlation? J Paleont 39/1:135–140

Johnson JG, Niebuhr WW (1976) Anatomy of an assemblage zone. Geol Soc Amer Bull 87:1693–1703

Johnston HC (1976) Cliques of a graph. Int J Computers & Inf Sci 5/3:209–238

Kauffman EG, Hazel JE (eds) (1977) Concepts and Methods of Biostratigraphy. Dowden, Hutchinson and Ross Inc, Stroudsburg, Pennsylvania, 658 pp

Kendall D (1969a) Incidence matrices,interval graphs and seriation in archeology. Pacific J Math 28:565–570

Kendall D (1969b) Some problems and methods in statistical archeology. World Archeology 1/1:68–76

Kendall D (1971) Seriation from abundance matrices. In: Hodson FR, Kendall DG, Tautu P (eds): Mathematics in the archeological and historical sciences. Edinburgh University Press, Edinburgh: 215–252
Kummel B (1966) The Lower Triassic formations of the Salt Range and Trans-Indus Ranges, West Pakistan. Bull Mus Comp Zool Harvard 134/10: 361–429
Kummel B, Teichert C (eds) (1970) Stratigraphic Boundary Problems, Permian and Triassic of West Pakistan. Univ Kansas, Dept of Geol Spec Publ 4: 474 pp
Lamb JL, Stainforth RM (1976) Unreliability of Globigerinoides Datum. Amer Assoc Petr Geol Bull 60/9: 1564–1569
Lang WD (1924) The Blue Lias of the Devon and Dorset Coasts. Proc Geol Assoc London 35: 169–185
Lekkerkerker CB, Boland JC (1962) Representation of a finite graph by a set of intervals on the real line. Fund Math 51: 45–64
Mallory VS (1959) Lower Tertiary biostratigraphy of the California Coast Ranges. Amer Assoc Petrol Geol, Tulsa: 416 pp
Malmgren BA, Kennet JP (1981) Phyletic gradualism in a Late Cenozoic planktonic foraminiferal lineage. Paleobiology 7/2: 230–240
Mamet B (1977) Foraminiferal zonation in the Lower Carboniferous: methods and stratigraphic implications. In: Kauffman EG, Hazel JE (eds) (1977: Concepts and methods of biostratigraphy. Dowden, Hutchinson and Ross Inc, Stroudsburg, Pennsylvania: 445–462
Martini E (1971) Standard Tertiary and Quaternary calcareous nannoplankton zonation. Proc IInd Plankt Conf. Edizione Technoscienza, Roma: 739–783
Martini E, Muller C (1986) Current Tertiary and Quaternary calcareous nannoplankton stratigraphy and correlations. Newsl Stratigr 16/2: 99–112
McCammon RB (1970) On estimating the relative biostratigraphic values of fossils. Bull Geol Inst Univ Uppsala (ns) 2: 49–57
Meister C (1986) Les ammonites du Carixien des Causses (France). Mém suisses Paléont 109: 209 pp
Merriam DF (ed) (1981) Computer applications in the Earth Sciences. Plenum, New York 385 pp
Millendorf SA, Brower JC, Dyman TS (1978) A comparison of methods for the quantification of assemblage zones. Comput Geosci 4/3: 229–242
Millendorf SA, Heffner T (1978) Fortran program for lateral tracing of time-stratigraphic units based on faunal assemblage zones. Comput Geosci 4/3: 313–318
Millendorf SA, Millendorf MT (1982) The conceptual basis for lateral tracing of biostratigraphic units. In: Cubitt JM, Reyment RA (eds): Quantitative stratigraphic correlation. John Wiley and Sons Chichester p 101–106
Miller TG (1965) Time in stratigraphy. Paleontology 8/1: 113–131
Miller FX (1977) The graphic correlation method in biostratigraphy. In: Kauffman EG, Hazel JE (eds) (1977: Concepts and methods of biostratigraphy. Dowden, Hutchinson and Ross Inc, Stroudsburg, Pennsylvani p 165–186
Moon JW, Moser L (1965) On cliques in graphs. Israel J Math 3: 23–28
Muller C (1978) Neogene clacareous nannofossils from the Mediterranean Leg 42-A of the Deep Sea Drilling Project. Init Repts DSDP 42/1: 727–751
Nakasawa K, Dickins JM (1985) The Tethys, her Paleogeography and Paleobiogeography from Paleozoic to Mesozoic. Tokai University Press, Tokyo, 317 pp
Oppel A (1856) Die Juraformation. Ebner and Seubert, Stuttgart, 857 pp
Pak DN (1984) Mathematical model for the construction of composite standards from occurrences of fossil taxa. Comput Geosci 10/1: 107–110
Palmer AR (1954) The faunas of the Riley Formation in central Texas. J Paleontol 28/6: 709–790
Palmer CP (1972) A revision of the zonal classification of the Lower Lias of the Dorset Coast. Newsl Stratigr 2/1: 45–54
Phelps, M (1983) A facies and faunal analysis of the Carixian-Domerian boundary beds in North-West Europe. Thesis, University Birmingham, 660 pp
Remane J (1963) Les Calpionelles dans les couches de passage Jurassique-Crétacé de la Fosse Vocontienne. Trav Lab Géol Fac Sci Grenoble 39: 25–82

Reyment RA (1980) Morphometric methods in biostratigraphy. Academic Press, New York, 175 pp
Roberts F (1976) Discrete mathematical models. Prentice Hall, New Jersey 559 pp
Roberts F (1978) Graph theory and its application to problems of society. NFS-CBMS monograph 29, Philadelphia 122 pp
Rubel M (1976) On biological construction of time in geology. Eesti NSV Tead Akad Toimetised, Keemia Geol 25/2:136–144 (in russian)
Rubel M (1978) Principles of construction and use of biostratigraphic scales for correlation. Comput Geosci 4/3:243–246
Rubel M (1982) The succesion of brachiopods as a tool for correlation. Ecostratigraphy of the East Baltic Silurian. Valgus, Tallin:51–62
Rubel M, Pak DN (1984) Theory of stratigraphic correlation by means of ordinal scales. Comput Geosci 10/1:97–105
Savary J, Guex J (1991 a) BioGraph: un nouveau programme de construction des corrélations biochronologiques basées sur les associations unitaires. Bull Géol Lausanne 313: 317–340
Savary J, Guex J (1991 b) Sériations archéologiques et associations unitaires: comparaison de deux algorithmes. Bull Géol Lausanne (in prep)
Schaaf A (1985) Un nouveau canevas biochronologique du Crétacé inférieur et moyen: les biozones à Radiolaires. Bull Sci Géol Strassbourg 38/3:227–269
Schindewolf OH (1950) Grundlagen und Methoden der paläontologischen Chronologie. Borntraeger, Berlin, 152 pp
Schindewolf OH (1953) Uber die Faunenwende von Paläozoikum zum Mesozoikum. Zt Dtsch Geol Ges 105/2:154–183
Schindewolf OH (1970) Stratigraphie und Stratotypen. Akad Wiss Math-Naturwiss Kl 1 Abh2 Mainz, 134 pp
Schlatter R (1980) Biostratigraphie und Ammonitenfauna des Unter-Pliensbachium im Typusgebiet (Pliensbach, Holzmaden und Nürtingen; Württemberg, SW-Deutschland). Stuttg Beitr Naturkd B/65:1–261
Scott GH (1974) Essay review: stratigraphy and seriation. Newsl Stratigr 3:93–100
Shaw AB (1964) Time in Stratigraphy. McGraw Hill, New York 365 pp
Siesser WG (1984) Paleogene sea level and climates: USA eastern Gulf coastal plain. Paleogeogr Paleoclim Paleoecol 47:261–275
Sneath PHA (1975 a) Quantitative method for lateral tracing of sedimentary units. Comput Geosci 1/3:215–220
Sneath PHA (1975 b) Clarification on a quantitative stratigraphic correlation technique. Comput Geosci 1/4:353–354
Snelling NJ (ed) (1985) The chronology of the geological record. Geol Soc Mem 10:343 pp
Southam JR, Hay WW (1978) Correlation of stratigraphic sections by continuous variables. Comput Geosci 4/4:257–260
Southam JR, Hay WW, Worsley TR (1975) Quantitative formulation of reliability of stratigraphic correlation. Science 188/4186:357–359
Spath LF (1924) The ammonites of the Blue Lias. Proc Geol Assoc 35:186–211
Spath LF (1942) The ammonite zones of the Lias. Geol Mag London 79:264–268
Sullivan FR (1964) Lower Tertiary nannoplankton from California Coast Ranges I: Paleocene. Univ Calif Publ Geol Sci 44:163–227
Sullivan FR (1965) Lower Tertiary nannoplankton from California Coast Ranges II: Eocene. Univ Calif Publ Geol Sci 53:1–74
Sweet WC (1970) Uppermost Permian and Lower Triassic Conodonts of the Salt Range and Trans-Indus Ranges. In: Kummel B, Teichert C (eds) (1970: Stratigraphic Boundary Problems, Permian and Triassic of West Pakistan. Univ Kansas, Dept Geol, Spec Publ 4:207–275
Taylor D (1986) The Hettangian-Sinemurian boundary (Early Jurassic): Reply to Bloos (1983). Newsl Stratigr 16/2:57–67
Thaler L (1972) Datation, zonation et mammifères. Mém B. R. G. M. 77/1:411–424
Tozer ET (1967) A standard for Triassic time. Bull Geol Surv Can 156:1–103
Tozer ET (1971) Triassic time and ammonoids: problems and proposals. Can J Earth Sci 8/8:989–1031

Tozer ET (1984) The Trias and its ammonoids: the evolution of a time scale. Geol Surv Can Misc Repts 35:1–171

Trümpy R (1971) Stratigraphy in mountain belts. Quart J Geol Soc London, 126:293–318

Waagen W (1895) Fossils from the Ceratite formation. Paleontol Indica 13/2:1–323

Wilkinson BM (1974) Techniques of data analysis-seriation theory. Archaeo-Physika 5:1–142

Worsley TR, Jorgens ML (1977) Automated biostratigraphy. In (Ramsay A (ed): Oceanic Micropaleontology, II. Academic Press, London: 1201–1229

Subject Index

GPSR Compliance

The European Union's (EU) General Product Safety Regulation (GPSR) is a set of rules that requires consumer products to be safe and our obligations to ensure this.

If you have any concerns about our products, you can contact us on ProductSafety@springernature.com

In case Publisher is established outside the EU, the EU authorized representative is:

Springer Nature Customer Service Center GmbH
Europaplatz 3
69115 Heidelberg, Germany

Batch number: 10704771

Printed by Printforce, the Netherlands